Society 5.0 and the Future of Emerging Computational Technologies

Demystifying Technologies for Computational Excellence: Moving Towards Society 5.0

Series Editors: Vikram Bali and Vishal Bhatnagar

This series encompasses research work in the field of Data Science, Edge Computing, Deep Learning, Distributed Ledger Technology, Extended Reality, Quantum Computing, Artificial Intelligence, and various other related areas, such as natural- language processing and technologies, high-level computer vision, cognitive robotics, automated reasoning, multivalent systems, symbolic learning theories and practice, knowledge representation and the semantic web, intelligent tutoring systems, AI and education.

The prime reason for developing and growing out this new book series is to focus on the latest technological advancements—their impact on society, the challenges faced in implementation, and the drawbacks or reverse impact on society due to technological innovations. With these technological advancements, every individual has personalized access to all services and all devices connected with each other communicating amongst themselves, thanks to the technology for making our life simpler and easier. These aspects will help us to overcome the drawbacks of the existing systems and help in building new systems with the latest technologies that will help society in various ways, proving Society 5.0 as one of the biggest revolutions in this era.

A Step Towards Society 5.0
Research, Innovations, and Developments in Cloud-Based Computing Technologies
Edited by Shahnawaz Khan, Thirunavukkarasu K., Ayman AlDmour, and Salam Salameh Shreem

Computing Technologies and Applications
Paving Path Towards Society 5.0
Edited by Latesh Malik, Sandhya Arora, Urmila Shrawankar, Maya Ingle, and Indu Bhagat

Reinvention of Health Applications with IoT
Challenges and Solutions
Edited by Dr. Ambikapathy, Dr. Shobana, Dr. Logavani, and Dr. Dharmasa

Healthcare and Knowledge Management for Society 5.0
Trends, Issues, and Innovations
Edited by Vineet Kansal, Raju Ranjan, Sapna Sinha, Rajdev Tiwari, and Nilmini Wickramasinghe

Society 5.0 and the Future of Emerging Computational Technologies
Practical Solutions, Examples, and Case Studies
Edited by Neeraj Mohan, Surbhi Gupta, and Chuan-Ming Liu

For more information on this series, please visit: www.routledge.com/Demystifying-Technologies-for-Computational-Excellence-Moving-Towards-Society-5.0/book-series/CRCDTCEMTS

Society 5.0 and the Future of Emerging Computational Technologies

Practical Solutions, Examples, and Case Studies

Edited by
Neeraj Mohan, Surbhi Gupta,
and Chuan-Ming Liu

CRC Press
Taylor & Francis Group
Boca Raton London

CRC Press is an imprint of the
Taylor & Francis Group, an **informa** business

First edition published 2022
by CRC Press
6000 Broken Sound Parkway NW, Suite 300, Boca Raton, FL 33487–2742

and by CRC Press
2 Park Square, Milton Park, Abingdon, Oxon, OX14 4RN

CRC Press is an imprint of Taylor & Francis Group, LLC

Library of Congress Cataloging-in-Publication Data
Names: Mohan, Neeraj, editor. | Gupta, Surbhi, editor. | Liu, Chuan-Ming, editor.
Title: Society 5.0 and the future of emerging computational technologies : practical solutions, examples, and case studies / edtied by Neeraj Mohan, Surbhi Gupta, and Chuan-Ming Liu.
Description: First edition. | Boca Raton : CRC Press, 2022. | Includes bibliographical references and index.
Identifiers: LCCN 2021039112 (print) | LCCN 2021039113 (ebook) | ISBN 9781032026039 (hbk) | ISBN 9781032026053 (pbk) | ISBN 9781003184140 (ebk)
Subjects: LCSH: Industry 4.0. | Society 5.0. | Artificial intelligence—Industrial applications—Case studies.
Classification: LCC T59.6 .S644 2022 (print) | LCC T59.6 (ebook) | DDC 658.4/0380285574—dc23/eng/20211027
LC record available at https://lccn.loc.gov/2021039112
LC ebook record available at https://lccn.loc.gov/2021039113

ISBN: 978-1-032-02603-9 (hbk)
ISBN: 978-1-032-02605-3 (pbk)
ISBN: 978-1-003-18414-0 (ebk)

DOI: 10.1201/9781003184140

Typeset in Times LT Std
by Apex CoVantage, LLC

Contents

Preface

This book provides modern, technological aspects, such as machine learning (ML), artificial intelligence (AI), and Internet of Things (IoT), for the implementation of Society 5.0. The book aims to provide the reader with a firm foundation and understanding of the recent advances in various domains, such as data analytics, neural networks, computer vision, and robotics. The book will provide a state-of-the-art look at these domains as well.

This book will cover various emerging technologies like AI, ML, Data Science, IoT, and big data applied for the realization of Society 5.0. The key feature of this book will be the presentation of practical solutions to existing problems in the fields of health care, banking, infrastructure management, smart cities, and more.

The book may be used by research scholars in the engineering domain who wish to contribute to a modern and secure society. IT professionals may refer to it for detailed examples, which enable them to be more productive in the implementation of ideas. University faculty may propose it to postgraduate students to refer to it for technological advancements.

The book content will be oriented toward Society 5.0 challenges, technology applications, and concrete implementations using ML, AI, big data, IoT, etc. The correlation between these technologies will also be explained by examples and case studies.

While we concentrate on practical applications in the fields of health care, image processing, networking, green cities, security, and others, we do so without glossing over the important formal details and theories necessary for a deeper understanding of the topics. The theoretical part is also important for understanding a topic completely.

Editor Biographies

Neeraj Mohan is an assistant professor in the Computer Science & Engineering Department in I.K. Gujral Punjab Technical University, Kapurthala (Punjab) India. He has a rich and quantitative academic experience of 19 years at various positions. He received his doctoral degree from I.K. Gujral Punjab Technical University, Kapurthala (Punjab) India in 2016. He is an active researcher with more than 50 research papers in reputed journals and conferences. His research interest areas are network traffic management and image processing. He has guided one Ph.D. thesis and 17 M.Tech. theses to date.

Surbhi Gupta holds a B.Tech. degree and Ph.D. from I.K. Gujral Punjab Technical University, Punjab, India. She received a merit for her master's degree at Punjab Agricultural University, Punjab, India. She is presently an assistant professor in the Department of Computer Science at Punjab Agricultural University, Ludhiana, India. She is involved in research on applications of image analysis using machine learning. She has authored over 40 international journal and conference papers and contributed as a reviewer for reputed journals including *Journal of Visual Communication and Image Representation* (Elsevier), *Imaging Science* (Taylor & Francis), and *Journal of Electronic Imaging* (SPIE).

Chuan-Ming Liu is a professor in the Department of Computer Science and Information Engineering (CSIE), National Taipei University of Technology (Taipei Tech), Taiwan, where he was the Department Chair from 2013 to 2017. Dr. Liu received his Ph.D. in Computer Science from Purdue University, West Lafayette, IN, USA, in 2002 and joined the CSIE Department in Taipei Tech in the spring of 2003. In 2010 and 2011, he held visiting appointments with Auburn University, Auburn, AL, USA, and the Beijing Institute of Technology, Beijing, China. He has services in many journals, conferences and societies and has published more than 100 papers in many prestigious journals and international conferences. Dr. Liu was the co-recipient of the ICUFN 2015 Excellent Paper Award, the ICS 2016 Outstanding Paper Award, the MC 2017 Best Poster Award, the WOCC 2018 Best Paper Award, and the MC 2019 Best Poster Award. His current research interests include Big Data management and processing, uncertain data management, data science, spatial data processing, data streams, ad-hoc and sensor networks and location-based services.

Contributors List

Arup Abhinna Achary
KIIT University
Bhubaneswar, India

Brijesh Bakariya
IK Gujral Punjab Technical University
India

Pooja Chadha
Guru Nanak Dev University
Amritsar, India

Ashoka D V
JSS Academy of Technical Education
Bengaluru, India

A. Deepa
Chandigarh Engineering College
Landran, India

Gittaly Dhingra
Thapar University
Patiala, India

Siddhartha Ghosh
Mahatma Gandhi Central University
Bihar, India

B. Gomathi
Hindusthan College of Engineering and Technology
Coimbatore, India

Rajni Goyal
Sh. LBS Arya Mahila College
Barnala, India

Vishal Goyal
Sh. LBS Arya Mahila College
Barnala, India

Surbhi Gupta
Punjab Agriculture University
Ludhiana, India

Neenu Juneja
Chandigarh Group of Colleges
Landran, India

Kamini
Chandigarh Group of Colleges
Landran, India

Sarabjeet Kaur
Chandigarh Group of Colleges
Landran, India

Priyanka Kaushal
Chandigarh Engineering College
Landran, India

Nimranjeet Kour
Guru Nanak Dev University
Amritsar, India

C. Gopala Krishnan
GITAM University
Bengaluru, India

Pavnesh Kumar
Mahatma Gandhi Central University
Bihar, India

Raushan Kumar
Mahatma Gandhi Central University
Bihar, India

Sachin Kumar
University of Jammu
India

Vinod Kumar
IK Gujral Punjab Technical University
India

Yogesh Kumar
Uttarakhand Technical University
Dehradun, India

Vibhakar Mansotra
University of Jammu
India

Neeraj Mohan
IK Gujral Punjab Technical University
Kapurthala, India

Sanjukta Mohanty
KIIT University
Bhubaneswar, India

A.H. Nishan
Francis Xavier Engineering College
Tirunelveli, India

Basamma Umesh Patil
SJB Institute of Technology, BGS
Health and Education City
Bengaluru, India

Ajay Prakash B V
SJB Institute of Technology, BGS
Health and Education City
Bengaluru, India

Chetan R
SJB Institute of Technology, BGS
Health and Education City
Bengaluru, India

Seema Rani
Ambala College of Engineering &
Applied Research
Ambala, India

Laki Sahu
College of Engineering & Technology
Bhubaneswar, India

Geetika Sharma
Guru Nanak Dev University, Amritsar,
India

Khushboo Sharma
Guru Nanak Dev University
Amritsar, India

Prince Sharma
Guru Nanak Dev University
Amritsar, India

Sandeep Sharma
Chang Gung University
Taoyuan, Taiwan

Sourabh Shastri
University of Jammu
India

Chamkaur Singh
Chandigarh Group of Colleges
Landran, India

Gurdit Singh
Punjab Agriculture University
Ludhiana, India

Harpal Singh
Chandigarh Engineering College
Landran, India

Kuljeet Singh
University of Jammu
India

Mandeep Singh
Guru Nanak Dev University
Amritsar, India

Tanpreet Singh
GNDEC
Ludhiana, India

Prasannavenkatesan Theerthagiri
GITAM University
Bengaluru, India

J Uma
Hindusthan College of Engineering and
Technology
Coimbatore, India

Chandan Veer
Mahatma Gandhi Central University
Bihar, India

Shashi Kant Verma
Govind Ballabh Pant Institute of
Engineering and Technology
Pauri Garhwal, India

Amit Wason
Ambala College of Engineering &
Applied Research
Ambala, India

1 Liquefied Petroleum Gas Level Monitoring and Leakage Detection Using Internet of Things

D V Ashoka, Basamma Umesh Patil, Chetan R, and Ajay Prakash B V

CONTENTS

1.1 INTRODUCTION

In 1910, Dr. Walter Snelling invented liquefied petroleum gas (LPG). LPG is produced by the combination of commercial butane and commercial propane. It includes both saturated and unsaturated hydrocarbons. As LPG has a flexible nature, it can be used for a variety of purposes, like domestic fuel, industrial fuel, motor fuel, residential heating, hot water systems and in agriculture. Because of its versatile nature, demand for the gas is increasing day by day. For cooking purposes, LPG is used as a standard fuel in India and other countries for convenience and economic reasons. In 2016, almost 8.9 million tonnes of LPG were utilized for cooking in a period of six months. A recent survey shows that due to the COVID-19 situation in India, LPG consumption increased rapidly in 2019 and 2020. India has become the world's second-largest consumer of LPG.

DOI: 10.1201/9781003184140-1

In 2019, India imported 14.538 million megatonnes of LPG, up 19.22% on the year, as requirement surpassed production. Consumption of LPG in 2019 was 26.944 million megatonnes, up 7.93% on the year, while production over the same year grew at a slower pace of 0.28%, to 12.77 million megatonnes.

Great interest in emerging technology comes from a competitive advantage, a point of view and a point of social need. Emerging technologies are helping to solve the toughest problems of the 21st century, including questions regarding water, health, energy and food environmental concerns. Many people from developed and developing countries alike seek to embrace emerging technologies as a solution to the problem. They fear the continued use of traditional technologies, which often create or contribute to the challenges facing the world today. Emerging technology can be a solution in the field of LPG monitoring.

In India, there are many monitoring systems that check the daily usage of petroleum products, but there is no efficient system that measures the domestic usage of LPG. This has given the motivation of building a system that could monitor the usage of LPG and intimate the user to book a new cylinder when it is about to get empty.

In the existing system, LPG users cannot predict the usage of the cylinder. They get to know that the cylinder is empty only when the flame goes off. The situation becomes difficult if the customer doesn't have an extra cylinder. To measure the natural gas or methane, the volume is calculated in cubic meters or cubic feet. At home, LPG gas is measured in terms of kilograms. The LPG gas in the gas bottles is weighed to check the weight of the gas. While producing or transporting, gas is measured in terms of tonnes, such as 1,000 kg or 2,000 kg, etc. Home LPG can be measured in liters while delivering using the tanker truck. In general, LPG is measured and sold in kilograms. Different sizes of LPG bottles are available, but gas is commonly measured in 45kg bottles. Due to the increase in the number of LPG users, there is an increase in the demand for LPG. Factors such as awareness of the usage of LPG cylinder and the provision for providing subsidies by the governments have increased the demand for LPG in India. This has ceased the usage of wood and its by-products, which, in turn, reduces the greenhouse effect. The low cost of LPG has also led to its adoption. Figure 1.1 shows the LPG coverage from April 2014 to April 2019.

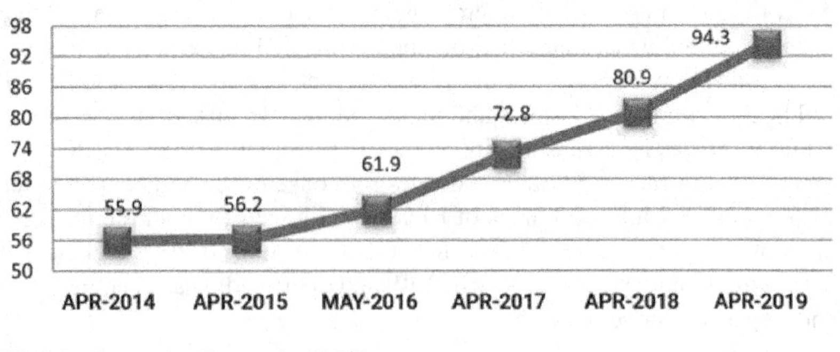

FIGURE 1.1 Increasing Demand of LPG.

One more problem of LPG is supply and demand. Due to the increase in demand, supply also needs to be increased. Many LPG providers are still encountering a delivery backlog to bulk LPG customers. This is due to issues with supply from LPG terminals in the winter and early spring which were then exacerbated by extreme weather conditions experienced across the country.

When the demand for cylinders increases, there will, in turn, be a delay in the supply of cylinders, which is one of the major consumer complaints: that cylinders are not delivered on time. In India, there are systems that monitor the consumption of petroleum products, but there are no proper techniques that monitor the usage of LPG.

In several scenarios, we have seen that leakage of LPG has caused severe threats to the people and also to the society. So, an efficient system is needed to detect the leakage of the gas. Apart from the safety problems, it is very difficult to monitor the remaining quantity of gas in the gas cylinders. Therefore, the system proposes reliable gas-leakage detection and monitoring the amount of gas present in the cylinder.

In the proposed system, when the cylinder is almost empty, an alert message is triggered to the consumer's mobile device via SMS. The undetermined usage of the LPG can be easily monitored so that the problem of supply and demand can be minimized. This brings transparency to the process of monitoring LPG. The proposed system has been developed to create awareness regarding the reduction in the weight of the gas in the gas container. The gas weight reduction is measured using a weight sensor, which is interfaced with Raspberry Pi. The proposed system measures and displays the level of LPG and can be employed for domestic purposes. It is an Internet of Things (IoT)-based system in which a weight sensor is used to monitor the level of LPG. The main objective of the proposed system is to monitor the gas level in the LPG cylinder from the beginning stage. Monitoring is done at the time of delivery from the distributor until the cylinder becomes empty. The weight of the cylinder is compared with certain threshold weights at regular intervals, and when the weight of the cylinder goes below the threshold value, it sends a message to the user to book the delivery of a new cylinder.

1.2 RELATED WORK

A system that performs automatic LPG booking, gas leakage detection and measurement of real-time gas levels in the cylinder, makes use of ATMega16A as a microcontroller, GSM module (SIMCOM-300), a gas sensor (MQ6) to detect leakages in LPG cylinders, a weight sensor to detect the weight of the cylinder, an LCD display to display weight and an exhaust fan. The objective of this work is to detect LPG leakages. It also has an additional feature of automatic gas booking. The system sends respective alert messages to the user (B. D. Jolhe *et. al.*, 2013).

The monitoring and controlling system for LPG leakage detection uses ATMega328p-16bit as a microcontroller, GSM module (SIMCOM-300), a gas sensor (MQ6) to detect leakages in LPG cylinders, a weight sensor to detect cylinder weight, an LCD display to display the weight, a DC motor to control the knob of the cylinder and a level sensor to detect the LPG level in the cylinder. The objective of this system is to detect LPG leakages. It also has an additional feature of automatic gas booking. The system sends respective alert messages to the user by making use of an alarm module (Soundarya and Anchitaalagammai, 2014)

The LPG gas-monitoring system uses a 16F877A microcontroller, GSM module (SIMCOM-300), a gas sensor (MQ6) to detect leakages in LPG cylinders, a weight sensor to detect cylinder weight and an LCD display to display the weight. The work detects LPG leakages. It has an automatic gas booking feature. The system sends respective alert messages to the user by alarm module. The system uses cost-efficient sensors compared to gas detectors (Arun Raj *et. al.*, 2015)

The system with WSN, GSM and embedded web server architecture that monitors internet-based kitchens, uses ARM1176JZF-S as a microprocessor, GSM module, a gas sensor (MQ6) to detect leakages in LPG cylinders, a weight sensor to detect the weight of the cylinder, a temperature sensor (DS1820) to monitor temperature, a light sensor for detecting the presence of light in the kitchen, a motion sensor to monitor movement in the kitchen, a fire sensor to detect explosions and an LCD display to display the weight. The goal of this system is to monitor the entire kitchen using various sensors. The system is implemented with an embedded web server and also a Zigbee network to monitor the system remotely (Sahani *et. al.*, 2015).

A smart IoT system to monitor gas levels, book gas cylinders and detect gas leakages makes use of a microcontroller and ARM processor, RF TX & RX modules to send alert messages using GSM technique, a gas sensor (MQ2) to detect leakages in LPG cylinders, a load cell to detect the weight of the cylinder, a temperature sensor (LM-35) to monitor temperature, a buzzer that is activated during LPG leakages and a siren (60db) that is triggered when there are any changes in any of the sensors. The main purpose of this work is to detect LPG leakages. The system sends respective alert messages to the user by Wi-Fi (Keshamoni and Hemanth, 2017).

A smart LPG monitoring and detection technique with NodeMCU to notifies people about gas leakages using different techniques, such as buzzers, SMS alerts and web servers. The purpose of this work is to monitor gas leakages (Mahfuz *et. al.*, 2020).

An IoT gas management system used for gas cylinders detects gas leakages and sends SMS alerts to the user. It is also capable of booking a cylinder automatically when it is empty (Shrestha *et. al.*, 2019).

An IoT system using an MQ6 sensor detects the level of the gas in the cylinder and automatically books a new cylinder when the LPG level is low (Tamizharasan *et. al.*, 2019). In the LPG gas monitoring and detection system based on microcontroller, the hardware devices, such as exhaust fans and switches, are integrated with the system, which uses an android application for controlling the hardware devices using the ZigBee module. The ZigBee module only works for short ranges (Rahul Kurzekar *et. al.*, 2017).

An industrial plant system (Kodali *et. al.*, 2018) using low-cost IoT for gas-leakage detection used MQ4, MQ6 and MQ135 for detecting different types of gases, such as LPG, benzene and methane. The data collected from the sensors are sent to the cloud, using the ESP 32 Wi-Fi module. A micro electro-mechanical interdigital sensor (Kodali *et. al.*, 2018) is used for detection of gas leakages. The domestic gas leakage detection employs the photolithography technique (Nag *et. al.*, 2016) where interdigital sensors are fabricated using single crystal silicon. The gas leakage detection is performed through electrochemical impedance analysis. LPG monitoring system

(Ali Ahsan *et. al.*, 2020; Infant Augustine *et.al.*, 2016), using IoT along with safety protocols for sensor nodes for different developing countries, has been proposed. The system uses a robust model where PCB is designed based on custom requirements, which facilitates the devices to be adjusted with the cylinder. This system comes with mobile application that is connected to a central server for communication between users and IoT devices.

An automatic cylinder booking method using wireless sensor networks (Rekha and Ashoka, 2019) is an efficient method of LPG leakage detection that continuously checks the quantity of gas available in the cylinder and automatically books a new cylinder by sending an SMS message to the distributor (Bairagi and Saikia, 2020).

LPG leakage can be efficiently monitored by using sensors and microcontrollers that can transmit data via LoRa technology from residential locations to the control center. Gas leakage warnings can avoid greater fire disasters by timely indications to the concerned authorities. This model uses Message Queuing Telemetry Transport (MQTT) protocol for the fast transfer of data (Siregar *et. al.*, 2020). An ARM7-based automated high-performance system (Rajashekar *et. al.*, 2014) was proposed for LPG refill booking and leakage detection that leverages LPC2148 and GSM technologies to book LPG cylinders. It delivers SMS alerts to customers when the gas begins to leak. It is a system based on the IoT (Srivastava *et. al.*, 2019) that analyzes several characteristics of an LPG cylinder and keeps the consumer informed via a mobile application. Depending on the consumer's consumption habit, the algorithm calculates the average number of days for which the LPG cylinder will last. The features of the system include displaying the proportion of LPG consumed, measuring the real-time weight of the cylinder and one-click refill booking through mobile application. The safety element of this home automation system is that when a gas leakage is discovered, the consumer is notified via a mobile application and an in-built buzzer. In this section, different techniques have come across in the area of gas monitoring and gas-leakage detection using IoT devices, microcontroller and NodeNCU. Thus, several research works have been carried out on LPG monitoring and leakage detection. In the next section, we are going to discuss the proposed methodology used for gas measurement using the load cell, MQ6 sensor and Raspberry Pi device for better and accurate gas-level monitoring.

1.3 METHODOLOGY

In this section, the proposed system is discussed for monitoring the gas level using weight techniques for measuring the cylinder and sending the user an SMS for booking a new cylinder with the distributor and using an MQ6 sensor for the detection of gas leakages. The proposed architecture diagram is given in Figure 1.2. The system architecture contains the load cell, Raspberry Pi kit, MQ6 sensor and GSM module. The load cell module is used for measuring cylinder weight. The Raspberry Pi kit is used for remote monitoring. The MQ6 sensor is used for leakage detection. The GSM module is used for sending the SMS to an Android mobile phone. The proposed system constantly observes the level of LPG in the cylinder by making use of the load cell. If the level is less than the threshold, the system informs the user by both SMS and on an LCD display. Thus, the users can have an idea about how long the LPG lasts.

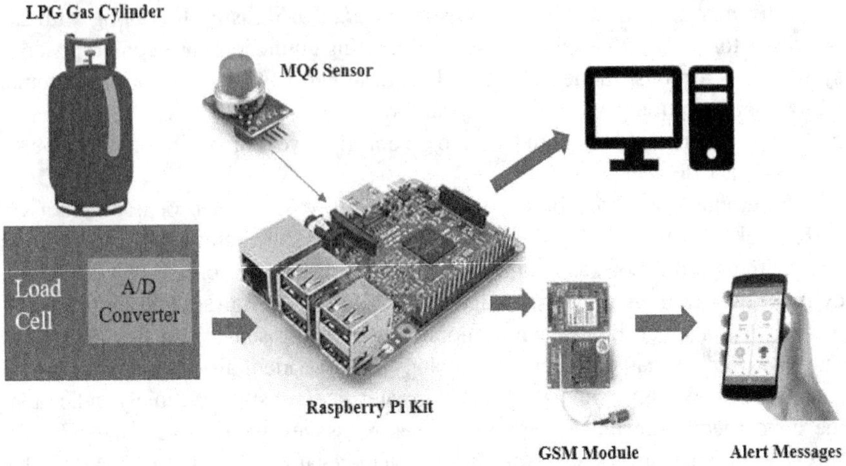

FIGURE 1.2 Proposed Architecture for Gas Monitoring and Leakage Detection Using Raspberry Pi.

The gas cylinder is placed on the load cell to measure the weight of the cylinder continuously. The load cell internally consists of an A/D converter, which converts analog signals to digital form. The load cell is then connected to Raspberry Pi, which is interfaced with a monitor, input devices (keyboard and mouse) and a continuous power supply. The GSM module is connected to Raspberry Pi, which accepts the cylinder weight from the load cell sensor, and the amount of gas present in the cylinder is measured and sent to the user's phone. The MQ6 sensor is used for sensing the gas leakage and sending this leakage information to the GSM module for an SMS alert to the user.

1.3.1 Process Flow Chart

In this section, the process flow chart is discussed for measuring the weight of the cylinder. Figure 1.3 illustrates the flow chart that depicts the working of the system for cylinder weight measurement. The weight of the cylinder is measured at regular intervals, starting from the time of delivery by the distributor.

The value of the cylinder's weight is measured and accepted. This value is compared with the threshold values that are fixed by the programmer according to the respective cylinders.

If the cylinder is missing from its position (i.e., if the cylinder is not placed on the weight sensor), then a message saying "Cylinder misplaced" is displayed. If the cylinder's weight is same as mentioned at the time of delivery during connection, then the statement "Cylinder is full" is displayed. If the cylinder connected is utilized and 50% of the gas is used up, then the statement "Cylinder is half filled" is displayed. If the cylinder connected is utilized 25% and the gas is about to run out then the statement "Cylinder is quarterly filled" is displayed. When the cylinder is completely used up, the load cell shows the weight of the empty cylinder and a "Cylinder is

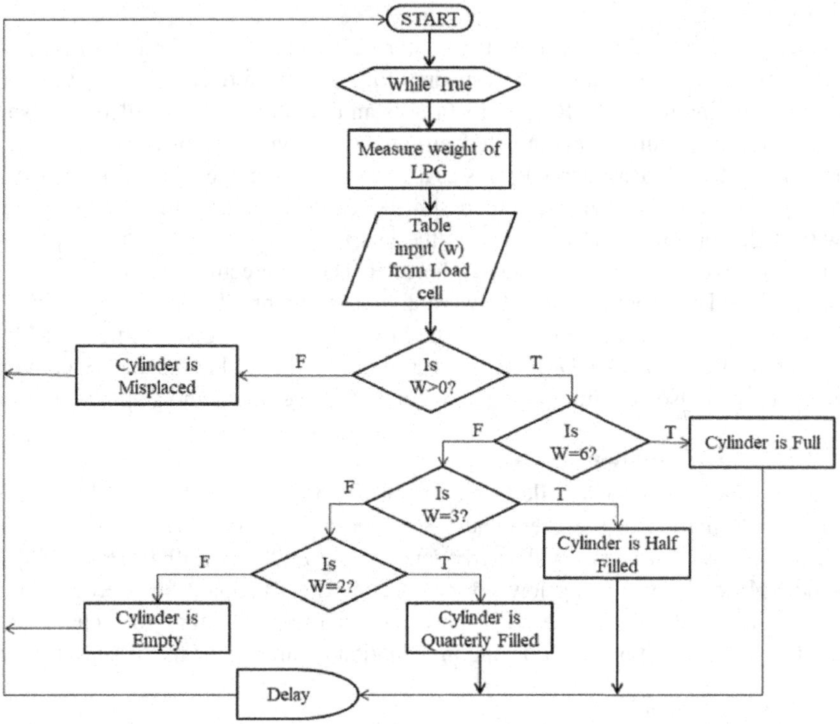

FIGURE 1.3 Flow Chart for Measuring Weight of LPG.

empty" message will be displayed. The value of full, half and quarterly filled is set by the programmer, as per the specification of the cylinder being used.

The prototype that has been used in this work is a water container that weighs around 1kg when it is empty and its fixed maximum threshold is 6kg, the condition checked for half-filled is at threshold 3kg, and the threshold for quarterly filled is 2kg.

The water in the proposed system prototype symbolically depicts the LPG gas in the cylinder. The container depicts the metallic part of the cylinder. In this prototype, 6kg has been considered, which tells us that 1kg is the weight of cylinder and 5kg is the amount of gas present in the cylinder. At half-filled, the weight of the LPG is 2.5~3kg. Therefore, the threshold for half-filled is $3+1=4$kg. If quarterly filled, 1kg is the weight of LPG, and 1kg is the weight of the cylinder. Therefore, the threshold for quarterly filled is $1+1=2$kg. If empty, there is 0kg of LPG and 1kg of cylinder weight. Therefore, the threshold for empty is $0+1=1$kg. If the cylinder is moved from the weighing scale, then no weight will be on the scale and it will display 0kg.

1.3.2 Hardware Description

This section describes Raspberry Pi, the MQ6 sensor, the GSM Module and the load cell.

1.3.2.1 Raspberry Pi

To design a module for LPG remote monitoring, there is a need for a processing unit. After exploring the features of Raspberry Pi, it has been decided to use this board as a processing unit. The Raspberry Pi uses an operating system called as Raspian, which is open-source software. It facilitates the interfacing of sensors, the GSM module and so on. Raspberry Pi acts as a portable system that can be easily operated using a keyboard and mouse. Figure 1.4 depicts a Raspberry Pi–embedded board with different units. The processor is the heart of Raspberry Pi. This chip is a 32 bit, 700 MHz system and is assembled on ARM11 architecture. It does not have any hard drive. The programs and software are stored on an SD card with a minimum 8GB capacity. A USB port may be of the Model B with two USB 2.0 ports or Model A with a single port. A DDMI port is responsible for digital audio and video output. Raspberry Pi also contains analog audio output, power input and a number of pins.

1.3.2.2 GSM Module

The Global System for Mobile Communications (GSM) module is used to communicate information to the user. The GSM module used in the proposed system is GSM800C, which works with a frequency of 850/900/1800/1900MHz. GSM can send SMS and voice using low power consumption. The key feature of the GSM module is the SIM card. The SIM (subscriber identity module) is a detachable smart card that contains a phone book and information regarding the user's subscription.

1.3.2.3 Load Cell

The main feature of any weighing method is the load cell. Load cells are not exciting to watch, but they are extremely precise transducers that can give the user

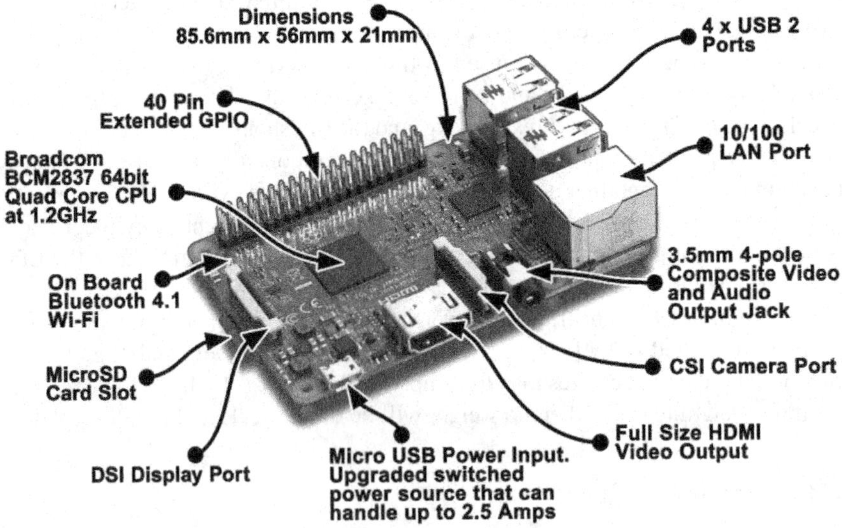

FIGURE 1.4 Raspberry Pi–Embedded Board.

information that is not generally obtainable by other technology due to some commercial factors.

1.3.2.4 MQ6 Sensor

The MQ6 gas sensor is responsible for detecting gases such as LPG, butane and carbon monoxide. This sensor comes with a digital pin so it can operate even without a microcontroller. It should be fixed 1–2 meters away from the cylinder.

1.4 IMPLEMENTATION

The system is partitioned into two phases, such as the measuring phase and the GSM phase. Figure 1.5 depicts the measuring phase of the proposed system. The weighing scale consists of a load cell and an A/D converter. The load cell is an op-Amp in which the positive voltage sent into an op-Amp produces a positive voltage output (i.e., positive weight), and if a negative voltage is sent to the op-Amp, it produces a negative output (i.e., negative weight).

As shown in Figure 1.6, the A/D converter converts the analog signal into digital form and displays the digital value on the LCD display of the weighing scale. The various steps are described next.

The output is taken from the weighing scale through the RS232 serial port. The serial port output is then sent to the converter chip, in which the value is changed to the form as accepted by Raspberry Pi. The output is then sent to the Raspberry Pi. In Raspberry Pi, there is one UART (universal asynchronous receiver and transmitter) port, to which the output of the converter is connected. The UART pins are TXD0 (pin 14), which is a transmitter, and RXD0 (pin 15), which is a receiver. The input from the UART receiver is taken and displayed on the monitor. The values are compared with the threshold values (i.e., fixed by the programmer at the beginning) and, in fulfillment of certain conditions, some statements are exclusively displayed. For example, if the cylinder is missing from its position (i.e., if the cylinder is not placed on the weight sensor), then the message "Cylinder misplaced" is displayed. The value

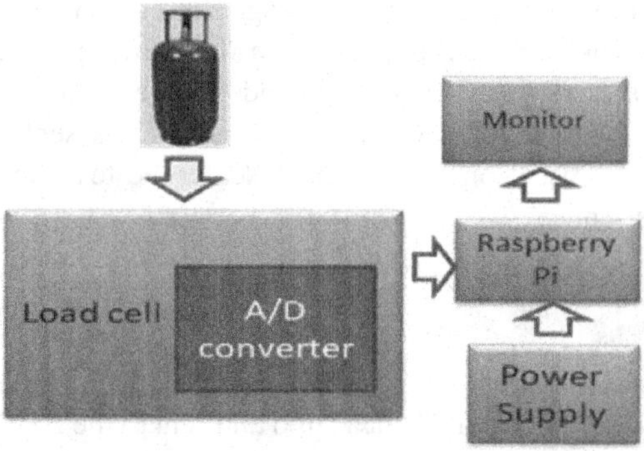

FIGURE 1.5 Phase 1: Measuring Phase of the System.

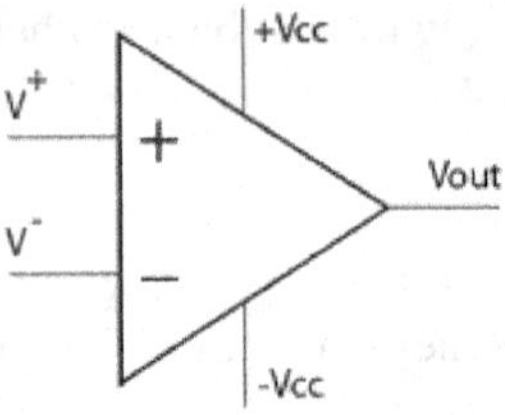

FIGURE 1.6 Op-Amp for Converting Analog Signals into Digital Form.

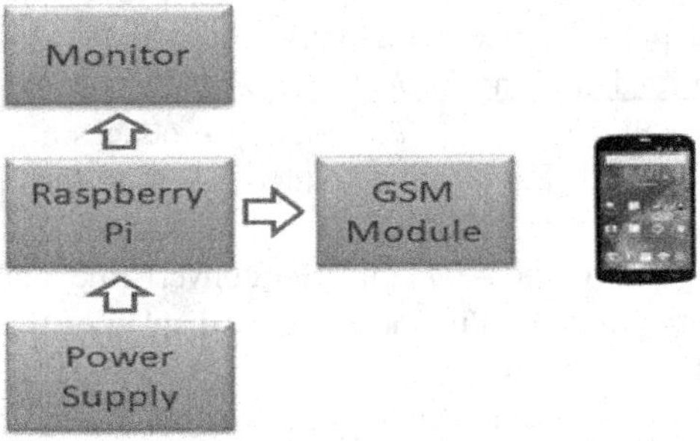

FIGURE 1.7 Phase 2: GSM Phase.

of full, half and quarterly filled is set by the programmer as per the specification of the cylinder being used.

The GSM phase is depicted in Figure 1.7. The Raspberry Pi has only one UART port, which has already been used for connecting with the weighing scale. Hence, an external UART port is connected.

The external UART that is connected is a USB-TTL board. The TTL board acts as a communication link between the Raspberry Pi and GSM module. The GSM module is powered from Raspberry Pi pin 2, which gives a power supply of 5V. The GSM module sends SMS for certain conditions to a particular user's mobile. When the program is executed, a text message saying "SYSTEM START" is sent. When the cylinder is full (i.e., as printed on the cylinder), a text message saying "CYLINDER FULL" is sent. When the cylinder content is reduced to 25%, the text message "CYLINDER IS GOING TO BE EMPTY" is sent. When the cylinder is relocated or not present on the weighing scale, the text message "CYLINDER MISPLACED" is sent.

1.5 RESULTS

Figure 1.8 illustrates the proposed system to monitor LPG. The cylinder is placed on the weighing device. The weight is measured and sent to the Raspberry Pi. The water in the proposed system's prototype symbolically depicts the LPG gas in the cylinder. The container depicts the metallic part of the cylinder.

FIGURE 1.8 The Set-Up of the System.

Note: The weight of the cylinder is weighed at regular intervals and the weight is displayed. In this prototype, the container filled with water is measured every two minutes.

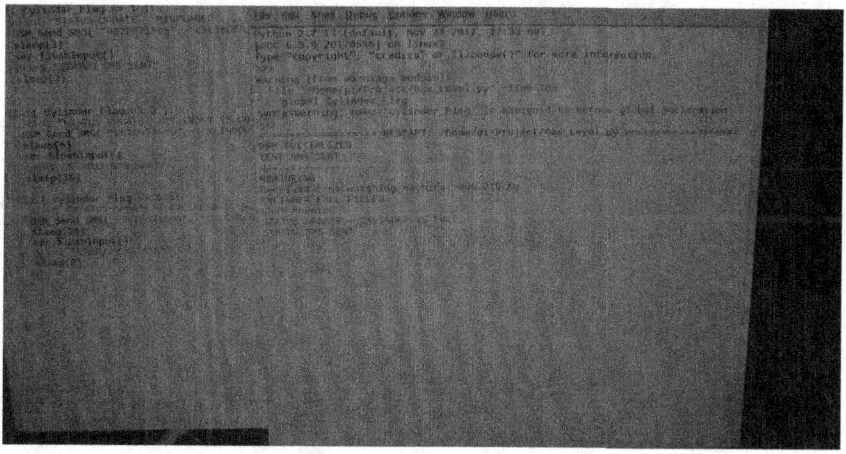

FIGURE 1.9 Weight Measurement of the Cylinder When It Is Full.

Figures 1.9 and 1.10 show the weight measurement of the cylinder when it is full, half-filled, quarterly filled and when the cylinder's weight is the same as mentioned at the time of delivery. In our prototype, we have taken 6kg, which tells that 1kg is the weight of the cylinder and 5kg is the amount of gas present in the cylinder.

The system displays appropriate messages based on the cylinder weights, as discussed in the flowchart shown in Figure 1.3. When cylinder is missing from its position (i.e., if the cylinder is not placed on the weight sensor), then a message status "Cylinder is misplaced" will be displayed. If the cylinder is relocated or removed from the weighing scale, there is no weight on the scale and it shows as 0kg.

Figure 1.11 shows a screenshot from a digital device. The system is designed to send an alert via SMS to a predefined mobile number during certain conditions. The

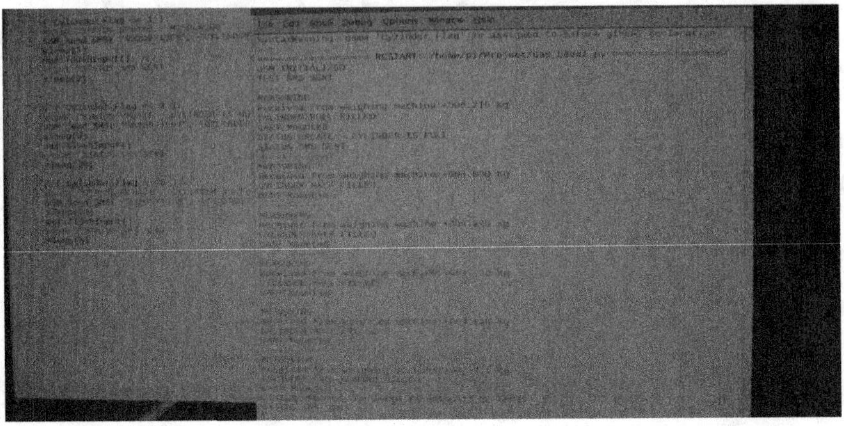

FIGURE 1.10 Weight Measurement of the Cylinder When It Is Half-Filled, Quarterly Filled and Empty.

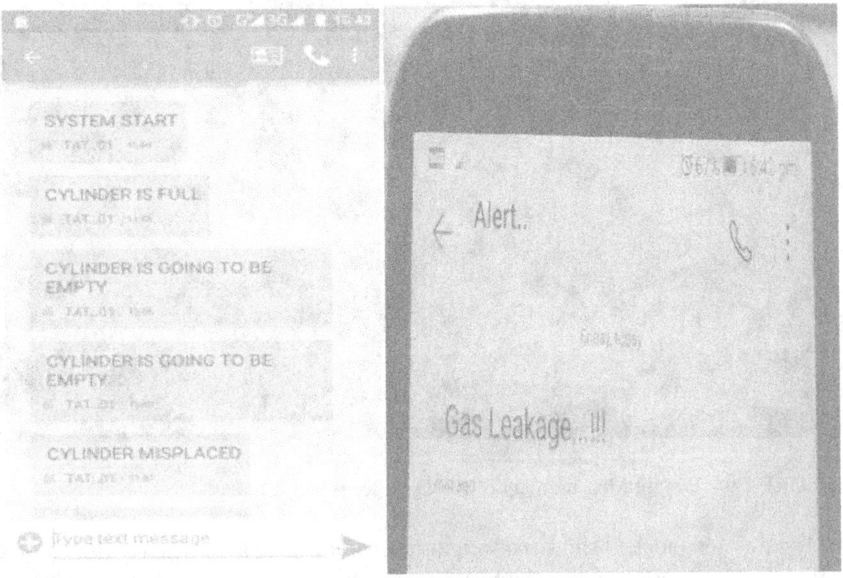

FIGURE 1.11 Mobile Screenshot for the LPG Monitoring Alerts and Gas Leakage Alert Message.

first SMS "system start" is sent when the program is executed. When a cylinder is placed on the weighing scale, if the weight is as predefined by the agency, then the message "Cylinder is full" is sent. When the cylinder is left with only 25% of its contents and is close to being empty, then the message "Cylinder is going to be empty" is sent. When the cylinder is misplaced or removed, then the message "Cylinder

TABLE 1.1
Comparison of Proposed and Existing Techniques

Sl No	Proposed Technique	Existing Technique
1	Raspberry Pi kit has clock speed of 1.2GHz.	Arduino kit has clock speed of 16Mhz.
2	Raspberry Pi kit can be connected to more sensors.	Arduino kit can connect to fewer sensors.
3	Python IDLE, Eclipse IDLE and others can be used for programming.	Arduino IDLE should be used for programming.
4	MQ6 Gas Sensor- Detect 200–10000ppm Concentration of gas.	MQ5 Gas Sensor- Detect 200–5000ppm Concentration of methane gas.

misplaced" is sent. The figure also shows the SMS warning message for the detection of an LPG leakage.

The comparison of the proposed methodology and the existing one show that the proposed technique is safer and the response time is less, compared to existing approaches. Table 1.1 shows a comparison of the proposed technique and existing techniques.

The advantage of the automated system for gas detection is its fast latency and correct detection of weights for booking the cylinder compared to manual and other existing approaches.

1.6 CONCLUSION & FUTURE WORK

The interfacing of sensors with Raspberry Pi and its configuration detects the weight of the cylinder. The interfacing of Raspberry Pi and the GSM module is to send messages to relevant users depending on the comparison of cylinder weights with the threshold weights. In the proposed system, a prototype module is developed for demonstration purposes. An MQ6 sensor is used to detect the leakage of LPG. A limitation of the system is that the gas cylinder must always be kept on the load cell for continuous monitoring of the gas cylinder, otherwise, whenever the user wants to measure the gas level in the cylinder, he has to place it over the load cell device. In the future, the work can be extended to the product level. To implement this work in a user-friendly and robust way, compacted and cost-effective systems should be implemented. The size of the system can be reduced by embedding most of the units on a single board with some change in the technology. The proposed work can be extended to further uses, such as automated gas refill booking and the entire kitchen monitoring using different sensors interfaced with Raspberry Pi. The work can also be extended using cloud-computing techniques through which we can monitor the usage of the cylinder on a daily basis by developing a user interface.

REFERENCES

Arun Raj, Athira Viswanthan, Atul, T. S. 2015. LPG Gas Monitoring System. *International Journal of Innovative Technology and Research (IJITR)*. Vol. 3, No. 2, pp. 1957–1960.

Ali, Ahsan, Lslam, M. Z., Siddiqua, R., Rhaman, M. K. 2020. *An IoT Based Interactive LPG Cylinder Monitoring System with Sensor Node Based Safety Protocol for Developing Countries*. IEEE Region 10 Symposium (TENSYMP). Dhaka, Bangladesh, pp. 398–401.

Bairagi, P. P., Saikia, L. P. 2020. *Development of a LPG Monitoring and Automatic Cylinder Booking System Based on Wireless Sensor Network*. 4th International Conference on Inventive Systems and Control (ICISC). Coimbatore, India, pp. 382–386.

Infant Augustine, A., Subramaniam, Kamalraj. 2016. LPG Consumption Monitoring and Booking System. *International Journal of Applied Engineering Research*. Vol. 11, pp. 3993–3998.

Jolhe, B. D., Potdukhe, P. A., Gawai, N. S. 2013. Automatic LPG Booking, Leakage Detection and Real Time Gas Measurement Monitoring System. *International Journal of Engineering Research & Technology (IJERT)*. Vol. 2, No. 4, pp. 1192–1195.

Keshamoni, K., Hemanth, S. 2017. *Smart Gas Level Monitoring, Booking & Gas Leakage Detector over IoT*. 2017 IEEE 7th International Advance Computing Conference (IACC). Hyderabad, pp. 330–332.

Kodali, R. K., Greeshma, R. N. V., Nimmanapalli, K. P., Borra, Y. K. Y. 2018. *IOT Based Industrial Plant Safety Gas Leakage Detection System*. 4th International Conference on Computing Communication and Automation (ICCCA). Greater Noida, India, pp. 1–5.

Mahfuz, N., Karmokar, S., Rana, M. I. H. 2020. *A Smart Approach of LPG Monitoring and Detection System Using IoT*. IEEE 2020 11th International Conference on Computing Communication and Networking Technologies (ICCCNT). Kharagpur, India, pp. 1–4.

Nag, A., Zia, A. I., Li, X., Mukhopadhyay, S. C., Kosel, J. 2016. Novel Sensing Approach for LPG Leakage Detection: Part I—Operating Mechanism and Preliminary Results. *IEEE Sensors Journal*. Vol. 16, No. 4, pp. 996–1003.

Rajashekar, P., Shyamaladevi, S., Rajaramya, V. G., Sebastin Ashok, P. 2014. Arm7, Based Automated High Performance System for LPG Refill Booking & Leakage Detection. *International Journal of Engineering Research and Science and Technology (IJERST)*. Vol. 3, No. 2, pp. 305–309.

Rahul, Kurzekar, Arora, H., Shrestha, R. 2017. *Embedded Hardware Prototype for Gas Detection and Monitoring System in Android Mobile Platform*. IEEE International Symposium on Nanoelectronic and Information Systems (iNIS). Bhopal, pp. 6–10.

Rekha, B., Ashoka, D. V. 2019. SRHM: Sustainable Routing for Heterogeneous Adhoc Environment in IoT-Based Mobile Communication. *Compusoft*. Vol. 8. pp. 3003–3010.

Sahani, M., Nayak, A., Agrawal, R., Sahu, D. 2015. *A GSM, WSN and Embedded Web Server Architecture for Internet Based Kitchen Monitoring System*. International Conference on Circuits, Power and Computing Technologies [ICCPCT-2015]. Nagercoil, pp. 1–6.

Shrestha, S., Anne, V. P. K., Chaitanya, R. 2019. *IOT Based Smart Gas Management System*. 3rd International Conference on Trends in Electronics and Informatics (ICOEI). Tirunelveli, India, pp. 550–555.

Siregar, B., Pratama, H. F., Jaya, I. 2020. *LPG Leak Detection System Using MQTT Protocol on LoRa Communication Module*. 4th International Conference on Electrical, Telecommunication and Computer Engineering (ELTICOM). Medan, Indonesia, pp. 215–218.

Srivastava, A. K., Thakur, S., Kumar, A., Raj, A. 2019. *IoT Based LPG Cylinder Monitoring System*. IEEE International Symposium on Smart Electronic Systems (iSES) (Formerly iNiS). Rourkela, India, pp. 268–271.

Soundarya, T., Anchitaalagammai, J. V. 2014. Control and Monitoring System for Liquefied Petroleum Gas (LPG) Detection and Prevention. *International Journal of Innovative Research in Science, Engineering and Technology*. Vol. 3, No. 3, pp. 696–700.

Tamizharasan, V., Ravichandran, T., Sowndariya, M., Sandeep, R., Saravanavel, K. 2019. *Gas Level Detection and Automatic Booking Using IoT*. 5th International Conference on Advanced Computing & Communication Systems (ICACCS). Coimbatore, India, pp. 922–925.

2 An In-Depth Analysis of Convolutional Neural Networks for Agricultural Purposes

Dr. Gurdit Singh and Er. Tanpreet Singh

CONTENTS

In computer vision, convolutional neural networks (CNNs), influenced by neuroscience, have gained recognition and success. The approach is one of the most powerful techniques for data processing. In reality, artificial neural networks (ANNs) inspired the CNNs, where the entire process is performed on multiple hidden layers: the convolutional layer, the non-linearity layer, the rectification layer, rectified linear units, the pooling layer, and the fully connected layer. A simplified architecture, such as visual detection, computer vision, and natural-language processing, addresses the

DOI: 10.1201/9781003184140-2

complex machine-learning problems with greater accuracy and prediction. Another field of CNNs is that of solving complicated agricultural issues. The key challenges are identifying crops, diseases, seeds, weeds, and pests. The problems have been solved with high precision and prediction through CNNs. In agriculture, CNNs have also been used to forecast environmental conditions, such as temperature, humidity, water usage, and soil conditions. Also, the calculation of vegetative plant and reproductive development is another functional area of CNN. In this article, we attempt to illustrate deep neural networks' usage in agriculture by describing CNNs' operations and layers. We have also established the benefits and drawbacks of utilizing CNN in agriculture and eventually concluding the future of deep learning in agriculture.

2.1 INTRODUCTION

Smart farming (Tyagi 2016) is crucial for addressing agricultural production's various challenges, including productivity, environmental impact, food protection, and sustainability (Gebbers and Adamchuk 2010). With the abrupt rise in world population (Kitzes et al. 2008), it is possible to see a major increase in food demand (FAO 2009). This is accompanied by the restoration of natural environments through the use of sustainable agricultural practices. When food is transported across the globe, it must hold a strong nutritional value (Carvalho 2006) because there are many issues associated with food from preparation to distribution, such as ambient temperature, humidity, weather factors, and delivery length. These issues can be resolved as a result of previous experience with dynamic, multivariate, and volatile agricultural environments. Via background, circumstance, and position understanding, ICT (information and communications technology) for large-scale environmental observation will facilitate and enhance governance and decision-/policymaking. ICT technologies, such as remote sensing (Bastiaanssen, Molden, and Makin 2000), Internet of Things (IoT) (Weber and Weber 2010), big data analytics (Chi et al. 2016; Kamilaris, Kartakoullis, and Prenafeta-Boldu´ 2017), and cloud computing (Hashem et al. 2015), are critical for understanding diverse agricultural ecosystems.

Large-scale photos of the rural environment are taken using satellites, remote-controlled rockets, aircraft, and drones. Remote sensing data may be collected on a routine basis through vast geographic areas, including those unavailable to human inquiry. The IoT collects and measures different parameters in a specific area using advanced sensor technologies. Finally, the combination of big data analytics and cloud computing allows real-time, large-scale exploration of cloud data (Waga 2013; Kamilaris et al. 2016). Via a greater understanding of climatic conditions and forecasts, we are able to establish cutting-edge software and services capable of agricultural production and food demand. Remote sensing and IoT record a plethora of documents, including tens of thousands of images of agricultural fields (covering a sizeable geographical area).

Furthermore, it will aid in the resolution of a number of issues that occur in agricultural fields (Liaghat, Balasundram et al. 2010; Ozdogan et al. 2010). Image processing is a significant field of study in the agriculture sector. Intelligent analysis methods aid us in image recognition, anomaly detection, and other tasks in a range of agricultural applications (Teke et al. 2013; Saxena and Armstrong 2014).

However, satellite imaging, which uses multi- or hyper-spectral scanning, remains the most commonly employed sensing tool. These use anti-radar, thermal, and near-infrared cameras to track and insert tiny particles into canned goods (Ishimwe, Abutaleb, Ahmed et al. 2014). Image-processing methods include machine learning; wavelet-based scanning; vegetation indices, such as the NDVI; and regression analysis (Saxena and Armstrong 2014). Aside from the techniques previously mentioned, deep learning (LeCun, Bengio, and Hinton 2015) is a relatively new technology that is gaining popularity through the different contexts in which it has been implemented (Wan et al. 2014; Najafabadi et al. 2015). It is a branch of artificial intelligence that is somewhat similar to ANNs (Schmidhuber 2015). A deep-learning process facilitates data representation by reorganizing the neural network's emphasis by hierarchical encoding.

The research investigates the issues associated with CNNs, also known as deep feed ANNs, a subset of deep learning (DL). CNNs are an extension of the classical ANN that adds depth to the network, allowing for hierarchical data representation (Schmidhuber 2015; LeCun, Bengio et al. 1995). They have been successfully incorporated into a number of visual imagery-related contexts (Szegedy et al. 2015). Since CNN is now the most widely used agriculture science technique, image-processing problems are intertwined. However, there are a few broad surveys available (Deng and Yu 2014; Wan et al. 2014; Najafabadi et al. 2015) that present and evaluate alternative fields of study and applications (Kamilaris and Prenafeta-Boldu´ 2018). The current study aims to establish CNN as a promising and high-potential approach for addressing a wide range of agricultural challenges.

The remaining part of the chapter is structured in the following way: In Section 2.2, we present the role of CNN and deep learning in agriculture. The functional architecture of CNN is thoroughly clarified in Section 2.3, including its current developments. Then, we describe the implementations of CNN in agriculture, including the various areas, different data sources, and the output indicators in Section 2.4. Section 2.5 deals with the benefits and drawbacks of using CNN in the agricultural sector. Section 2.6 discusses the futuristic dimensions of the usage of deep learning in agriculture. The conclusion is then provided in Section 2.7.

2.2 DEEP LEARNING

By translating the data and viewing it hierarchically, deep learning brings depth to traditional machine learning. Automatic task extraction, also known as feature learning, is a direct benefit of DL. Due to the more dynamic models, deep learning can solve more complicated problems more efficiently than other methods. Such models may be used to solve classification problems as long as there are ample big databases to include sufficient information about the situation. DL is comprised of many modules that differ in size and function depending on the network design being used. The layers in deep-learning models are hierarchical and perform exceptionally well at classification and estimation, which enables them to be scalable and adaptable to a broad variety of complex problems. When applied to media, DL is capable of working with a wide range of data forms, including film, photos, and audio. Additionally, natural language, population data, and more are explanations of how to use DL.

The removal of the need for feature engineering (FE) is one of the most significant benefits of using DL in image processing. Traditionally, picture classification practices relied heavily on hand-engineered features, the efficacy of which had a significant influence on the final outcome. FE is a time-consuming, dynamic framework that must be adjusted as the meaning or dataset changes. Thus, FE is a costly endeavor that is reliant on the expertise of specialists and does not properly generalize (Amara, Bouaziz, and Algergawy 2017). On the other hand, DL does not employ FE, preferring to discover significant features by preparation.

The downside of DL is that it usually needs more preparation time. The training time, on the other hand, is usually shorter than for other machine learning–based approaches (Chen et al. 2014). Other disadvantages include concerns that may occur as pre-trained models are used on limited or substantially different datasets, optimization issues caused by the model's sophistication, and hardware constraints.

In Section 2.5, we address the benefits and drawbacks of DL as described in the surveyed publications.

2.2.1 Role of Computer Vision in Agriculture

According to the authors' output measures, the latest study indicates that the bulk of related work exists. CNN consistently outperforms GMM (Gaussian mixture modeling) in terms of precision (Reyes, Caicedo, and Camargo 2015; Santoni et al. 2015). Although the sample size in this study is small, satisfactory precision has been observed in the majority of agricultural challenges, especially when compared to other techniques used to solve the same problems. Specifically, the areas of diagnosing plant and leaf diseases, tree detection, ground-cover classification, fruit counting, and weed identification also apply to the categories with the highest degree of precision. CNN-based models were trained using sensory data obtained from the field (Kuwata and Shibasaki 2015) and a mixture of static and dynamic environmental variables (Song et al. 2016). In both cases, the performance (i.e., root mean square error) was greater than that of the other methods under consideration. The same laboratory conditions are used to compare research using CNN, including datasets and performance metrics, model parameters, precision, and quality. Of the 23 articles, 15 (65%) made explicit, right, and reliable comparisons between CNN and other commonly employed strategies studied in related work.

2.2.2 The Importance of Deep Learning in Agriculture

Deep analysis is a new agricultural technique that has already been extended successfully to other disciplines. Deep learning in agriculture is to use more complex algorithms to detect and predict the output with more accuracy. There are several complex algorithms, like AlexNet, Caffe, Theano, and various others, that can find the minute details with better accuracy than traditional CNN. The difference between CNN and DL is that DL follows a more complex structure where the number of hidden layers is increased compared to CNN. Various methods used for detecting diseases are elaborated upon in Table 2.1.

TABLE 2.1
Agriculture-Related Uses for Deep Learning.

Ref.	Agricultural Area	Dataset Used	Deep Learning Models
(Sladojevic et al. 2016)	Leaf disease detection	Database containing 4,483 images	CaffeNet
(Mohanty, Hughes, and Salath´e 2016)	Plant disease detection	PlantVillage dataset of 54,306 images	AlexNet, GoogleNet
(Amara, Bouaziz, and Algergawy 2017)	Plant disease detection	Dataset of 3,700 images of banana diseases	LeNet
(Chen et al. 2014)	Land-cover Classification	A natural transition region with a dense vegetation site on the KSC and an agricultural site in Pavia	Hybrid of PCA, Autoencoder and logistic regression
(Luus et al. 2015)	Land-cover Classification	UC Merced land-use dataset	Author-defined
(Lu et al. 2017)	Land-cover Classification	Aerial images from drone of agriculture area	Author-defined
(Kussul et al. 2017)	Crop type Classification	Landset-8 and Sentinel-1A RS satellite images	-
(Reyes, Caicedo, and Camargo 2015)	Plant recognition	LifeCLEF 2015 plant dataset	AlexNet
(Reyes, Caicedo, and Camargo 2015)	Plant recognition	MalayaKew (MK) Leaf dataset	AlexNet
(Grinblat et al. 2016)	Plant recognition	INTA Argentina produced a total of 866 leaf photographs	Author-defined
(Douarre et al. 2016)	Soil and root Segmentation	X-ray tomography of soil	Author's created CNN with SVM for classification
(Kuwata and Shibasaki 2015)	Crop yield estimation	Maize yields from 2001 to 2010	Author-defined
(Rahnemoonfar and Sheppard 2017)	Fruit counting	Author-created database of 24,000 synthetic images	Inception-ResNet
(Chen et al. 2017)	Fruit counting	Orange and apple images	CNN and linear regression
(Bargoti and Underwood 2017)	Fruit counting	Apples, almonds, and mangoes	Authors created faster region-based CNN with VGG16 model
(Sa et al. 2016)	Fruit counting	122 pictures in RGB and near-infrared (NIR)	A more efficient CNN with a region-dependent approach based on the VGG16
(Christiansen et al. 2016)	Obstacle detection	A total of 437 images	AlexNet
(Christiansen et al. 2016)	Obstacle detection	48 photos were used as background data and 48 images were used	AlexNet and VGG
(Xinshao and Cheng 2015)	Crop identification	3,980 photographs featuring 91 different forms of weed seeds	PCANet + LMC classifiers

(Continued)

TABLE 2.1
(Continued)

Ref.	Agricultural Area	Dataset Used	Deep Learning Models
(Dyrmann, Karstoft, and Midtiby 2016)	Identification of weeds	BBCH 12–16: 10,413 photographs	Variation of the VGG16
(Sørensen et al. 2017)	Identification of weeds	A total of 4500 images	DenseNet
(Song et al. 2016)	Soil moisture content prediction	Soil data collected from an irrigated corn field	Macroscopic cellular automata focused on deep belief networks (DBN-MCA)
(Santoni et al. 2015)	Cattle race Classification	1,300 images were collected	CNN Matrix with Grey Level Co-Occurrences (GLCM-CNN)

2.3 WORKING OF CNN/DEEP NEURAL NETWORK

CNN usually performs better when big, meaning when they have broader, deeply intertwined layers of neural networks (Andri et al. 2018). The biggest downside to these architectures is the expense of computing. Thus, high CNNs are usually impractically low, particularly for built-in IoT devices (Canziani, Paszke, and Culurciello 2016). The latest research efforts focus on reducing the expense of computing deep-learning networks for daily usage while preserving the quality of predictions (Hasanpour et al. 2018). To explain the implementation of state-of-the-art CNN architectures in agricultural systems, we checked the related literature's precision and computational criteria, including recent network updates, as seen in Figure 2.1. Classical state-of-the-art deep-network architectures include: LeNet (LeCun et al. 1998), AlexNet (Krizhevsky, Sutskever, and Hinton 2017), NIN (Lin, Chen, and Yan 2013), ENet (Paszke et al. 2016), ZFNet (Zeiler and Fergus 2014), GoogleLeNet (Szegedy et al. 2015), and VGG 16 (Simonyan and Zisserman 2014). Modern architectures include: Inception (Szegedy et al. 2016), ResNet (He et al. 2016), and DenseNet (Huang et al. 2017). ANNs perform very well in the machine-learning field. They have been used to classify various things like image, video, audio, and text. Each area of classification requires a different type of neural network. For instance, when predicting the sequence of words, we may need RCNN or LSTM. For image classification, we require CNNs. The CNNs mostly have three layers, namely the input layer, the hidden layer, and the output layer.

An image is fed to the neural network for processing and classification. The output from each layer in the CNN is called forward propagation. When the outcome is provided again to the network from backwards, the process is called backward propagation. During the backward propagation, error and loss have been computed to reduce their rate in order to produce accurate results. The overall architecture of the CNN can be described in Figure 2.1.

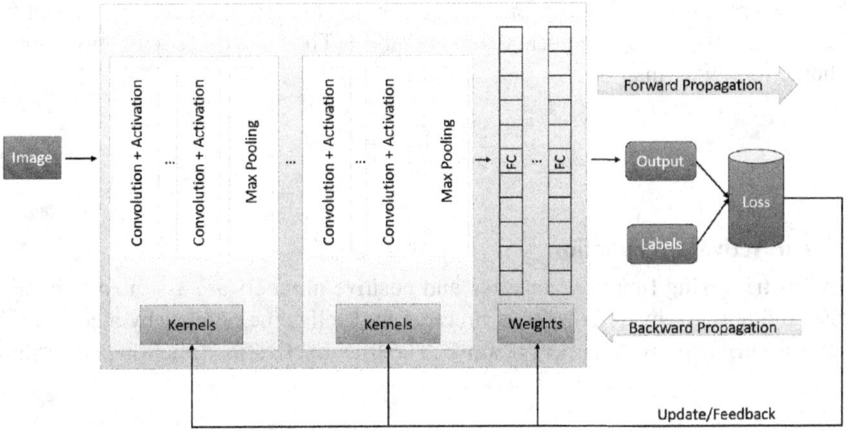

FIGURE 2.1 Convolutional Neural Network Architecture.

Taking an example of a color image, it has three channels: RGB (red, green, blue), each channel varies from 0 to 255. The smaller the value, the darker the color. For example, if RGB is (0, 0, 0), the color will be black, and when RGB is (255, 255, 255), it will be white. In a color image, each pixel has three channels, and we denote those values in a matrix form of RGB, respectively. Those matrices are ready to pass to the CNN layers, which are as follows:

2.3.1 Convolutional Layers

A convolutional layer in CNN takes the collection of learnable filters. These filters are used to identify low-level features or trends in the background. Consequently, the matrix on this layer can be expressed as (NxNx3), and the filter matrix can be defined as (MxMx3).

This filter is then convoluted (slid) over the whole picture to get an activation map. A new selection of filters creates a different set of features maps, and they're ready to move to the next layer of CNN.

2.3.2 Non-Linearity Layer

The non-linearity layer of a CNN is made up of an activation function that takes the feature map given by the previous convolutional layer and outputs the activation map. The activation function is a volumetric processing of the input volume on an element-by-element basis. The dimensions of the input and output are equivalent. In a CNN, many different types of activation functions are used.

Binary Step Function

In this activation function, the input values are standardized between the values 0 and 1. In Binary Step Function, where the input is below zero, the value will be set to 0, whereas if the input is greater than or equal to zero, the value will be set to 1. In this way, the values are normalized to 0 and 1. The binary step activation function should look like this

$$x = \begin{Bmatrix} 1; x >= 0 \\ 0; x < 0 \end{Bmatrix} \tag{2.1}$$

Linear Activation Function

In this triggering function, negative and positive numbers are assumed to be negative and positive by default. This function multiplies the number by a constant but doesn't turn negative numbers positive. The linear activation function can be shown as:

$$f(x) = ax \tag{2.2}$$

Non-linear Activation Function

Non-linear activation functions are those that are capable of understanding more complicated behaviors, including the capacity to understand images, video, and audio. For certain non-linear simulations, the dataset is of high dimensionality. The non-linear activation function contain the following:

Sigmoid/Logistic

Sigmoid is the most widely used activation mechanism of neural networks. The function's scale is from 0 to 1. If the value occurs on the opposite side of the X axis, 0 is allocated. If the value falls on the positive side, 1 is assigned. This is also regarded as a logit function. The sigmoid function simply computes values from 0 to 1. The sigmoid or logistic activation function can be shown as:

$$f(x) = \frac{1}{1 + e^{-x}} \tag{2.3}$$

TanH/Hyperbolic Tangent

Tanh is a feature that is identical to the sigmoid feature. The only distinction is that it is symmetrical around the root. In this case, the possible values vary from -1 to 1. The tanH function can be shown as:

$$tanh(x) = \frac{2}{1 + e^{-2x}} - 1 \tag{2.4}$$

ReLU (Rectified Linear Unit)

Because of its popularity in the neural network world, the ReLU activation function has gained notoriety. The name "Relu" stands for "rectified linear unit." Since the ReLU mechanism does not induce all neurons at the same moment, individual neurons may be activated more rapidly. As a result, neurons are only deactivated when the linear transformation contribution is less than zero. The ReLU function can be shown as:

$$f(x) = max(0, x) \tag{2.5}$$

Leaky ReLU

A leaky ReLU functionality is a more advanced form of ReLU. As we can see, the ReLU function suppresses neurons below a certain threshold but activates neurons above that threshold. The aim of leaky ReLU is to resolve this issue. We classify ReLU as a minor linear variable of x, rather than as 0 for negative x values. The leaky ReLU function can be shown as:

$$f(x) = \begin{Bmatrix} 0.01x; x < 0 \\ x; x >= 0 \end{Bmatrix} \tag{2.6}$$

Parametric ReLU

Parametric ReLU is different from the ReLU feature intended to deal with the issue of gradients being negative in the left half of the axis. The parameterized ReLU adds a new parameter to the derivative to generate negative values. The parameterized ReLU function can be shown as:

$$f(x) = \begin{Bmatrix} ax; x < 0 \\ x; x >= 0 \end{Bmatrix} \tag{2.7}$$

Exponential Linear Unit

An exponential linear unit (ELU) is a modified version of ReLU that raises the slope of the answer function's negative component. Unlike leaky ReLU and parametric ReLU, ELU describes negative values with a log curve rather than a straight line.

The ELU can be shown as:

$$f(x) = \begin{Bmatrix} x; x >= 0 \\ a(e^x - 1); x < 0 \end{Bmatrix} \tag{2.8}$$

Softmax

A number of sigmoids are equal to the softmax function. We realize sigmoid curves are percentages and they can reflect the likelihood of a data point belonging to a certain class. Since sigmoid is commonly used for classification and regression problems, the softmax function is useful for multi-class classification problems. The function returns the sample's data distribution.

The softmax function can be shown as:

$$\sigma(z)_j = \frac{e^{z_j}}{\sum_{k=1}^{K} e^{z_k}} \tag{2.9}$$

Swish

Swish is an activation mechanism first discovered by researchers at Google (which is lesser known). Using swish is as computationally effective as ReLU but greatly boosts efficiency over ReLU. The swish values vary from negative infinity to infinity. The swish activation function can be shown as:

$$f(x) = \frac{x}{1+e^{-x}} \tag{2.10}$$

2.3.3 Pooling Layer

This layer conducts compression and reduction of data in the network. It decreases the volume of overfitting by the gradual reduction of the network's capacity. To properly grasp how the pooling principle functions, we must first understand stride and padding.

Stride

Stride is the number of columns that a matrix is broader throughout. When a stride value is 1, we pass the filters one pixel at a time. As we work, we increment the filters by two pixels at a time, and so on. Figure 2.2 lays out how convolution will work when done over a stride of 2.

Padding

Often, the filter does not match on the picture completely. We have two choices:

- Pad the image with zeros, such that the picture matches (Figure 2.3).
- Delete the portion of the picture where the filter wasn't really applied. This is called validity padding, which only validates part of the picture.

Padding are of two types:

1. **Valid padding**
   In this, there is no padding at the edges of the image. The input image is left in its valid/unaltered shape.

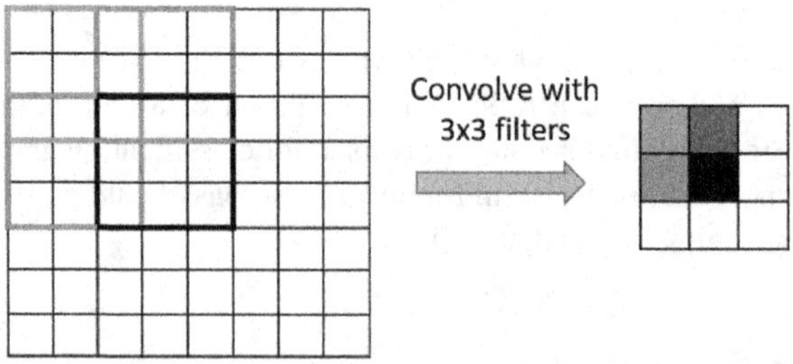

FIGURE 2.2 Stride of Two Pixels.

0	0	0	0	0	0	0	0	0
0								0
0								0
0								0
0								0
0								0
0								0
0								0
0	0	0	0	0	0	0	0	0

FIGURE 2.3 Padding by Zero at the Edges of Image.

2. **Same padding**
   In this, we pad the image by adding pad layers. Usually denoted by "p." So, if we pad it by 1, the following equation holds;

$$[(n+2p)\times(n+2p)image]*[(f\times f)filter]\rightarrow[(n\times n)image] \tag{2.11}$$

where "n" is the number of pixels in the row and column of an image, "p" is the padding size input, and "f" is the filter size. From equation (2.11), we can conclude:

$$p=\frac{(f-1)}{2} \tag{2.12}$$

So, the pooling layers are of two types:

1. **Average pooling (Figure 2.4)**
   In this, the average or mean is computed from the pool.
2. **Maximum pooling (Figure 2.5)**
   In this, the maximum number from the pool is selected and the rest are dropped.

One of these approaches may be used throughout the network to greatly reduce the number of parameters and amount of calculation needed. The adjustment in the dimensions of a graph would not change the underlying network. After total pooling, we are able to conclude the following:

$$Maxpool=\frac{(N-F)}{S}+1 \tag{2.13}$$

where "N" is the number of pixels in an image, "F" is the filter size, and "S" is the stride input size.

2.3.4 Flatten

In this layer, all the matrices are unrolled to form a one-dimensional matrix of size Nx1.

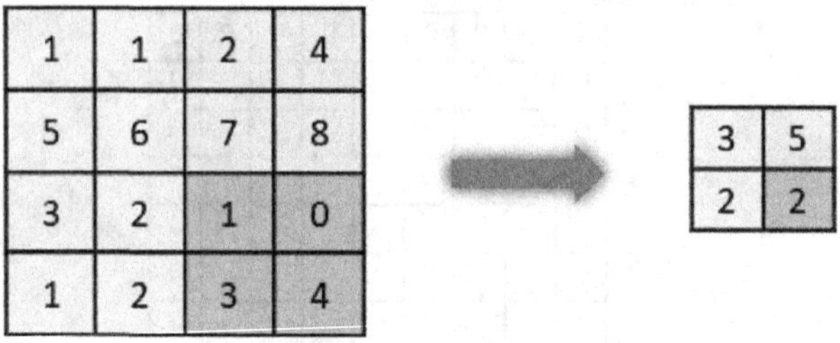

FIGURE 2.4 Average Pooling

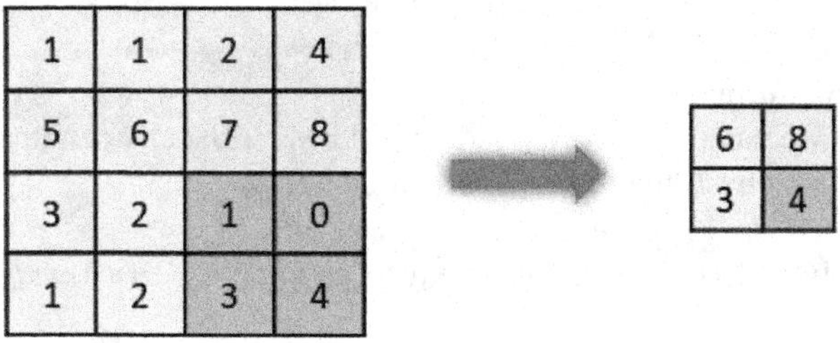

FIGURE 2.5 Max Pooling.

2.3.5 Fully Connected Layer

Neurons in this layer provide all the details from the previous layers (Figure 2.7). Their activations may be calculated by multiplying a matrix with a bias, then adding a bias reduction. This is the last step of CNN's takeover. The CNN is composed of several layers and layers of links. The completely linked portion of the CNN network calculates its own weight values through backward propagation. The neurons use the necessary weight to give the most important tag. Finally, the neurons "vote" on each of the labels to decide the final classification.

2.3.6 The Network's Training

When training a network, the kernel on top of the convolutional layer and the weight on top of the fully connected layer are chosen to minimize the disparity between predicted and real labels on the training dataset. Backward propagation, which is used in gradient descent, is the most often used technique for training neural networks. Forward propagation on a training dataset is used to assess a model's output, and learnable parameters, such as kernels and weights, are adjusted based on the loss value using an optimization algorithm, such as backward propagation or gradient descent (as seen in Figure 2.1).

Loss function: A loss function, also known as a cost function, is used in backward propagation to assess the precision of the neural network's output predictions and marks. Though cross entropy is often used to quantify classification (Figure 2.6)

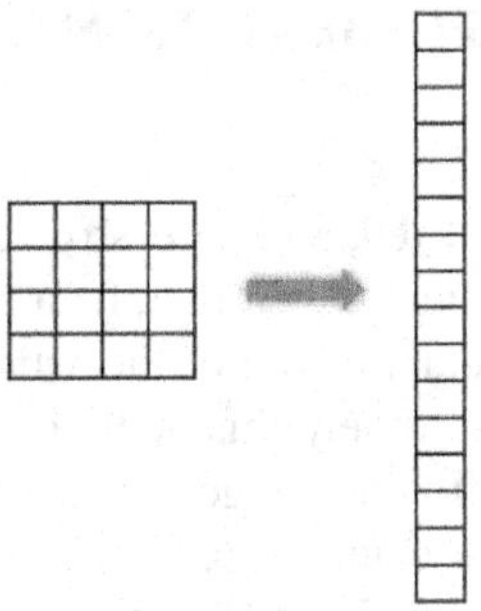

FIGURE 2.6 Flattening Layer.

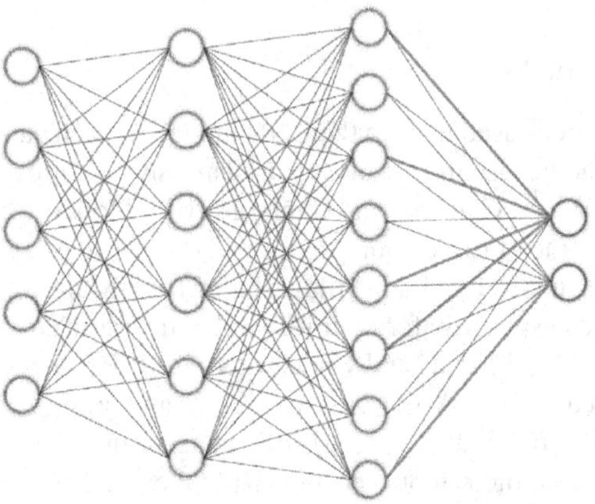

FIGURE 2.7 Fully Connected Layer.

accuracy, the most common regression approximation is mean squared error. One of the hyperparameters that must be calculated depending on the form of data specified is the type of loss function to be used.

Gradient descent: Gradient descent is a popular optimization algorithm that uses backward propagation to minimize loss in a neural network. The gradient of the loss function describes the path in which the function stretches the fastest, and each learnable parameter is set in the opposite direction of the gradient using an arbitrarily large step size defined by a hyperparameter called learning rate.

The learning rate is a critical hyperparameter that must be set prior to training. To optimize estimation, the gradients of the loss function with respect to the parameters are determined using a subset of the training dataset called mini-batch and then applied to the parameter adjustments. This method is regarded as stochastic gradient descent, and a mini-batch scale is one of its hyperparameter values. Other gradient descent algorithms, such as SGD, RMSprop, and Adam (Qian 1999; Kingma and Ba 2014; Ruder 2016), have been suggested and are in use, but the details of these algorithms are outside the scope of this chapter.

2.4 AGRICULTURE IMPLEMENTATION OF CNNS

2.4.1 Area of Application

There were a total of 12 fields listed, with disease detection utilizing plants and leaves, ground-cover classification, plant recognition, fruit counting, and weed identification being the most important. Notably, both of the articles were written in the past two years, showing how novel and timely this method is in the area of agriculture. The majority of these papers dealt with image recognition and field identification, such as obstacle detection and fruit counting (Christiansen et al. 2016; Steen et al. 2016; Rahnemoonfar and Sheppard 2017), while others concentrated on forecasting possible values, including maize yield (Kuwata and Shibasaki 2015) and soil moisture content in the area (Song et al. 2016). From another angle, the bulk of papers concentrated on maize, with just a few covering ground cover and livestock farming.

2.4.2 Sources of Data

When the data sources used to train the CNN model for each paper were investigated, it was revealed that the researchers mostly used massive picture datasets totaling thousands of images (Reyes, Caicedo, and Camargo 2015; Mohanty, Hughes, and Salath´e 2016). Although some of these photographs and datasets came from well-known and easily accessible sources, like PlantVillage, LifeCLEF, MalayaKew, and UC Merced, others were created especially for their research purposes (Xinshao and Cheng 2015; Sladojevic et al. 2016; Bargoti and Underwood 2017; Rahnemoonfar and Sheppard 2017; Sørensen et al. 2017). Articles on land cover and crop-type classification relied on a small number of photos (e.g., 10–100 images) obtained by unmanned aerial vehicle (UAV) or satellite remote sensing (Kussul et al. 2017). Douarre et al. (2016) provide a useful essay on root and soil segmentation using X-ray tomography images. Furthermore, several programs have documented historical text details using SVR (Kuwata and Shibasaki 2015) over corn field (Song et al. 2016). In general, the greater the complexity of the case, the more data is needed.

2.4.3 Overall Accuracy and Performance Measurements

A number of metrics of impact were used by the different researchers, the most frequent being the percentage of accurate forecasts on the date range. The performance metrics used also included RMSE, F1 Score, QM (Douarre et al. 2016), RFC (Chen et al. 2017), and LC (Reyes, Caicedo, and Camargo 2015). The comparable work done by CA (correct predictions) is normally strong (i.e., over 90%) and implies that CNN has been successfully implemented in a number of agricultural issues. In Chen et al. (2017), Lee et al. (2015), and Steen et al. (2016) with at least 98% accuracy, the best CA statistics were identified.

2.4.4 Techniques Used

Various complex techniques have been used by various authors in the field of agriculture. The key aim of the writers is to classify or diagnose different diseases linked to plants and seeds. The precision of the algorithm to discover problems in the

TABLE 2.2
Various Networks Used in the Agricultural Field.

S. No.	Networks
1	CNN
2	AlexNet
3	Fully convolutional network
4	DNN
5	Mask R-CNN
6	YOLOv3
7	ResNet
8	GoogLeNet
9	VGG
10	Inception
11	LSTM
12	Faster-CNN, R-FCN, SSD
13	DenseNet

agriculture sector improves when a large amount of information becomes available. So, to increase the accuracy, Table 2.2 lists the various known or complex algorithms available for agricultural purposes.

2.5 COMPARISON OF CNN MODEL WITH USAGE

Apart from the precision improvements observed in the classification/estimation problems of the surveyed works, CNNs provide several benefits for image processing. Historically, conventional methods produced acceptable outcomes with the use of hand-engineered features that had a significant impact on the outcome. Feature engineering is a dynamic process that evolves in response to improvements in the problem or data collection. Thus, FE is a time-consuming and expensive approach that is highly dependent on the skills of specialists and does not generalize well (Amara, Bouaziz, and Algergawy 2017). On the other hand, CNN would not need MRI because it is capable of identifying critical features without guidance. The model was trained to identify and label each fruit it encountered during fruit counting (Rahnemoonfar and Sheppard 2017). Experience enhances the output of neural networks (Pan and Yang 2009). They remain robust in terms of adversity, including poor illumination, a complex background, the size and orientation of the images, and varying resolutions (Amara, Bouaziz, and Algergawy 2017). Their primary disadvantage is that CNN occasionally fails to train on a timely basis. After training, the performance of this method is significantly greater than that of other machine-learning systems, such as KNN and SVM (Chen et al. 2014; Christiansen et al. 2016). Another significant disadvantage is the need for large datasets (hundreds of thousands of images) and their accurate annotation, which is often a delicate process requiring domain-specific information. A new analysis of CNN data revealed that insufficient data labeling is a problem that can have a significant impact on accuracy and precision. Additionally, there are disadvantages associated with utilizing pre-trained

TABLE 2.3
Advantages and Disadvantages of Deep Neural Networks.

Framework	Year	Advantages	Disadvantages
Tensorflow	2015	Multi GPU parallel computing Automatic derivation Flexible portability parallel design	Code complexity Frequent interface change Bad input channels
Caffe	2014	Robust and fast Component modularization	Not flexible enough Insufficient scalability
Keras	2017	Rich variety of models	Slow training speed
Theano	2008	Closely integrated with Numpy Usage of GPU Speed and stability optimizations	Low level framework Patchy support for pre-trained models
Torch	2002	Support convolutional network	Not flexible enough
PyTorch	2017	Dynamic neural network Flexible expansion	Server architecture lacks flexibility

models on close and limited datasets, as well as optimization problems associated with the models' measurements and hardware limitations. Table 2.3 compares the advantages and disadvantages of the most common open-source systems. To accelerate training, TensorFlow and Keras were used for discovery and experimentation.

2.6 FUTURE OF AGRICULTURAL PROFOUND LEARNING

It will be interesting to see CNN responding to agricultural issues, such as plant phenology, seed identification, consistency of soil and leaf phosphorus, irrigation, plant water stress detection, pests, and grain cultivation deficiencies. Given that many of the areas of the research referred to earlier use CNN-related data processing techniques (such as linear and logistical regression, SVM, KNN, K-mean clustering, wavelet filtering, and Fourier transformation), the applicability of CNN to this issue would need to be investigated. CNN includes the use of aerial imagery (by drones, for instance) to track seed performance; to improve the consistency of wine output by grapes harvesting at an optimum level of maturity; and to monitor animals and activity to evaluate general health, to detect diseases, and various other circumstances. The research did not have complex models, such as recurring neural networks or long-term memory, but included some complex models. An illustration may be an assessment of the growth rate of plants, trees, and even insects on the basis of past consecutive measurements, an estimated harvest, a determination of their survival rate, or an insect deterrent from spreading a harmful insect. These models may be used to further explain climate change, to predict weather conditions and activities, and to calculate the environmental impacts of physical or artificial operations.

2.7 CONCLUSION

The current paper analyzes CNN-based study practices in the agricultural field. It details how the region and challenge were selected, the technical specificities of the models were defined, the overall precision/accuracy achieved, and the data sources used. In terms of output comparisons, neural networks were linked to other existing methods based on the authors' implementations. The research shows that CNNs are advantageous in a wide variety of applications and outperform commonly used image-processing techniques. The ability to correctly approximate too complex problems and avoid the need for an FE beforehand are the primary benefits. After applying CNN to the task of locating lost vegetation, the current authors discovered that CNN is extremely reliant on the scale and accuracy of the training data collection. Future research will apply CNN's fundamental concepts and best practices to other agricultural areas that have not yet been completely used. A plethora of these areas have been discussed. The current survey aims to motivate researchers to experiment with deep learning and CNNs, as well as to apply them to agricultural computer vision and image-processing problems and to analyze the results. CNN's cumulative benefits are encouraging, particularly with their plans to further secure their food supply and encourage organic farming.

REFERENCES

Amara, Jihen, Bassem Bouaziz, and Alsayed Algergawy (2017). "A deep learning-based approach for banana leaf diseases classification". In: *Datenbanksysteme fu¨r Business, Technologie und Web (BTW 2017)-Workshopband,* Gesellschaft für Informatik e.V.

Andri, Renzo et al. (2018). "Hyperdrive: A systolically scalable binary-weight CNN inference engine for mW IoT end-nodes". In: *2018 IEEE Computer Society Annual Symposium on VLSI (ISVLSI)*. IEEE, pp. 509–515.

Bargoti, Suchet and James Underwood (2017). "Deep fruit detection in orchards". In: *2017 IEEE International Conference on Robotics and Automation (ICRA)*. IEEE, pp. 3626–3633.

Bastiaanssen, Wim G. M., David J. Molden, and Ian W. Makin (2000). "Remote sensing for irrigated agriculture: Examples from research and possible applications". In: *Agricultural Water Management* 46.2, pp. 137–155.

Canziani, Alfredo, Adam Paszke, and Eugenio Culurciello (2016). "An analysis of deep neural network models for practical applications". In: *arXiv preprint arXiv:1605.07678*.

Carvalho, Fernando P. (2006). "Agriculture, pesticides, food security and food safety". In: *Environmental Science & Policy* 9.7–8, pp. 685–692.

Chen, Steven W. et al. (2017). "Counting apples and oranges with deep learning: A data- driven approach". In: *IEEE Robotics and Automation Letters* 2.2, pp. 781–788.

Chen, Yushi et al. (2014). "Deep learning-based classification of hyperspectral data". In: *IEEE Journal of Selected Topics in Applied Earth Observations and Remote Sensing* 7.6, pp. 2094–2107.

Chi, Mingmin et al. (2016). "Big data for remote sensing: Challenges and opportunities". In: *Proceedings of the IEEE* 104.11, pp. 2207–2219.

Christiansen, Peter et al. (2016). "Deep Anomaly: Combining background subtraction and deep learning for detecting obstacles and anomalies in an agricultural field". In: *Sensors* 16.11, p. 1904.

Deng, Li and Dong Yu (2014). "Deep learning: Methods and applications". In: *Foundations and Trends in Signal Processing* 7.3–4, pp. 197–387.

Douarre, Cl´ement et al. (2016). "Deep learning based root-soil segmentation from X-ray tomography images". In: *bioRxiv*, p. 071662.

Dyrmann, Mads, Henrik Karstoft, and Henrik Skov Midtiby (2016). "Plant species classification using deep convolutional neural network". In: *Biosystems Engineering* 151, pp. 72–80.

FAO, U (2009). "How to feed the world in 2050". In: *Rome: High-Level Expert Forum* 732, p. 733.

Gebbers, Robin and Viacheslav I. Adamchuk (2010). "Precision agriculture and food security". In: *Science* 327.5967, pp. 828–831.

Grinblat, Guillermo L. et al. (2016). "Deep learning for plant identification using vein morphological patterns". In: *Computers and Electronics in Agriculture* 127, pp. 418–424.

Hasanpour, Seyyed Hossein et al. (2018). "Towards principled design of deep convolutional networks: Introducing simpnet". In: *arXiv preprint arXiv:1802.06205*.

Hashem, Ibrahim Abaker Targio et al. (2015). "The rise of 'big data' on cloud computing: Review and open research issues". In: *Information Systems* 47, pp. 98–115.

He, Kaiming et al. (2016). "Deep residual learning for image recognition". In: *Proceedings of the IEEE Conference on Computer Vision and Pattern Recognition*, IEEE, pp. 770–778.

Huang, Gao et al. (2017). "Densely connected convolutional networks". In: *Proceedings of the IEEE Conference on Computer Vision and Pattern Recognition*, IEEE, pp. 4700–4708.

Ishimwe, Roselyne, K. Abutaleb, F. Ahmed et al. (2014). "Applications of thermal imaging in agriculture—a review". In: *Advances in Remote Sensing* 3.3, p. 128.

Kamilaris, Andreas, Andreas Kartakoullis, and Francesc X. Prenafeta-Boldu´ (2017). "A review on the practice of big data analysis in agriculture". In: *Computers and Electronics in Agriculture* 143, pp. 23–37.

Kamilaris, Andreas and Francesc X. Prenafeta-Boldu´ (2018). "Deep learning in agriculture: A survey". In: *Computers and Electronics in Agriculture* 147, pp. 70–90.

Kamilaris, Andreas et al. (2016). "Agri-IoT: A semantic framework for internet of things- enabled smart farming applications". In: *2016 IEEE 3rd World Forum on Internet of Things (WF-IoT)*. IEEE, pp. 442–447.

Kingma, Diederik P. and Jimmy Ba (2014). "Adam: A method for stochastic optimization". In: *arXiv preprint arXiv:1412.6980*.

Kitzes, Justin et al. (2008). "Shrink and share: Humanity's present and future ecological footprint". In: *Philosophical Transactions of the Royal Society B: Biological Sciences* 363.1491, pp. 467–475.

Krizhevsky, Alex, Ilya Sutskever, and Geoffrey E. Hinton (2017). "Imagenet classification with deep convolutional neural networks". In: *Communications of the ACM* 60.6, pp. 84–90.

Kussul, Nataliia et al. (2017). "Deep learning classification of land cover and crop types using remote sensing data". In: *IEEE Geoscience and Remote Sensing Letters* 14.5, pp. 778–782.

Kuwata, Kentaro and Ryosuke Shibasaki (2015). "Estimating crop yields with deep learning and remotely sensed data". In: *2015 IEEE International Geoscience and Remote Sensing Symposium (IGARSS)*. IEEE, pp. 858–861.

LeCun, Yann, Yoshua Bengio, and Geoffrey Hinton (2015). "Deep learning". In: *Nature* 521.7553, pp. 436–444.

LeCun, Yann, Yoshua Bengio et al. (1995). "Convolutional networks for images, speech, and time series". In: *The Handbook of Brain Theory and Neural Networks* 3361.10, p. 1995.

LeCun, Yann et al. (1998). "Gradient-based learning applied to document recognition". In: *Proceedings of the IEEE* 86.11, pp. 2278–2324.

Lee, Sue Han et al. (2015). "Deep-plant: Plant identification with convolutional neural networks". In: *2015 IEEE International Conference on Image Processing (ICIP)*. IEEE, pp. 452–456.

Liaghat, Shohreh, Siva Kumar Balasundram et al. (2010). "A review: The role of remote sensing in precision agriculture". In: *American Journal of Agricultural and Biological Sciences* 5.1, pp. 50–55.

Lin, Min, Qiang Chen, and Shuicheng Yan (2013). "Network in network". In: *arXiv preprint arXiv:1312.4400*.

Lu, Heng et al. (2017). "Cultivated land information extraction in UAV imagery based on deep convolutional neural network and transfer learning". In: *Journal of Mountain Science* 14.4, pp. 731–741.

Luus, Francois P. S. et al. (2015). "Multiview deep learning for land-use classification". In: *IEEE Geoscience and Remote Sensing Letters* 12.12, pp. 2448–2452.

Mohanty, Sharada P., David P. Hughes, and Marcel Salath´e (2016). "Using deep learning for image-based plant disease detection". In: *Frontiers in Plant Science* 7, p. 1419.

Najafabadi, Maryam M. et al. (2015). "Deep learning applications and challenges in big data analytics". In: *Journal of Big Data* 2.1, p. 1.

Ozdogan, Mutlu et al. (2010). "Remote sensing of irrigated agriculture: Opportunities and challenges". In: *Remote Sensing* 2.9, pp. 2274–2304.

Pan, Sinno Jialin and Qiang Yang (2009). "A survey on transfer learning". In: *IEEE Transactions on Knowledge and Data Engineering* 22.10, pp. 1345–1359.

Paszke, Adam et al. (2016). "Enet: A deep neural network architecture for real-time semantic segmentation". In: *arXiv preprint arXiv:1606.02147*.

Qian, Ning (1999). "On the momentum term in gradient descent learning algorithms". In: *Neural Networks* 12.1, pp. 145–151.

Rahnemoonfar, Maryam and Clay Sheppard (2017). "Deep count: Fruit counting based on deep simulated learning". In: *Sensors* 17.4, p. 905.

Reyes, Angie K., Juan C. Caicedo, and Jorge E. Camargo (2015). "Fine-tuning deep convolutional networks for plant recognition." In: *CLEF (Working Notes)* 1391, pp. 467–475.

Ruder, Sebastian (2016). "An overview of gradient descent optimization algorithms". In: *arXiv preprint arXiv:1609.04747*.

Sa, Inkyu et al. (2016). "Deepfruits: A fruit detection system using deep neural networks". In: *Sensors* 16.8, p. 1222.

Santoni, Mayanda Mega et al. (2015). "Cattle race classification using gray level co- occurrence matrix convolutional neural networks". In: *Procedia Computer Science* 59.2015, pp. 493–502.

Saxena, Lalit and Leisa Armstrong (2014). "A survey of image processing techniques for agriculture". In: *Proceedings of Asian Federation for Information Technology in Agriculture*, Australian Society of Information and Communication Technologies in Agriculture, pp. 401–413.

Schmidhuber, Ju¨rgen (2015). "Deep learning in neural networks: An overview". In: *Neural Networks* 61, pp. 85–117.

Simonyan, Karen and Andrew Zisserman (2014). "Very deep convolutional networks for large-scale image recognition". In: *arXiv preprint arXiv:1409.1556*.

Sladojevic, Srdjan et al. (2016). "Deep neural networks based recognition of plant diseases by leaf image classification". In: *Computational Intelligence and Neuroscience* 2016, https://doi.org/10.1155/2016/3289801.

Song, Xiaodong et al. (2016). "Modeling spatio-temporal distribution of soil moisture by deep learning-based cellular automata model". In: *Journal of Arid Land* 8.5, pp. 734–748.

Sørensen, Ren´e A. et al. (2017). "Thistle detection using convolutional neural networks". In: *2017 Efita Wcca Congress*, Montpellier Supagro, p. 161.

Steen, Kim Arild et al. (2016). "Using deep learning to challenge safety standard for highly autonomous machines in agriculture". In: *Journal of Imaging* 2.1, p. 6.

Szegedy, Christian et al. (2015). "Going deeper with convolutions". In: *Proceedings of the IEEE Conference on Computer Vision and Pattern Recognition*, IEEE, pp. 1–9.

Szegedy, Christian et al. (2016). "Rethinking the inception architecture for computer vision". In: *Proceedings of the IEEE Conference on Computer Vision and Pattern Recognition*, IEEE, pp. 2818–2826.

Teke, Mustafa et al. (2013). "A short survey of hyperspectral remote sensing applications in agriculture". In: *2013 6th International Conference on Recent Advances in Space Technologies (RAST)*. IEEE, pp. 171–176.

Tyagi, Avinash C. (2016). "Towards a second green revolution". In: *Irrigation and Drainage* 4.65, pp. 388–389.

Waga, Duncan (2013). "Environmental conditions' big data management and cloud computing analytics for sustainable agriculture". In: *Available at SSRN 2349238*.

Wan, Ji et al. (2014). "Deep learning for content-based image retrieval: A comprehensive study". In: *Proceedings of the 22nd ACM International Conference on Multimedia*, Springer, pp. 157–166.

Weber, Rolf H. and Romana Weber (2010). *Internet of Things*. Vol. 12. Springer.

Xinshao, Wang and Cai Cheng (2015). "Weed seeds classification based on PCANet deep learning baseline". In: *2015 Asia-Pacific Signal and Information Processing Association Annual Summit and Conference (APSIPA)*. IEEE, pp. 408–415.

Zeiler, Matthew D. and Rob Fergus (2014). "Visualizing and understanding convolutional networks". In: *European Conference on Computer Vision*. Springer, pp. 818–833.

3 Implementation of Blockchain Technology in Indian Banking Sector
A Descriptive Study

Dr. Pavnesh Kumar and Siddhartha Ghosh

CONTENTS

3.1 INTRODUCTION TO BLOCKCHAIN TECHNOLOGY

Blockchain has garnered great attention in the last decade, but the technology behind it is older. Stuart Haber and W. Scott Stornetta first described a cryptographically secured chain of data or blocks in 1991 (Stuart Habe, 1991).

Their work was the base for computer scientist Nick Szabo, who used the technology for a digital decentralized currency in 1998 called 'Bit gold' (Institute of Chartered Accountants in England and Wales, 2021). These earlier breakthroughs, though, remained in obscurity until 2008, when a developer operating under the pseudonym Satoshi Nakamoto published a white paper that introduced the model for the blockchain. Just one year later, in 2009, Nakamoto used the published model for introducing blockchain-based currency called bitcoin (Nakamoto, 2009).

DOI: 10.1201/9781003184140-3

In the initial years, blockchain and bitcoin were indistinguishable, but in 2014, the blockchain technology (which wasn't a proprietary technology, i.e., it was open source) was considered for other applications as well. It opened the doors for blockchains to be implemented in various fields. A simple analogy to understand this is a wheel. A wheel has just one property, it can roll on a surface smoothly; but just this one basic feature makes it apt for applications in transportation, machinery, toys, or even pulling our luggage. The same can be said about blockchain; it can be applied to everything involving computers and the internet.

Blockchain, in its essence, can be compared to a security system functioning like a database. Just like databases store a ton of information electronically—all while making it possible for the users to navigate across the data—blockchains achieve the same. While databases are used for limited purposes, blockchain can be used for a host of purposes. The difference between the two lies in the smallest unit of what they are made up of. While databases can be compared to an elaborate spreadsheet containing only text, blockchain is made up of small data units called "blocks" arranged in a structure similar to a chain, i.e., each block is connected digitally with a minimum of two more blocks. These blocks can contain data that is much more diverse in nature than simple text and hence gives blockchain technology its versatility.

The blockchain gets bigger and bigger as new blocks are added to the chain with a timestamp that is irreversible in nature. Any tampering to a specific block is not possible since blockchain isn't stored in any specific server or location. It is a distributed system spread across a peer-to-peer network and is hence decentralized in nature. Simply put, the blocks in the chain cannot be altered by compromising any specific computer system or server. Also, each user existing on the blockchain is issued a digital signature. This makes the network resilient to attacks since any digital attack is generally routed through an army of computers, all working in synchronization to take down their target. Any "unusual amount of traffic" in the blockchain can hence be easily detected with a proper record of each computer system due to their digital signature and hence genuine works over the blockchain, like transactions or record keeping, can be easily identified from malicious acts of system attacks or sabotage.

3.2 USES OF BLOCKCHAIN TECHNOLOGY IN BANKING

While we have discussed the inception of the current state of blockchain technology, it will be essential to cover the areas in banking where the benefits of blockchain technology can be reaped to provide better and, at times, even new services that would have not have been possible earlier. Traditionally, the banking sector in India has been laden with paperwork to satisfy all the stringent rules laid out by the regulatory authorities for performing due diligence on the part of the banks. This elaborate paperwork used to make routine banking transactions and operations a time-consuming affair. However, things improved with the advent of the internet, with mobile applications and internet banking. Internet banking, though, had its own set of challenges. The primary one being lack of security features that can be tied to it. Credit and debit cards were and still are prone to fall prey to malicious activity. Internet banking passwords can be taken from the actual users in a phishing attack.

Some products, like contactless debit or credit cards, can easily be used for unauthorized transactions.

It is here that blockchain technology can be implemented for bolstering up the security as well as effectivity of banking. Some of which is discussed briefly here:

1. Foreign Transactional Payments—Foreign payments have two aspects attached to them, the first being of number of documents put up for scrutiny for each transaction, and the other being the time taken to execute the transaction after genuineness of the documents is ensured. Blockchain technology can be used for offering a seamless documentation solution tied up with the actual payment being made. Currently, such transactions require third-party authorization, where payments are actually processed through means like SWIFT transactions. Since blockchain doesn't depend on any such outside parties, transactions can be performed much faster and more economically since transactional fees can be bypassed.
2. Share Trading in Stock Exchanges—Blockchain technology can also be used for executing share trading orders. All of share trading involves at least three parties viz. stock exchange, brokers and users. Any normal transaction starts by the user prompting an order initiation by using a banking interface that is then communicated to the exchange by the banking backend software. The trade is then executed. All of this involves many steps, which lengthens the transaction process. Also, once the trading is finished, the record has to be maintained in the demat account of the user. An efficient way to share trades can be a blockchain-based technology that guides the banking system to carry out trading smoothly and securely. Since blockchain technology has a lot of security built into its core, it can provide an added layer of security to the customers and can also attract more people into share trading given its robust security measures.
3. For Lending Purposes—Currently, bank lending is a lengthy process involving physical forms to be filled out by the requesting party with signatures that are then scrutinized by the bank along with documents like proof of income or a profit & loss statement, balance sheet, ITR returns and other such documents. All this makes the process tedious and time consuming. Since a lot of guidelines are involved in this, the process can't be completely done away with in its current form. Blockchain technology can easily be used for record-keeping purposes with a tamper-proof time-stamp feature. It effectively means that documents can be shared over a blockchain-based platform where authenticity can be verified easily. Once the genuineness of the documents is ascertained, all the bank has to do is decide to sanction the loan and then transfer the amount to the appropriate account.
4. Other Necessary and Auxiliary Activities—Apart from selling products, a lot of activities of a bank or financial institution involves record keeping, accounting and time-to-time audits. Though digitalization has reduced the amount of paperwork for these activities, digitalization alone cannot solve all the woes of a bank. Even after digitalization, banking frauds happen, errors are made at the bank personnel's end and, at times, sabotages are orchestrated

by collusion between a bank's customers and its employees. Blockchain brings with it the necessary checks and balances that can provide safety in undertaking these operations. It might help the auditing parties to verify the records seamlessly while also checking which bank employee has been accessing the records and for what purposes. This can be done by the personal signature the employees' computers leave on the blockchain interface. Not only can a blockchain-based system save time at the bank's end, it can also provide a level of security much more preferable to the top management.

3.3 BLOCKCHAIN TECHNOLOGY IMPLEMENTED BY INDIAN BANKS—AN OVERVIEW

In the previous sections, we discussed blockchain technology and its potential uses in the banking domain. In this section, we will be discussing a few of the practical applications and solutions being deployed by banks in India using blockchain technology.

Reserve Bank of India (RBI) is the apex banking body of the country and regulates all other public- and private-sector banks. RBI has been testing blockchain-based technology since at least 2017. RBI's research arm, known as Institute for Development and Research in Banking Technology (IDRBT), has partnered with the New York company MonetaGo as a technology partner (Shetty, 2017). IDRBT has also published a white paper titled "Applications of blockchain technology in banking and financial sector in India", where they have acknowledged blockchain technology as disruptive in nature with a potential to revamp and propel the financial sector of India to new heights (Institute for Development and Research in Banking Technology, 2017).

Yes Bank too has warmed up to the idea of using blockchain technology. It has initially decided to onboard 32 vendors, including Bajaj Electrical, to an invoice-financing blockchain platform. The invoicing procedure currently in practice takes around four days to complete and includes verification, presenting, recording and reconciling of the invoices before disbursement of any working capital loan. All these steps take time, and the planned blockchain platform can cut redundancy in the whole process, facilitating speedy disbursals of loans (Yes Bank, 2017).

Axis Bank has gone ahead with deploying blockchain technology for a solution aimed at smoothing inward remittances. They have created a platform called RAKBank, which will service their retail clients in Gulf countries along with a partnership forged with Standard Chartered Bank (Singapore); they will also cater to businesses having exposure to corporate trade remittances. A Ripple Inc. blockchain platform is being used for setting up this global network for making transactions (Axis Bank, 2017).

—Federal Bank too has partnered with Ripple to offer a global payments system. Ripple uses blockchain technology for facilitating global remittances, and the partnership between the two is expected to result in a blockchain-based platform that can process transactions securely. It should be noted that a large segment of Federal Bank customers are NRIs (non-resident Indians) from the southern part of India, based in the Gulf, Europe and the USA. These customers regularly send remittances back home to India and a blockchain-based solution will definitely suit Federal Bank to offer a seamless remittance solution (pymnts.com, 2019).

ICICI Bank has also joined the blockchain bandwagon by partnering with Emirates NBD. ICICI Bank became the first bank in India to successfully carrying out a transaction involving international trade finance and remittance (Bloomberg Quint, 2016). In 2016, the bank used a blockchain-based solution to authenticate documents pertaining to imports and exports, such as purchase orders, invoices and goods insurance. The blockchain-based solution simplified the whole process and was executed in a few minutes, while traditional methods involve a lot of paperwork and require third-party services, like couriers, which adds to the time slag in each transaction. The solution was designed to be real-time visible to both importers and exporters and enables them to track and authenticate transactions at various levels (cio.com, 2016; Edgeverve.com, 2017).

JP Morgan has launched a blockchain-based platform that has been joined by seven Indian banks, namely ICICI Bank, Axis Bank, Yes Bank, Union Bank of India, Federal Bank and Canara Bank (Bhatia, 2019). The new platform will facilitate faster international payments and remittances, as the blockchain platform will have to execute fewer steps to complete the transaction while also being not dependent on external parties. The blockchain-based platform is expected to reduce costs of transactions while also adding extra layers of security. The platform is called Interbank Information Network (IIN) (Economictimes.com, 2019).

The State Bank of India has partnered with JP Morgan to accelerate international transactions through the US bank's blockchain technology.

According to reports, the partnership is expected to lower transaction costs and payment processing times for SBI customers. They stated that the time required to resolve cross-border payment-related inquiries can be reduced from up to a fortnight to only a few hours. This enables beneficiaries to receive cross-border payments more quickly and with fewer measures.

JP Morgan's blockchain technology, dubbed Liink, is designed to operate as a peer-to-peer network, with financial institutions, corporations and fintech firms worldwide subscribing to it. This allows users to perform reliable and peer-to-peer data transfers at a higher rate and with greater power. Additionally, it reduces the risk associated with international transactions (Economictimes.com, 2021).

3.4 OBJECTIVES OF THE STUDY

The objectives of this research study are as follows:

- To examine the genesis of blockchain technology and what makes it stand apart from regular computer programs and digital methods of processing data.
- To understand how blockchain technology is being implemented in the banking industry.
- To suggest the model of disruption possible in accordance with regulations present in the sector.
- To suggest how the model fits into the case of blockchain technology's implementation in the banking sector.

3.5 RESEARCH METHODOLOGY

A thorough literature review regarding the title and correlated concepts was done. Secondary data of both qualitative as well as quantitative nature was analyzed. Latest information was sought from technology journals, research publications, news reports, books, magazines and various websites. Libraries and corporate reports were also consulted while writing this chapter. The collected literature was cross-checked and properly validated to provide the latest information.

3.6 UNDERSTANDING WHAT DISRUPTIVE INNOVATION MEANS

The theory of disruptive innovation was first suggested in 1995 by Clayton Christensen, who used the model to describe the way a new entrant to the market disrupts the business using unconventional techniques. This theory describes "disruption" as a process where a small company with fewer resources at its disposal can start challenging big, established players in the market. Mostly, the company bringing the disruption targets niche segments of the market to not catch the attention of established players who are busy maximizing profits rather than refining their products or services. After gaining a foothold, the entrant company can offer their disruptive services to a wider audience in the market, effectively altering the way people consume or use products or services in that specific segment (Clayton et al., 2015). People often misunderstand some degree of innovation or success for a company to be disruptive in nature, though it is not necessarily the case (Clayton and Christensen, 1995).

The components of this theory can be summed up in the following points:

- Every innovation is not disruptive in nature, suggesting that just finding newer ways to perform the same task doesn't necessarily mean disruption is at work.
- Disruptions originate in either the low end of a market, where major players are busy selling to or servicing the top tier of customers who bring in major profits, or a new market altogether.
- Disruptive innovations do not become an instant hit but, like a process, need time to polish their offerings in competitive quality with that to major companies in the market.
- Disruptive innovations do not appeal to all customer segments at the beginning but, rather, only to a select few. These select segments might have some unique requirement to which the innovation can exclusively cater.

3.7 IS BLOCKCHAIN ACTUALLY A DISRUPTION?

If we carefully analyze blockchain technology with the tenets of the theory of disruptive innovation, it becomes clear that it isn't actually a disruption. Even before blockchain was implemented, many aspects of banking operations and transactions used digital mediums for processing and completing tasks. And although blockchain is a revolutionary technology that has the potential to reshape the banking industry, it isn't a disruption in spirit. Blockchain, with its implementation in the banks, isn't altering the business model in any way. At best, it offers a better and secure way to do those tasks or transactions that are already happening.

3.8 UNDERSTANDING DISRUPTIONS AND REGULATIONS

While we have covered how disruptions work and how blockchain isn't a disruption in spirit, it is time we also cover a very important aspect of any newer technology-driven revolution that may or may not qualify as a disruption. Regulations are lawful guidelines that various institutions must follow. They can be in the form of dos and don'ts, best practices, mandatory duties and other rules and guidelines that can be implemented over time. Some sectors of the economy, like banks, are tightly regulated, while others, like the automobile industry, have fewer regulations.

There is an inverse relationship between regulations of a sector and the disruption possible in it. Tighter regulations leave lesser scope for third parties to be involved, the kind of data that can be shared with them, the requisite approval needed, the subsequent paperwork for such collaborations and other legal formalities.

3.9 MODEL OF DISRUPTION POSSIBLE IN ACCORDANCE WITH REGULATIONS

In the model suggested in Figure 3.1, we can see that three elements come to play:

- The type of industry itself;
- The kind of regulations over it; and
- The pressure exerted by the regulations over the industry.

Tighter regulations leave lesser scope for disruptions. Sectors such as hospitality or retail are subjected to far fewer regulations and hence have less pressure to comply and have a wider scope for blockchain to be introduced as a disruption bringing radical changes to the existing business model.

It is notable to mention, though, that these are subjected to change. Any changes in regulations can shift that specific sector right or left. With a rightward shift, the sector will have fewer rules to follow and more field open for advancements. A leftward shift will have the opposite effect.

FIGURE 3.1 Relationship between Regulations of a Sector and Possible Disruption.

3.10 WHERE DOES BANKING LIE (ON THIS MODEL)?

Banking, as we can see, lies to the extreme left, suggesting that the industry has lesser scope for blockchain-based disruptions. Blockchain can still be implemented, but it will be used to perform the same tasks more efficiently. The business model of the banks functioning in India is unlikely to change.

However, there has been a renewed interest at the administrative level to formalize the rules and guidelines regarding blockchain and its possible uses for banking, a lot needs to change at the legislature level. We have lately seen various banks exploring with blockchain technology, deploying it for their operations and bringing about some changes in their product offerings. A faster trade and finance dashboard for performing global payments and a platform for verifying the credit worthiness of business entities are a couple examples. But all these collaborations on the bank's part still depend on a foreign blockchain initiative.

Actual changes in regulations will be seen with an advent to an Indian blockchain initiative where Indian banks will be more comfortable sharing data and integrating blockchain in their network. Something of this sort was seen in 2012, when National Payments Corporation of India launched RuPay, an international card-payment service, as a homegrown alternative to the existing Visa and Mastercard offerings. A rightward shift of the banking sector is on its way but will need some more time to bear fruit in terms of actual innovation and newer product offerings and services.

3.11 FINDINGS

The findings of the study have been listed here:

- Blockchain technology is indeed a new revolutionary technology that has the potential to change banking as we know it in India and elsewhere.
- Blockchain technology is not exactly disruptive in spirit. We are yet to see any new business model spring up in the banking sector with the deployment of blockchain.
- Widescale disruption can only happen when adequately allowed by regulations imposed on the sector. Currently, the banking sector is under many regulations, which doesn't allow for widescale and radical changes through blockchain at the product level.
- We will gradually see a rightward shift of the banking sector, meaning fewer regulations with which to comply in terms of technological collaboration, meaning a wider field available to integrate blockchain, offer newer products and cater to newer customers. An example can be offering micro credit services to the poor via establishing their identity quickly by deploying a blockchain-based platform.

3.12 SUGGESTIONS

We have the following suggestions for blockchain implementation in the banking sector in India:

- Concrete but flexible legislation is required to bring into immediate effect for giving the banks a greater degree of freedom in choosing blockchain-based solutions for their products and services.
- Since blockchain is still in its nascent stage, proper monitoring of recent developments in the area should be done, and any possible risk to the banking institutions or people using blockchain-based solutions should be nipped in the bud.
- India should think of an indigenous blockchain-based umbrella institution that can seamlessly integrate all the banks in India with blockchain-based solutions.
- Such a development will also safeguard India's security interest in the future since depending on foreign companies for blockchain-based digital platforms jeopardizes our country's financial health, as banks form any country's financial backbone.
- Entrepreneurs should think about springing up newer business models around blockchain to harness its true potential and bring transformational changes to the banking sector.

3.13 CONCLUSION

Only time will tell what lies ahead for blockchain-based banking in India. But that being said, one thing is clear: Blockchain is here to stay and not some technological fad or buzzword that might fall to obscurity in the near future. The slow process of disruption in the banking sector might well be on its way, but it will still take some time to deliver radically different products or services and spring up newer business models. India has a large population with a major chunk of it still barely using formal banking channels. Blockchain might very well change this by reaching out to the last-mile customer, even those living in the rural interiors of the country.

REFERENCES

Axis Bank. (2017). Axis Bank launches ripple-powered instant payment service for retail and corporate customers. Retrieved April 5, 2021, from www.axisbank.com/docs/default-source/press-releases/axis-bank-launches-ripple-powered-instant-payment-service-for-retail-and-corporate-customers.pdf?sfvrsn=6

Bhatia, R. (2019). JP Morgan succeeds where Facebook failed; Grows blockchain network to India. *Analytics India Magazine*. Retrieved from https://analyticsindiamag.com/jp-morgan-blockchain-network-banks-india/

Bloomberg Quint. (2016). India's first banking transaction on blockchain. Retrieved from www.bloombergquint.com/business/indias-first-banking-transaction-on-blockchain

cio.com. (2016). ICICI Bank executes India's first blockchain-based transaction. Retrieved April 5, 2021, from https://cio.economictimes.indiatimes.com/news/strategy-and-management/icici-bank-executes-indias-first-blockchain-based-transaction/54824471

Clayton, M. and Christensen, J. L. (1995). Disruptive technologies: Catching the wave. *Harvard Business Review*, 43–53.

Clayton, M. C., Raynor, M. R. and McDonald, R. (2015). What is disruptive innovation? *Harvard Business Review*. Retrieved from https://hbr.org/2015/12/what-is-disruptive-innovation

Economictimes.com. (2019). Seven Indian banks join JP Morgan's blockchain platform. Retrieved April 5, 2021, from https://m.economictimes.com/markets/stocks/news/seven-indian-banks-joins-jp-morgans-blockchain-platform/articleshow/71281642.cms.

Economictimes.com. (2021, February 23). SBI joins JPMorgan's blockchain-based payment network. Retrieved April 5, 2021, from https://m.economictimes.com/industry/banking/finance/banking/state-bank-of-india-joins-jpmorgans-blockchain-based-payment-network/articleshow/81157341.cms

Edgeverve.com. (2017). ICICI Bank—Reimagining the trade finance process with blockchain. Retrieved April 5, 2021, from www.edgeverve.com/finacle/client-stories/icici-bank-reimagining-trade-finance-process-with-blockchain/

ICAEW. (2021). Retrieved February 28, 2021, from www.icaew.com/technical/technology/blockchain/blockchain-articles/what-is-blockchain/history

Institute for Development and Research in Banking Technology. (2017). Applications of blockchain technology to banking and financial sector in India. Retrieved from www.idrbt.ac.in/assets/publications/Best Practices/BCT.pdf

Nakamoto, S. (2009). Bitcoin: A peer-to-peer electronic cash system. Retrieved April 27, 2021, from https://bitcoin.org/bitcoin.pdf

pymnts.com. (2019, March 29). Ripple, India's Federal Bank team on cross-border payments. Retrieved April 5, 2021, from www.pymnts.com/blockchain/2019/india-federal-bank-cross-border/

Shetty, M. (2017). *RBI arm tests tech behind Bitcoin*. Retrieved 17 February 2022, from https://economictimes.indiatimes.com/articleshow/56588911.cms

Stuart Habe, W. S. (1991). How to time-stamp a digital document. *Journal of Cryptology*, *3*(2), 99–111.

Yes Bank. (2017). Yes Bank implements multi-nodal blockchain solution in India. Retrieved April 5, 2021, from www.yesbank.in/media/press-releases/fy-2016-17/yes-bank-implements-multi-nodal-blockchain-solution-in-india

4 A Case Study of Trust Management for Authorization and Authentication in IoT Devices Using Layered Approach

Dr. Kamini, Dr. Chamkaur Singh, Dr. Neenu Juneja, and Ms. Sarabjeet Kaur

CONTENTS

DOI: 10.1201/9781003184140-4

4.1 INTRODUCTION

In modern scenarios, the development of web-based applications is rising rapidly. To build such interbed-based applications, the major role of Internet of Things (IoT) technology came into existence (Cicflowmeter 2019). In general, to connect various hardware devices, such as routers, bridges, and gateways, IoT provides interconnections. It also allows various objects to be controlled and behave from remote places from distributed network infrastructure. Along with advantages of reducing human efforts, IoT is a very good and intelligent technique to deal with and provide access to physical devices (Adware 2019). The advanced facility of autonomous control features the control number of devices without any human interaction.

"Things" in the IoT sense, is the combination of hardware, software, data, and services. "Things" can refer to a wide range of devices, such as DNA analysis devices to monitor the environment, electric clamps in coastal waters, Arduino chips in home automation, and many others. With the availability of existing technologies, these devices gather useful data to share on different distributed network devices (Abeshu and Chilamkurti 2018). This author has also presented with examples, including home automation systems, which use Wi-Fi or Bluetooth to exchange data between various devices in a home. One important feature is connectivity, which refers to establishing a proper connection between all the things of IoT to an IoT platform, may it be server or cloud. After connecting the IoT devices, it needs high-speed messaging between the devices and cloud to enable reliable, secure, and bidirectional communication. The real features of analyzing the data collected and use them to build effective business intelligence with the connection of relevant things (Bhowmick and Duchi 2018). There is a need to develop a smart system that requires good insight into the data gathered from all these things. For instance, if a coffee machine is about to run out of beans, the coffee machine itself would order the coffee beans of your choice from are tailer. There are sensor-based devices that merge with IoT technologies for detection and measuring any change in the environment and report on their status (Dangi 2012).

Authorization and authentication have a significant role in processing security in IoT devices. IoT is nowadays getting significant interest from the community and society. Individuals from around the globe are centered on pushing forward in endeavors to upgrade usability, maintainability, and security through standardization and improvement of best practices (Diro et al. 2017). In this contemporary era, security becomes an essential part of all devices because of its effect as one of the most constraining components to more extensive IoT appropriation (Bhagoji et al. 2019). Plenty of research regions exist in the security space, extending from cryptography to organize security mechanism. These domains range from home automation, environmental monitoring, healthcare, logistics, and smart homes. Nevertheless, IoT is confronting numerous security issues, for example, verification, key administration, recognizable proof, accessibility, protection, and trust management.

In fact, setting up trustful connections between various hubs in IOT give an underlying security achievement to have trusty frameworks that avoid different malicious hubs (Pajooh et al. 2021). This chapter provides a study of existing research applicable to IoT at each layer of IoT architecture for the use of identity management, authentication, and authorization. Moreover, comparison on existing security can be analyzed to identify the challenges to traditional security solutions and provide support for bringing the trust between the environment to IoT for better authorization and authentication (Dhir and Kumar 2020). Further, the chapter focuses on knowledge about the various trust-management models and techniques for better communication among devices and the outside environment.

These are the passive networks in IoT technology that bring active networks. It would not be possible to hold and maintain an effective IOT environment without sensors. In this way, IoT brings the connected technology, product, or services to actively participate between each other. Without making any failure in the system, it is extremely important to create end-to-end management of IoT system devices.

For example, it has been seen in the real-life example due to unavailability of people at home for a few days (Zhao et al. 2018). Due to this reasons, retailers did not provide beans in cases where the coffee machine ordered them. This reason can lead to the failure of the IoT system. So, there is a need for endpoint management. The major contribution of this work is to create a relationship between the client node and devices at the application layer. Comparison for trust management at different layers is also included in this chapter. The security challenges and various network attacks are also discussed in comparison to traditional protocols. The opnet-based simulation surveys the devices connected with the trusted node of networks.

This chapter is organized as follows. Section 4.2 is focused on IoT-based security challenges related to device, user authentication, insufficient awareness, and security-based attacks. Section 4.3 discusses the cyberattacks on each layer of IoT architecture related to identification, sensing, communications, services, and web semantics. With the popularity of IoT devices in the world, the threat of cyberattacks has increased. Section 4.4 focuses on trust management in various different layers of computer networks. Here, the response results from an opnetbased simulation have stated about the devices which are connected to servers accessing the application services, and their traffic response time for uploading and downloading data from the server is recorded. Section 4.5 is based on IOT security challenges as compared to traditional security protocols. The validations plan for the analysis of authorization and authentication mechanisms in IoT have also been discussed. Section 4.6 contemplates the writing of the validation plans for IoT. The assessment relies upon the multi-criteria course of action subject to the IoT application zones. Smart network is taking over customary force frameworks because of its productivity and adequacy, yet there are security issues, which are discussed. The proposed mechanism for providing secure roaming to users without authentication is possible for group signatures that allow telnet users to gain the access of any web service at any place. Handshaking and the verification process for the protocol are only between the client and server, not between end-to-end devices.

4.2 IOT SECURITY CHALLENGES

When we work on our computers on the internet, we think it's only us. Without our knowledge, there are "things" that are communicating continuously with the internet. For instance, a vehicle sending messages to the mechanic that it needs service

(Abadi et al. 2016). Although IoT is amazing, it's not totally safe; the technology is still immature. In the current state, not only the manufacturers but also the users have to overcome many security issues, like manufacturing different standards, updation in management, physical concreting, less knowledge, and awareness for users.

Fewer Complaints on the Part of IoT Manufacturers: IoT manufacturers are less compliant. New IoT devices are vulnerable because security is not taken into consideration and time and resources spent on the misless (Zhang et al. 2019). For example, most of the fitness trackers who are using After the initial pairing, Bluetooth remains visible, and if the devices used for fitness tracker have the MAC addresses are the same as the padlock device's, then a smart fingerprint padlock can be accessed with a Bluetooth key that has the same, and a smart refrigerator can reveal Gmail login credentials. One of the many security issues with IoT is that it lacks in universal security standards, and manufacturers do not cease creating these poorly secured devices (Abeshu and Chilamkurti 2018).

Insufficient Awareness and User Knowledge: People who use the internet regularly have started to learn to overcome problems like phishing, viruses, spam emails, and have started using stronger passwords to secure their Wi-Fi networks. Because IoT is a relatively new technology, not many people are aware of it. Even though the security issues mostly occur from the manufacturers, there are bigger issues that can be created by the users (Alom and Taha 2017). Thus, the users' lack of awareness of the functioning of IoT poses the biggest risk. For example, hackers often trick people to gain access to a network. Social engineering attacks are by far an overlooked security risk. An example of social engineering was the Stuxnet attack in 2010 that happened at a nuclear facility in Iran when a worker plugged a USB flash drive into one of the internal computers (Adware 2019). Directed to PLCs (programmable logic controllers), it led to an explosion of the plant by corrupting 1,000 centrifuges.

Update Management for IoT Security Problems: Insecurity in the software or firmware is another IoT risk. Although the device can be sold with updated software, new issues are bound to arise. It is important to maintain security on IoT devices through updates as soon as the issues are found. Few IoT devices can be used continuously without updating them in comparison to smartphones or computers, which are updated automatically. It is easy for a hacker to hack the system, as, while updating the device, it will be down for a short span of time. If there is an unencrypted connection and unprotected update files, it can become simpler for the hacker to get ahold of sensitive information.

Lack of Physical Hardening: IoT security issues can also be caused due to lack of physical hardening. Some IoT devices do not need a user to operate, but they need to be secured from outside threats. It is, however, a challenging task for the manufacturers to ensure security in low-cost devices. Maintaining IoT devices securely is in the hands of the users as well. It is easy to tamper with a house-used smart motion sensor or a video camera if it's unprotected.

Botnet Attacks: Collectively, many malware-infected devices are a threat. It takes an army of bots to infect the devices and hack them (Bhowmick et al. 2018). The main concern is that all IoT computers are vulnerable to malware attacks. These computers lack the ability to update software protection on a daily basis, like a machine does. As a result, the computers are easily transformed into infected zombies that can be used as tools to transmit massive quantities of data. Due to their vulnerability to malware attacks, IoT devices lack security. Moreover, electrical grids, telecommunication systems, transportation systems, water-cleaning treatment facilities that many people use, and more are at high risk of attack by a botnet (Bonawitz et al. 2019).

Industrial Espionage and Eavesdropping: Privacy invasion is one of the many important IoT security issues. The user's sensitive data may be used, thus making spying a real problem. Many IoT devices store critical user information, which can be bare to the hacker. Some specific IoT devices with security problems have been banned in some countries (Briland et al. 2017). For example, Germany banned the My Friend Cayla doll after labeling it an espionage device.

Hijacking the IoT Devices: Out of the many malwares, ransomware is a nasty one. Instead of destroying the sensitive files, this malware encrypts them and blocks them. To unlock the files, the hacker can demand a ransom fee. Around 70% of the surveillance cameras in Washington, DC, were hacked and infected with ransomware just before the inauguration speech given by former president Donald Trump, The police were unable to record for several days. It is rare, however, for IoT devices to be infected with ransomware (Chu et al. 2008). It is a very popular concept nowadays, though, and there might be a risk of smart devices to be infected in the future.

Data Accuracy Risks of IoT Security in Health Care: Data is being moved on IoT as it is collected from the outside world and is sent to storage without being encrypted. A hacker can control and alter the data and can transmit false information, which in turn may lead to risking patients' lives. For example, healthcare devices, like pacemakers, are at risk of being hacked and giving altered pacing or shocks or even depleting the battery (Cicflowmeter 2019).

IoT Bots for Cryptomining: For cryptocurrency mining, colossal CPU and GPU resources are needed. Infected botnets may not cause damage, but they do mine cryptocurrencies. The open-source cryptocurrency Monero was mined with a video camera. IoT miners have a lot of potential to attack, flood, and disrupt the entire crypto market with a single attack. At first, people's realization after the Mirai attack that any IoT device is a resource for the bot army, was the beginning (Dai 2017). The security risks at present and in the coming years are inevitable. IoT security problems will become more complex with more variations of devices. There is a need to create universal IoT standards to better control security issues. Unlike earlier, where the smart devices never needed an internet connection, we are surrounded by objects that continuously collect our personal and sensitive information. Without proper security, the data that hackers can steal is unimaginable. The aforementioned IoT security threats are just a few out of many. Security needs to be of utmost importance in these devices (Dangi 2012).

4.3 POSSIBLE CYBERATTACKS ON EACH LAYER OF IOT ARCHITECTURE

4.3.1 Perception Layer

The perception layer is also called the sensor layer. It identifies things to collect the information out of it. There are many sensors available, such as temperature, pressure, proximity, accelerometer, optical, gas, and smoke. These sensors are attached to the objects and are responsible for collecting the information. These sensors are used as per the need of application. As the list specifies, these sensors collect information like temperature, gas leakages, changes in environment, any kind of motion, etc. Though these sensors add lots of functionality to our devices, they are targeted by many types of attacks (Dhar and Bose 2020). The attackers replace the sensor with their own to hack the information. Some of the commonly occurring attacks on the perception layer are discussed here:

- **Eavesdropping:** A real-time, unauthorized attack. In it, a hacker attacks communications privately made by telephone, text, fax, or video conference. The attacker tries to hack the information that is being exchanged or transferred over the network. Insecure data transmission leads to such attacks, and the hacker can misuse the information being transmitted.
- **Node capture:** The most dangerous attack that occurs in this layer. In this type of attack, a hacker completely controls the key as well as a gateway node. This may capture the whole communication of the sender or receiver. Also, since a key is involved, the protected records are stored in memory.
- **Fake node and malicious code:** This is a type of attack where an intruder indulges a node to the already existing system that gives unauthorized data as input. The main aim of this attack is to stop communicating real information. A node added by an intruder uses the energy of authorized nodes and destroys then network.
- **Replay attack:** This is also known as a playback attack. In this type of attack, a hacker secretly listens to a conservation between a sender and receiver and takes authorized information. The attacker then sends the accessed information to the victim by showing his identity proof. The data has been translated so that someone who receives it will recognize it as a legitimate request and respond as the attacker wishes.
- **Timing assault:** This attack is commonly found in computers of limited processing power. It allows an attacker to uncover insecurities and discover secrets about a system's protection by watching it, such as how long it takes the system to react to various queries. It also keeps track of the input and cryptographic algorithms.

4.3.2 Network Layer

Another name for the network layer is the transmission layer. This layer bridges the gap between application and the perception layer (Dhir and Kumar 2020). It transfers the information that is being collected through sensors deployed in physical objects. The transmission medium can be wireless or wired. It is also responsible for connecting the devices and the smart things to each other via network. It has the most critical security concerns concerning the confidentiality and authorization of information sent through the network. The following are some of the most important security threats and issues that exist in the network layer:

- **Denial-of-Service (DoS) attack:** Prevents a genuine user from accessing the network and its resources. This attack is basically about sending multiple unauthenticated requests to the server, keeping it busy for a long time. This makes the server unavailable to authenticated users, thus harming the whole network transmission.
- **Man-in-the-middle (MITM) attack:** In this type of attack, an attacker silently obstructs the information shared by a sender or receiver. They believe as if they are in direct communication. As the attacker gains the control of information, he can manipulate it. This may cause a hazardous threat to network security, which allows attackers to access and change real-time information.
- **Storage attack:** As we know, information is stored either on a storage device or the cloud. Both can be under security threat, and the information of the user can be accessed by an attacker who can change it before transmission. The duplicity of information provides the advantage to the attackers by giving more chances to the attackers for performing attacks.
- **Exploit attack:** This refers to any immoral or illegal assault carried out by machines, a group of data, or a sequence of commands. It happens due to security vulnerabilities present in an application, whole system, or in any hardware. It is done by getting control of any information or system over a network.

4.3.3 Application Layer

This layer has all the applications deployed on a server that uses IoT technology (Diro et al. 2017). Some IoT applications can be smart home, smart city, smart etc. It is responsible for providing services to a particular application. These services can be of any type and depends on the requirement of the application for which it is being used as well as on the type of information provided by the sensors. The application layer is the most sensitive with regard to security. Particularly, when IoT is being implemented for smart homes, it causes many threats that affect the application layer directly. Some of the security threats and issues that commonly occur in the application layer are:

- **Cross-site scripting:** This attack is also known as an "injection attack". It inserts some client-side script into the existing code of the trusted site. It allows an unauthorized user to fully modify the contents of the application to meet his needs and to use accepted information in an unauthorized manner.
- **Malicious code attack:** This is the injection of code into some section of software that can create unintended modifications and have an impact on the device. Antivirus software would not be able to interrupt or monitor this kind of attack.
- **The ability of dealing with mass data:** Due to a huge amount of data, network is not capable of dealing with data transmission as per specifications due to the hardware and a significant volume of data transfer between users. As a result, it leads to network noise and a loss of data.

4.4 TRUST MANAGEMENT IN APPLICATION NETWORK

4.4.1 Introduction

Presently, due to widely spread IoT networks, the framework design of a security for rising intelligence environments should be based on a modular architecture for providing adaptability (Table 4.1). First and the foremost, a common context of vibrant intelligence was started with home services, and with the passage of time it has extended to workspaces, open areas, and clinical environments. The proposed system design has a layered structure and includes at each level (hub, proxy server, and cloud) a component mandated to monitor and act accordingly if the module shows symptoms of attacks (Dong et al. 2020).

The structural design presented is based on the prototype centralized model but introduces decentralized framework elements by using the intermediate mode as a key element. The interconnectivity between the various nodes of networks must be monitored and managed by the proxy server while communication due to single mode of link is managed.

ALGORITHM 4.1 REPUTATION COMPUTATION

If Periodic Trust packets are received from network nodes then Combine trust values for every node to its reputation value
for All Nodes do
if Disbelief > bad_threshold then
Block the node as a bad node Notify the operator and network nodes end if
end for end if

ALGORITHM 4.2 DODAG UPDATE PROCESS IF MY PARENT IS A BAD NODE THEN

Wait for periodic DIO messages
Select a new parent that is not a Bad Node end if

4.4.2 Trust Attack Model

On-Off attacks (OOAs) can be viewed as a random attack. Just like its name, a malicious node could turn on or turn off its switch to useful services. With this mechanism, the node will not get low trust value and could perform an attack before the trust system is aware of it. With its feature of randomness, this attack is the hardest to detect (Enthoven and Al-Ars 2020).

Like the On-Off attack, the opportunistic service attack (OSA) could be good or bad at different times; in other words, a malicious node carrying OSA could perform a good service for one kind of service and misbehave for another service. This is also known as a selective behavioral attack on network technology.

TABLE 4.1
Trust Management Framework Architecture.

Layer	Key Components	Service Domain
Business process layer	Business process models, business ecosystems, price and cost model	Manages overall IoT system activities, builds business models, graphs, flow charts Transformation decision-making based on "Things", i.e., big data analysis and app.
Application layer	Reporting, analytics, and control models	E–health care, retail, military transportation, energy supply chain surveillance
Management service layer	Data abstraction, aggression, access models	Device modeling, configuration management, trust management, security Control, data element analysis
Transmission layer	Zigbee, Bluetooth, GPS, Wi-Fi, GSM, 3G/4G, Infrared, AMQP, DDS	Controlling and transformation
Network layer	Wireless LAN, PAN, IPV6/IP routing, 6LoWPAN, IEEE802.15.4	Communication and processing units
Perception layer	RFID, wireless sensors, acutators, embedded devices, machine to machine (M2M)	Wireless sensor networks, acutator network

A self-promotion attack (SPA) is one in which it seems like the services is bragging to gain more trust from its peers to be selected as a service point, but then they will perform bad services.

In a bad-mouthing attack (BMA), bad nodes make other well-behaved nodes lose their high reputation by providing them bad trust evaluations to decrease their chance of being selected as serving nodes. This mechanism would increase other bad nodes to be selected and mess up the normal services so that qualified services would not be done in an efficient way.

In a ballot-stuffing attack (BSA), bad nodes collide with each other and intentionally vote for one of their peers to increase the opportunity of the malicious node being selected as a recommended node.

4.4.3 Statistics Collection Through OPNET

The application demand allows the collection of the following statistics at the client and/or the servers (Figure 4.1).

At the requesting node:

1. Response time for each request in each application demands that it originate in this node (Figure 4.2).
2. Traffic sent and traffic received in bytes/sec and packets/sec for each application demand that it originate in this node (Figure 4.3).

At the responding node:

1. Traffic sent and Traffic received in bytes/sec and packets/sec for each demand of the node services.

4.5 CHALLENGES OF IOT TRUST MANAGEMENT

However, existing WSN and MANET are inadequate for IoT. IoT actually expands on WSN and VANET requirements, which could be concluded as wider architectures, more heterogeneous, inconstant resource capabilities and increased autonomy. The IoT trust authorities must have high security provisions in place for all IoT hubs at all times, from entity recognition to service provisioning, data acquisition to infrastructure governance. At the outset of an agency operation, security instruments should consider facilities. The following are the problems of IoT trust management (Feng 2017):

Unpredictability: With the IoT links and information transfers, the common network gateway aids in the delivery of a high ratio of goods and services. The hardware system, networking, I/O channels, and performance can all be used to differentiate such services. One explanation for these differences may be the hardware of the things, which could result in varying processing speeds, storage capabilities, and energy utilization (Geetha and Thilagam 2020). All connected with IoT can be interpreted as a network node that can be located, addressed, and interact using various

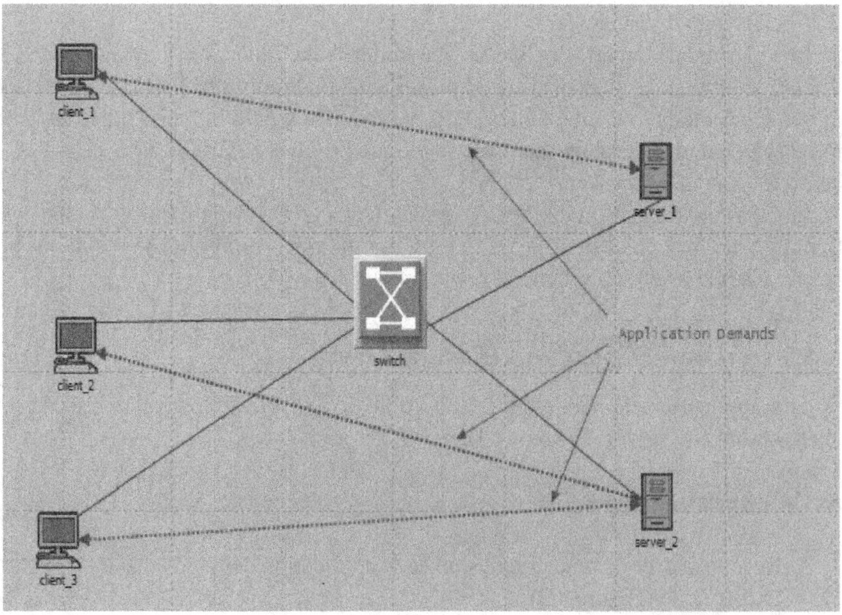

FIGURE 4.1 On-Demand IoT-Based Application Scenario Model Application.

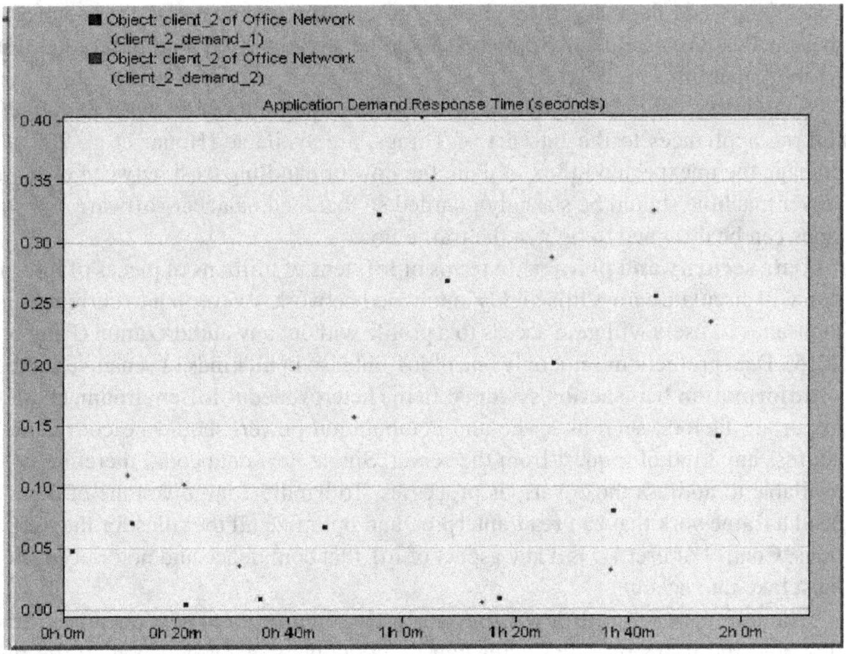

FIGURE 4.2 IOT-Application Response Time.

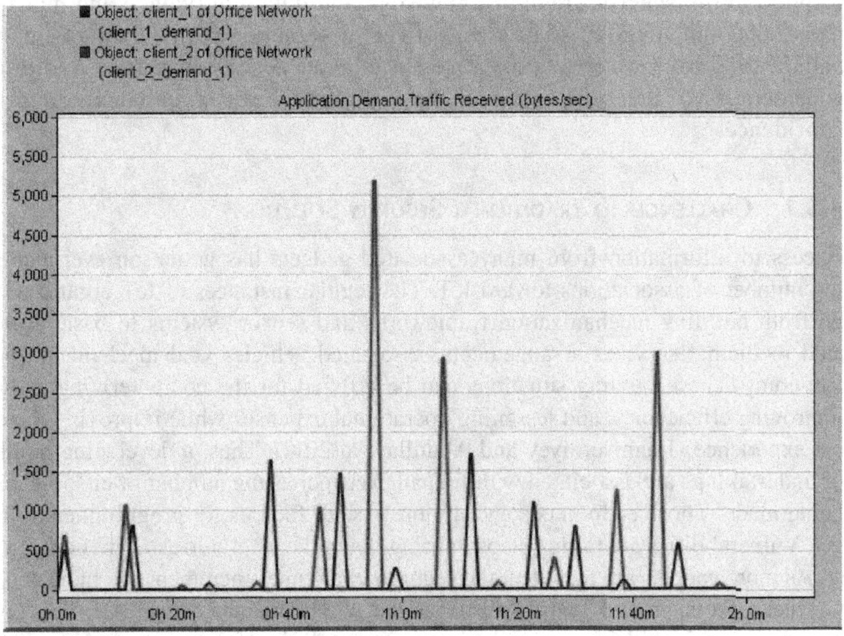

FIGURE 4.3 IOT Application Demand Traffic Received.

conventions and data structures. In this way, security agreements that connect all these nodes into an IoT environment should be slightly heavy and have good transferable capability.

Scalability: Numerous communications, knowledge transfers with the effect of linking appliances to the Internet of Things, are available (Honar et al. 2021). To manage the unexpected influx of data, the flow of handling trust between client and server machine should be strongly guarded so that load balancer software and strategies can be designed to lighten the traffic flow.

Data security and privacy: In terms of IoT, tens of millions of pieces of information will circulate across this highly automated network. According to current trends, thousands of users will gain access to a profile without any authorization (Pang et al. 2015). Data protection is not only one of IoT but also in all kinds of sensitive subjects.

Information transaction systems: In the heterogeneous IoT environment, all the important factors, such as space and computation power, should be considered to address any kind of request from the server. Single flow data could therefore not be available to address those various processes. To handle a large volume of data, we need a framework that can read, interpret, and optimize all the rules for the transactions. Control of user ID is a key aspect of IoT that confidence and notoriety systems must take into account.

The hub network in IoT will have a lasting identity and several more identities, and the hub's real character can be conceived. Approval is also the main subject of management character. Validation and clearance are simultaneously subject to certain open testing issues. The issues provide suggestions for how to proceed within the context of an integrated and appropriated strategy to respond to the specific trade of knowledge and data, confidence and management (Zhu and Jin 2020). The trust officer's interaction with existing demands and the range of resources is supported by a better-designed process. Additional considerations, such as governance, have shown interest in enhancing internet confidence.

4.5.1 Challenges to Traditional Security Solutions

Access to information from many associated gadgets has gotten an ever-increasing number of associations toward IoT. The regular instances of IoT organizations go from building mechanization frameworks and sensor systems to basic associated medicinal services arrangements, associated vehicles, and mechanical apply autonomy. These sending situations can be utilized for the computerizing gadget, improving efficiencies, and lessening operational expenses while improving the client experience (Imamverdiyev and Abdullayeva 2018). Thus, a developing number of undertakings are associated with IoT; an ever-increasing number of endpoints are being made. Those endpoints suggest more assault focuses for programmers to target. Vulnerabilities are taking steps to bring down the level of trust right now.

As more gadgets are included, the security challenges increment. As indicated by Gartner reports, around 30 billion IoT gadgets will be associated by 2021. This gives programmers 26 billion potential targets. These difficulties are unique in relation to customary cybersecurity since security difficulties of ongoing IoT systems can have extensive impacts on security. Lately, there has been an expansion of botnets among IoT gadgets. A botnet exists when programmers control web-associated gadgets remotely and use them for illicit purposes. Aventurecouldhavetheirgadgetsco-selectedasapieceofabotnetwithoutrealizing it.

The trouble is that an incredible number of associations don't have constant security answers to address this.

A couple of years prior, security experts were centered only around ensuring cell phones and PCs. Today, there is an expansion of IoT gadgets. There are more than 10 billion gadgets around, which number could increment to twenty billion by 2020. More IoT gadgets mean greater security liabilities, and it's a developing assignment for security experts. Notwithstanding the shortcomings of the IoT gadgets, the contrary stress is with interconnected heritage frameworks. In an endeavor with a rising number of IoT gadgets, heritage advances may sound strange. A rupture of an IoT gadget could likewise prompt the breaking of an inheritance framework that needs current security guidelines (Kumar et al. 2020).

Many IoT devices employ original default passwords that are weak. Although it's recommended that one should simply change the passwords, some IT leaders fail to employ this easy step. A weak, easy-to-guess password could make an IoT device susceptible to a brute force attack. This issue is so prevalent that California banned default passwords in 2018 (Liu and Lang 2019). Phishing is a security worry over each endeavor advances, and IoT gadgets speak to the most recent attack vector. Programmers can send a sign to an IoT gadget that triggers numerous difficulties. In spite of the fact that it is among the essential regular sorts of security assaults, and they are frequently halted, numerous associations can't suitably prepare their laborers about the most up-to-date phishing dangers.

Security experts ought to be proactive to stop IoT security ruptures before they happen. Notwithstanding, a few endeavors may have a nonattendance of a solid administration framework that would screen action and supply bits of knowledge into potential dangers. Without such an answer, a venture won't distinguish potential breaks in time. Programming refreshes is another strategy IT experts use to ensure that PCs and cell phones are secure. Some IoT gadgets come up short on the amount of programming refreshes that different advances may get. Additionally, undertakings battle to supply basic security updates to IoT gadgets inside the field. Undertakings must ensure client information (that goes for both an organization's outside and inner clients). This is a need on the grounds that numerous laborers are utilizing IoT gadgets gave by their bosses. At the point when a rupture happens, prompting bargain of individual information, an undertaking's notoriety would endure a tremendous shot, which is the reason it is one of the most noteworthy IoT security challenges that can't be overlooked (Liu and Lang 2019).

4.6 ANALYSIS OF AUTHORIZATION AND AUTHENTICATION MECHANISMS

This area contemplates the writing of the validation plans for IoT. The assessment relies upon the multi-criteria course of action subject to the IoT application zones.

A Merkle tree, likewise called a hash tree, is an information structure executed for information confirmation and synchronization. All the leaf hubs are at a similar profundity and are as far left as could be expected under the circumstances. It keeps up information honesty, utilizing hash capacities for this reason.

4.6.1 Smart-Networks Grid

Smart-network grids are taking over customary force frameworks because of their productivity and adequacy, yet security issues are present in such fields. To gather and collect the information and utilization for power each homes that has smart networks that can be used while sending the information by means of remote correspondence to the neighborhood gateway (NG).

4.6.2 RFID and NFC-Based Applications

Radio frequency identification (RFID) is a remote innovation that is comprised of labels that we can connect to any physical object or even people. The primary motivation behind RFID is the recognizable proof or discovery of the labeled items. RFID can be used in different fields, e.g., production networks, health care, climate detecting, and so on (Niknam et al. 2020). Lightweight validation convention for RFID labels depends on PUF (physically unclonable function). The convention is comprised of three exchanges: label acknowledgment, check, and update. In the primary exchange, the tag is perceived by the label reader (Sharma et al. 2020). The subsequent exchange is where the reader and the tag commonly confirm each other's legitimacy. In the last exchange (update), one should keep up the most recently utilized key for the following confirmation.

To protect the inventory network of associated gadgets, validation and detectable quality of the IoT gadgets is empowered through an RFID-based arrangement. The validation procedure consist of two levels: checking the network between the tag and the IoT gadget and afterward endorsing the detectable quality of the tag. In an IoTRFID-based framework, the RFID reader is associated with the internet forming an IoT end gadget (Osorio et al. 2017). On the other hand, it is related with the marked things through RFID correspondence conventions. The concept of IoT RFID-based applications in the fifth generation computer systems is for the reader to reserve and store the keys for the visited labels for the purpose of decreasing the expense of calculations and to increment the security capacity.

4.6.3 Vehicular Networks

Vehicles these days are allied with the internet or IoT to shape vehicular systems or Internet of Vehicles (IoV). Such availability is utilized in various circles. Providing traffic data to clients, sharing riding administrations, charging electrical autos, and so forth. The field of electrical vehicles (EV) has become a pattern, and vehicle verification is a difficult point in EV frameworks. Two-factor confirmation plot for EV, however, tends to be set up in various fields. The plan is a mix of exceptional logical elements. The vehicle is regarding the server through a remote availability and with the charger by a charging link, so it relies upon the physical network to check the character (Zhu and Jin 2020). A vehicle can check numerous messages simultaneously (Zhu and Jin 2020), and their marks can be amassed into one message, which diminishes the extra room required by a vehicle or an information gatherer. Broadcast validation plot called prediction-based authentication (PBA) that ensures against DoS assaults and opposes bundle misfortune. The convention depends on Merkle tree, which is a data structure used for the development of moment check of pressing messages, and a self-produced message authentication code (MAC) stockpiling instead of putting away all the getting marks. PBA as a communication validation conspire in VANETs (vehicular ad hoc networks).

4.6.4 Wireless Sensor Networks

Nowadays, sensors are embedded into different areas (appliances in homes, vehicles, grids, etc.) that have the features of wireless sensor networks (WSNs), which is the capability of adding the connectivity and sensing as per environment. A validation convention at the media get to control a sublayer called optimization of communication for ad-hoc reliable industrial systems (OCARI) in which they utilize a one-time shared key. This is a system reasonable for asset-compelled gadgets. Blom's key pre distribution scheme and the polynomial pattern are viewed as reasonable for being utilized as key administration conventions for some IoT use cases. The utilization of the BAN-logic (rationale of conviction and activity that guarantees one part of communication accepts that the key in verification is nice) guarantees a common authentication protocol known as E-SAP, which stands for efficient-strong authentication protocol for remote well-being care applications. It comprises various features: a two-factor verification (keen card and secret word), shared verification between sensors, symmetric cryptography for guaranteeing message privacy and the capability to alter passwords. The Biometric and symmetric crypto-framework applied For Novel shared confirmation and key administration for the Wireless Sensor Networks (WSN) frameworks. They contrasted their plan, and the plans given in the writing with respect to calculation and correspondence cost and security dangers utilizing both BAN-rationale and AVISPA (Automated Validation of Internet Security Protocols and Applications) devices.

4.6.5 General IoT Applications

The two-step verification for protocol is required for performing handshaking between client machine and server machine for providing the end-to-end security between the devices. For performing this activity, a password or a shared secret key for providing authentication and PUF is the another factor for authentication. To accomplish shared validation and secure key administration for resource-constrained devices, a gigantic number of gadgets needing to get to the system prompts an overburden for the confirming server and to accomplish common confirmation and secure key administration for asset-compelled gadgets, a gathering-based lightweight verification is utilized, and a key understanding plan called GLARM. With different mixes of MAC code, bunch-based key pair mix different gadgets to be validated for GLARM security stages (Liu et al. 2017). Light weight gadget validation convention for IoT frameworks called speaker-to-microphone (S2M) accomplishes separation verification between remote IoT devices. To check the experimental results, mobile phone–based (Pang et al. 2015) and PC-based tests have been conducted. A PUF uses the hardware fingerprint to authenticate IoT devices known as the new hardware-based approach (Paudel et al. 2019).

4.7 CONCLUSION AND FUTURE SCOPE

To conclude, this chapter has studied IoT with the existing technology for use in identity management, authentication, and trust between various devices. This chapter surveys the network layer for providing the end-to-end security between individual layers. In the future, a survey of IOT devices with society, communities, and social interaction can be concluded to resolve upcoming issues in trust management. Additionally, challenges to traditional security protocols and their support for trust between the IOT devices for better authorization and authentication has also been discussed.

To support the trust, opnet-based simulation results with various client and end-server devices have been shown for better communication in the environment.

REFERENCES

Abadi, M., Chu, A., Goodfellow, I., McMahan, H. B., Mironov, I., Talwar, K., &Zhang, L. (2016). Deep learning with differential privacy. In *Proceedings of the 2016 ACM SIGSAC Conference on Computer and Communications Security* (pp. 308–318). https://dl.acm.org/doi/abs/10.1145/2976749.2978318.

Abeshu, A., & Chilamkurti, N. (2018). Deep learning: The frontier for distributed attack detection in fog-to-things computing. *IEEE Communications Magazine*, 56(2), 169–175. doi: 10.1109/MCOM.2018.1700332.

Alom, M. Z., & Taha, T. M. (2017, June). Network intrusion detection for cyber security using unsupervised deep learning approaches. In *2017 IEEE National Aerospace and Electronics Conference (NAECON)* (pp. 63–69). IEEE. doi: 10.1109/NAECON.2017.8268746.

Android Adware Dataset. (2019). www.unb.ca/cic/datasets/android-adware. Last accessed 30 May 2019.

Bhagoji, A. N., Chakraborty, S., Mittal, P., & Calo, S. (2019, May). Analyzing federated learning through an adversarial lens. In *International Conference on Machine Learning* (pp. 634–643). PMLR.

Bhowmick, A., Duchi, J., Freudiger, J., Kapoor, G., & Rogers, R. (2018). Protection against reconstruction and its applications in private federated learning. *arXiv preprint arXiv:1812.00984.*

Bonawitz, K., Eichner, H., Grieskamp, W., Huba, D., Ingerman, A., Ivanov, V., Kiddon, C., Konecny, J., Mazzocchi, S., McMahan, H. B., Van Overveldt, T., Petrou, D., Ramage, D., & Roselander, J. (2019). Towards federated learning at scale: System design. In*SysML.*

Buczak, A. L., &Guven, E. (2016). A survey of data mining and machine learning methods for cyber security intrusion detection. *IEEE Communications Surveys and Tutorials*, 18(2), 1153–1176. https://doi.org/10.1109/COMST.2015.2494502

Cai, H., Zhu, L., & Han, S. (2018). Proxylessnas: Direct neural architecture search on target task and hardware. *arXiv preprint arXiv:1812.00332.*

Chu, F., Yuan, S., & Peng, Z. (2008). Machine learning techniques. *Encyclopedia of Structural Health Monitoring*, 967–974.https://doi.org/10.1002/9780470061626.shm184

Cicflowmeter. (2019). www.unb.ca/cic/research/applications.html#CICFlow Meter. Last accessed 30 May 2019.

Dai, J., et al. (2017). Deformable convolutional networks. In *Proceedings of the IEEE International Conference on Computer Vision* (pp. 764–773).

Dangi, C. S. (2012). Cyber security approach in web application using SVM. *International Journal of Computer Applications (0975–8887)*, 57(2), 30–34.

Dhar, S., & Bose, I. (2020). Securing IoT Devices Using Zero Trust and Blockchain. *Journal of Organizational Computing and Electronic Commerce (1091–9392)*, 31, 18–34. doi: 10.1080/10919392.2020.1831870

Dhir, S., & Kumar, Y. (2020). Study of machine and deep learning classifications in cyber physical system. In *2020 Third International Conference on Smart Systems and Inventive Technology (ICSSIT)* (pp. 333–338). doi:10.1109/ICSSIT48917.2020.9214237.

Diro, A. A., Chilamkurti, N., & Kumar, N. (2017). Lightweight cybersecurity schemes using elliptic curve cryptography in publish-subscribe fog computing. *Mobile Networks and Applications*, 22(5), 848–858.

Dong, Y., Chen, X., Shen, L., & Wang, D. (2020). EaSTFLy: Efficient and secure ternary federated learning. *Computers & Security*, 101824. doi:10.1016/j.cose.2020.101824

Elman, J. L. (1990). Finding structure in time. *Cognitive Sciences*, 14(2), 179–211.

Enthoven, D., & Al-Ars, Z. (2020). An overview of federated deep learning privacy attacks and defensive strategies. *arXiv preprintarXiv:2004.04676.*

Esposito, C., Tamburis, O., Su, X., & Choi, C. (2020). Robust decentralised trust management for the internet of things by using game theory. *Information Processing & Management*, 57(6), 102308.

Feng, C. (2017). A user-centric machine learning framework for cyber security operations center. *978-1-5090-6727-5/17/$31.00 ©2017 IEEE*, 173–175.

Geetha, R., & Thilagam, T. (2020). A review on the effectiveness of machine learning and deep learning algorithms for cyber security. *Archives of Computational Methods in Engineering*, 371–390. https://doi.org/10.1007/s11831-020-09478-2

Goodfellow, I. J., Pouget-Abadie, J., Mirza, M., Xu, B., Warde-Farley, D., Ozair, S., Courville, A., & Bengio, Y. (2014). Generative adversarial networks. *Communications of the ACM*, 63(11), 139–144.

Hitaj, B., Ateniese, G., & P´erez-Cruz, F. (2017). Deep models under the GAN: Information leakage from collaborative deep learning. *CoRR, abs/1702.07464.*

Honar Pajooh, H., Rashid, M., Alam, F., & Demidenko, S. (2021). Multi-layer blockchain-based security architecture for internet of things. *Sensors*, 21(3), 772.

Imamverdiyev, Y., & Abdullayeva, F. (2018). Deep learning method for denial of service attack detection based on restricted Boltzmann machine. *Big Data*, 6(2), 159–169.

Kumar, Y., Kaur, K., &Singh, G. (2020). Machine learning aspects and its applications towards different research areas. In *2020 International Conference on Computation, Automation and Knowledge Management (ICCAKM)* (pp. 150–156). doi: 10.1109/ICCAKM46823.2020.9051502.

Liu, H., & Lang, B. (2019). Machine learning and deep learning methods for intrusion detection systems: A survey. *Applied Sciences*, 9(20), 4396.

Liu, W., Wang, Z., Liu, X., Zeng, N., Liu, Y., & Alsaadi, F. E. (2017). A survey of deep neural network architectures and their applications. *Neurocomputing*, 234(11–26), 139.

Liu, Y., Ma, Z., Liu, X., Liu, J., Jiang, Z., Ma, J., . . . Ren, K. (2020). Learn to forget: Memorization elimination for neural networks. *arXiv preprintarXiv:2003.10933.*

Niknam, S., Dhillon, H. S., & Reed, J. H. (2020). Federated learning for wireless communications: Motivation, opportunities, and challenges. *IEEE Communications Magazine*, 58(6), 46–51. doi:10.1109/mcom.001.1900461.

Osorio, S., van Ackere, A., &Larsen, E. R. (2017). Interdependencies in security of electricity supply. *Energy*, 135, 598e609.

Pang, S., Peng, Y., Ban, T., Inoue, D., &Sarrafzadeh, A. (2015, September). A federated network online network traffics analysis engine for cybersecurity. In *Proceedings of the International Joint Conference on Neural Networks*. https://doi.org/10.1109/IJCNN.2015.7280563.

Paudel, R., Muncy, T., & Eberle, W. (2019, December). Detecting DoS attack in smart home IoT devices using a graph-based approach. In *2019 IEEE International Conference on Big Data (Big Data)* (pp. 5249–5258). IEEE.

Sharma, A., Pilli, E. S., Mazumdar, A. P., & Gera, P. (2020). *Towards Trustworthy Internet of Things: A Survey on Trust Management Applications and Schemes*. Computer Communications.

Zhang, X., Chen, J., Zhou, Y., Han, L., & Lin, J. (2019). A multiple-layer representation learning model for network-based attack detection. *IEEE Access*, 7, 91992–92008.

Zhao, J., Mili, L., & Wang, M. (2018, January). A generalized false data injection attacks against power system nonlinear state estimator and countermeasures. *IEEE Transactions on Power Systems*, 33(5), 4868e77.

Zhiqi, Bu, Dong, Jinshuo, Long, Qi, & Su, Weijie J. (2019). Deep learning with Gaussian differential privacy. *arXiv preprint arXiv:1911.11607*.

Zhu, H., & Jin, Y. (2020). Real-time federated evolutionary neural architecture search. *arXiv preprint arXiv:2003.02793*.

5 Enthralling Aspects in Automation of Face Recognition and Aging

A. Deepa

CONTENTS:

5.1 INTRODUCTION

The current era is the digital era, with remarkable growth in soft technologies and computing. The world has drawn its attention to the advantages of digital processing and data handling. The identification of face and age from images is used in various systems, such as safety measures, biometric access, smart gadgets, e-accounts and in similar processes (Lin et al., 2012; Eason et al., 1955; Zimbler et al., 2001; Shannaq & Elrefaei, 2019). The automatic inference and analysis of the data has become a mandatory enhancement in the intelligent systems. In these intelligent systems, the face plays a crucial role in identification and the estimation of the age of the person. The face remains the manifesting descriptor of a person. As per the discussion in the research by Lin et al. (2012), SIFT (scale-invariant feature transform) (Gupta et al., 2020) and MLBP (multi-level local binary pattern) features are used to find the identification of features of the input image and the age of the subject in the image. The multiple feature discriminant analysis (MFDA) is utilized to obtain the features. The identification of the face is obtained by various LDA-based classifiers. Eason (1955) uses the ANFIS (adaptive neuro-fuzzy inference system) method (Moghadam et al., 2013) to identify age by utilizing LBP (local binary pattern) and HOG (histogram of oriented gradients)

DOI: 10.1201/9781003184140-5

features. Zimbler et al. (2001) utilizes the geometric features to estimate age through the PCA (Hlaing et al., 2014; Hlaing et al., 2012) (principal components analysis) method. Alley (1998) uses the AGES (aging pattern subspace algorithm) method to calculate age. Shannaq utilizes artificial neural network (ANN) procedures to estimate age group. Age is classified into four age groups: childhood, adolescence, youth and elderly. Aging is a biological phenomenon thatoccurs in every human based on gender, habitat and physical and external factors (Lanitis, 2004). Face aging occurs with changes in the bones and muscles and becomes visible through the skin. At a younger age, the skin appears clear, as the tissues are tightly arranged. As one ages, the signs of aging become visible at the forehead, eyebrows, cheeks and near the jawline. The eyelids start to sag down, and dark circles form under the eyes. Though there are manifesting changes accomplished in this dominion, the pertaining entailment of digital image processing is on demand. In the concern of processing, there are several challenging aspects to be considered. Image processing with a still image of a face is an intricate process. The intricacy is faced with respect to different aging manners, pose and lighting effects. In addition, there is an unavailability of exact values for classification, makeup effects (Cunjian et al., 2014), facial hair and mannerism changes. The proposed system concentrates on processing the facial image, extracting the features from the input image to classify the age group of the input image. The process of identification of the features from the face consists of two major stages, such as extracting the features and classifying the face into the required category. Extraction of features and establishment of classifiers are the major considerations in the process of facial image processing and estimation of age from the classified results. The inclusion of geometrical values and the texture values together provide better estimation of age. Using the procedures of normalization first followed by face cropping and image processing to extract features enhances the accuracy of age estimation.

5.2 AGING

A person's age is most noticeable in the face, compared to other areas of the body. The face is preordained to express aging. The overall physical appearance can be managed to be rescinded, but the facial aging cannot be reversed. The major inference of age factor is observed in the region over eye, chin, cheeks and mouth part. Additionally, the absence of teeth can alter age estimation, and cosmetics added to the skin can soften the signs of aging.

5.2.1 Transition in the Face with the Progression of Age

The face, as noticed with the progression of age, reveals various changes that are visually perceptible, with remarkable variations in the facial features. In human vision, the changes on the face are noticed with minimal dissimilitude, but in actual digital analysis, the features show variations in color, texture, orientation, shape and key points. In Figure 5.1, the face of the same person at the ages of 3, 19, 24 and 29 is shown in a, b, c and d, respectively. The figure shows difference in all aspects, such as in the forehead, eyes, cheeks and cheekbones. In these four age groups, there are considerable changes in the size of the facial features. So, the analysis of the

[a] [b] [c] [d]

FIGURE 5.1 Transition of Aging.

age-estimation algorithm illustrates the result that the classification of age includes both shape and texture to be considered by considering the key values obtained from other facial parts. The angle of the face with respect to pose varies, and the gradient value also changes. So, these details are added in the finite estimation of age.

5.2.2 Aging in Men

The male face shows less transition in aging than the female face (Han, 2014). This results in a harder texture to the skin compared to females. There are various reasons for the transition away from the soft texture of a baby's face as the person grows. The presence of facial hair, exposure to external factors and hormonal changes contribute to the texture variation in males. Baldness is also a highly visible sign of aging. A person may look more aged due to several factors, such as lifestyle, consumption of alcohol, smoking and increased sugar intake. By reducing stress and moral practices, the aging can be delayed. As in Figure 5.2., the images depict age progression from 35 to 90 years old.

With respect to these features, there is need for an algorithm thatbalances various aspects in categorizing a male face and female face and hence providing a standard age-classification method.

5.2.3 Aging in Women

When considering the process of aging in women, as in Figure 5.3, it is visible more prominently in every age group. The face of a baby from 0 to 5 years looks almost similar in both the male and female faces. The changes begin only from the second age group, 6–12 years. The female face changes in appearance due to external changes in hairstyle and mannerism. The difference of gender becomes prominent on a face from the third age group 13–19. The face shows less variation in gender in the first two age groups. Changes are visible in the cheeks as well as the forehead due to the polarities in the ways males and females dress and groom themselves. But when analyzing the fiducial features, the values almost remain the same. In this system, the estimation is hence not only considered on the basis of texture values and

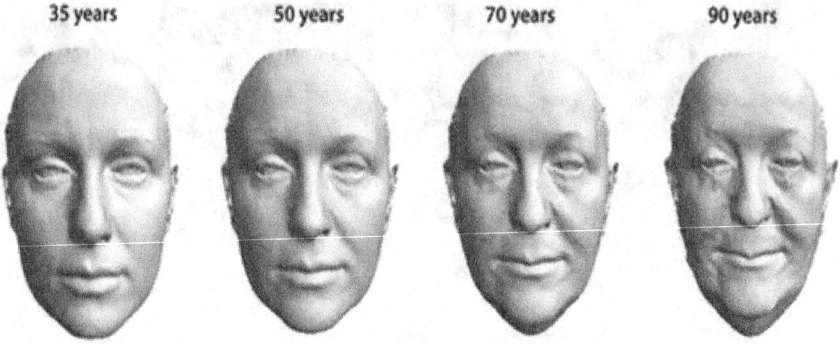

FIGURE 5.2 Progression of Age in Male Faces.

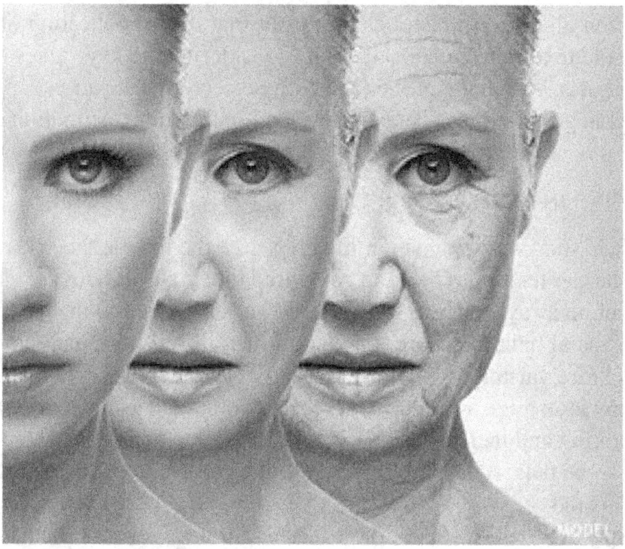

FIGURE 5.3 Progression of Age.

makeup effects but also on other various factors, such as the orientation of the face and the shape and texture of the face. The fiducial values extracted from certain key points of the face depict variation in the gradient value with which age information can be obtained. It is found that elasticity and skin tone start changing effectively beginning in the 35–39 age group, and these changes continue. Wrinkles, unevenness in skin tone and dry skin begin to appear on the face. These features can be analyzed from the fiducial points, the shape and the texture details.

5.3 PROPOSED MODEL

The appearance of the human face encounters different variations with the progress of aging (Ling et al., 2010). The major certitude in this deed is that estimation cannot

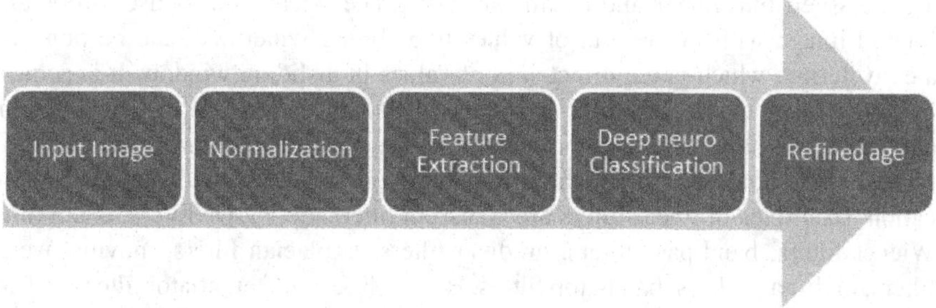

FIGURE 5.4 Proposed Method.

be provided with certainty. The fact is there are several hurdles in judging age, even by humans. In this deed there is a need for an automated system that can provide an age estimation that defines a specific age category based on the obtained features. The objective of the research is to normalize the input face image and to extract the required features from the normalized face image. Figure 5.4 depicts the procedure followed in the proposed methodology. In extracting the features from a facial image, the details of eye portion were determined, the fiducial points were located and the measures of the face were also determined with respect to the geometric ratios. The fractal directional code method (Costa et al., 2012) was used to extract the fractal features, and major information regarding the feature values was obtained. The features were extracted in terms of color and gradient, descriptors of the key points, orientation, shape and texture. With this regard, a total of 45 features, which included 10 values for orientation, 10 values for texture, 15 values for key point descriptor, 5 values for shape and 5 for color and gradient, were extracted. With these fractal values, local features, such as the geometric measures of the forehead, nose, mouth and height and width of face, were also considered. The facial angle, with respect to the angle developed between the eye points and the lip midpoint, was taken. The angle value provided the age range. This range and the fiducial value ratio were compared, and a final age group was identified. All the feature values and geometric measures were used to train the system (Cuixian et al., 2010). With proper training, the age was estimated for the testing image. The Otsu algorithm was used for performing normalization, and the classification was done using deep neural network (DNN). DNN (Jianyi et al., 2013) utilizes softmax normalization at the last stage of classification to maintain the boundary values and consists of three layers to train the set. The back propagation with scaled conjugate gradient methodology trained the system, and the classification was obtained in varying age groups (0–5, 6–12, 13–19, 20–24, 25–29, 30–34, 35–39, 40–44, 45–49, 50–55, 56–59, 60 and above). Concocting the features obtained from bio-features and fractal observations, a better accuracy is achieved. Hence, an ideal model has been proposed to retrieve a facial image, preprocess it and identify an approximate age group.

5.3.1 Filtering

Filtering is a method to change or intensify the image. The requirement is to identify the objective of the algorithm and the expectation of the proposed model. The filters are classified into linear and nonlinear filters. The linear filter is used to obtain a filtered image with linear sum of values in a sliding window. It can be done with a convolution, which is the direct sum of values in a sliding window. It is achieved by magnifying the spectrum by an image. This can be utilized in both frequency and spatial domain. Efficient computation is used with fast Fourier transform. The reverse is the nonlinear filter, which cannot provide an output with straight magnification. Examples of linear filters are Gaussian filters, derivative filters, mean filters, Wiener filters, band-pass filters, gradient filters, Laplacian filters, moving average filters, high-pass filters, band-stop filters, Hilbert filters, differentiator filters, list convolve and list correlate filters. The nonlinear filters (Suo et al., 2010; Wang et al., 2013) generally used are median filters, minimum filters, max filters, mean shift filters, entropy filters, Comer filters, ridge filters, Kuwahara filters and bilateral filters. A median filter is applicable in smoothening and enhancing an image but with some adverse effects too. This filter does not distinguish the noise and fine details. The minimal value pixel element is ignored, as it has less impact compared to the neighboring pixels. But in considering feature extraction for instance age estimation, the fine details are also equally considered. So, removing the fine details can adversely affect the accuracy of the algorithm. To nullify this effect, the adaptive median filter is used. This adaptive filtering provides better results in the uniform maintenance of image quality. The image can be subject to equalization, smoothing and sharpening after application of filtering, as shown in Figure 5.5. When the image is subject to processing, the facial features are to be maintained so as to obtain the data for estimating the age of a person from the input image. So, the requirements are the fiducial values, texture values and the edge details. When noise is present, median filtering is done, and when contrast issues are present, the image is sharpened and the values are extracted from the image. So, the adaptive filtering suits in obtaining a proper filtered image.

By performing analysis on various images and the filtered images, the matching filter, which is appropriate in several analyses, is listed in Table 5.1. Filtering is essential for facial image processing because only a processed image can provide a refined classified age group. The adaptive median filter applies the technique of spatial processing to retrieve the original image in the presence of noise. The identification of noise in the original image is done by segregating the pixel as noise and averaging the adjacent pixel values.

According to the required size and threshold value, the size of the neighborhood pixel is adjusted. The noisy pixel is identified by considering the pixel that has a value totally different from that of the neighboring pixel values. The pixel value possessing noise is superseded with the value of the pixel obtained by median processing. The removal of the impulse noise is achieved, which results in the smoothing of the image. The distortion and boundary values are reduced. The image is subject to filtering, which makes adaptation at the level of application of intensity to the input image.

To consider the level of intensity of the image, for the region of interest (ROI) the algorithm is applied.

1. If the neighboring pixels have intensity greater than the nearby pixels, the averaging is done at two levels and two numerals for intensity are calculated.

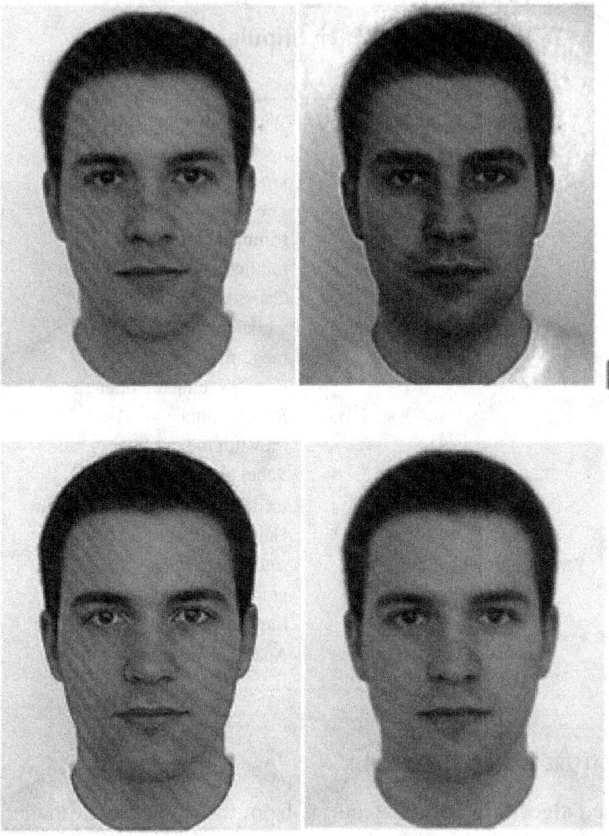

FIGURE 5.5 Application of Filters [a] Input Image [b] Equalized Image [c] Sharpened Image [d] Smoothened Image.

Intensity 1 is obtained as Level 1=I(ROI)-I(Min) and Intensity 2 as Level 2=I(ROI)-I(Max).

2. If Level 1 > 0 and Level 2 < 0, then find Intensity 2.
3. Else window size enlarged. In case the size of the window is lower than the size of the maximum intensity, Step 1 is repeated.
4. Else output the average intensity.
5. Intensity 2 is obtained from Level 1 and Level 2.
6. Level 1 is obtained by subtracting the minimum value from the average value and Level 2 is obtained by subtracting the maximum value from the average value. If Level 1 is greater than 0 and Level 2 is less than 0, the resultant value will be the average value, and the median value is provided as the output.

TABLE 5.1
Ideal Filter for Each Stipulation.

Stipulation	Ideal Filter
Image restoration	Wiener filter
Image smoothing	moving median filter
	k-neighbors filter
	Lee's filter
	trimmed mean filter
	minimum variance filter
	Gaussian filter
Edge detection	variance filter
	range filter
	Kirsch's template filter
	Roberts filter
	Prewitt gradient filter
	Sobel gradient filter
Removal of noise	median filter
	Gaussian filter
Improving contrast	sharpening filter
	gradient filter
	Laplacian filter
Facial image analysis	Adaptive mean median filter

5.3.2 Feature Extraction

The proposed algorithm obtains features from the input facial image. Facial features play a major role in estimating the accuracy of the system. A total of 45 features from the face are obtained and analyzed, and then deep neural classification is done to obtain the results. Along with extracted features, the angle value of the face is analyzed to exact the accuracy of estimation. The image is required to be segmented and filtered to enhance the feature retrieval from the input image. So, the process is made easy, as it performs independently, and the labeling also becomes easy. The Otsu algorithm is used to normalize the image before starting the extraction process. The fractals are obtained from the image. The features are then derived from the converted fractals. Using the threshold binary decomposition (TBD) algorithm and segmentation-based fractal texture analysis (SFTA) algorithm, the retrieval of features related to texture can be achieved from the image that is in gray scale. The feature vectors are retrieved, and they provide the texture details. The input color image is changed to grayscale to obtain good accuracy. Using local directional number (LDN) pattern, the texture structure and the transition of the intensity are obtained. The LDN renders the result as a six-bit binary code, which is provided to every pixel in the input image. The directional information of the facial image is encoded using LDN. The region over the face is categorized into several regions, and from each region the LDN feature is extracted. LDN is a pattern used to measure the variation in intensity of the face at certain pixel positions. The features are extracted from several pixel positions. These valuable descriptors provide the value in terms of six-bit code to ease the detection rate. These extracted features are combined to achieve a feature vector. This value aids as the descriptor of the face. The following are the five vital features that classify age.

5.3.2.1 Orientation

To extract the orientation values, edge values, certainty of corners and edge ratio are calculated. This value deteriorates with age. By considering the property of the input image, the assignment of axis to the fiducial points is achieved. The approach taken to find orientation is as given follows:

1. The scale of key point is used to select an input image (I), which is smoothened.
2. The gradient magnitude (M) is obtained by

$$M(x,y)=\sqrt{I(x+1,y+1)-I(x-1,y))\wedge 2+\left(I(x,y+1)-I(x,y-1)\right)\wedge 2} \quad (5.1)$$

3. Orientation of the histogram B is obtained from sample values of the gradient orientations.
4. The maximum value obtained in histogram is identified.
5. For certain points, multiple orientations are also considered.

5.3.2.2 Texture

Image texture analysis is exerted in the process of segmentation and classification. It provides information of the spatial arrangement of the intensity value and the color details in an image. For a better segmentation process, the details are obtained from the average gray level and spatial frequency. To analyze the details of the structure, structured approach and statistical approach are used. The edge detection of the image is used to find the texture details. The texture analysis is possible with the help and direction of edge points. The texture of the face shows severe changes with every age group. The number of age groups considered are 12, so with each group, there are noticeable differences in the texture values. The texture values are extracted from the regions, as per mentioned in Figure 5.6. These are the regions where the variation in texture is found in different age groups. The forehead, which appears puffy in the ages 0–5, becomes a little flat in the 6–12 age group. From 13 to 19, the forehead becomes flatter, and then aging starts to progress with visible transmutation in the texture. Similarly, the texture shows changes around the eyes. The fat around the eye at younger ages drops off, with a drooping effect and fine lines becoming more prominent as one ages, resulting in impressions and dark circles around the eyes. In the area beneath the eyes over the upper cheeks, a similar transmutation is noted. The fatty skin becomes flat, clear, even loosened, wrinkly, droopy and shrunken. A similar effect is noticed in the mid region of the cheeks, too. The corners of the lips are also an important region where the signs of aging, such as wrinkles, are noted. The texture of the region beneath the lips, that is, the chin, remains varied as a result of age progression. These ten regions in the face render the texture value, which acts as a lucrative feature in evaluating the age of a person.

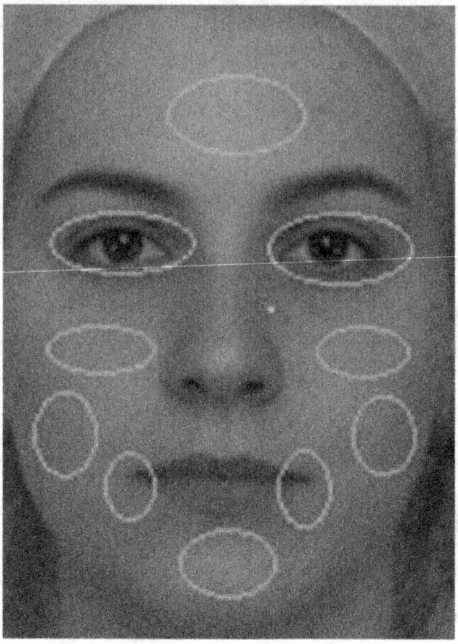

FIGURE 5.6 Texture Extracted from These Regions.

5.3.2.3 Key Point Descriptor

The key point is the region in the image with associated orientation. It is explained by four parameters in the geometric frame. Key point coordinator x and y plot scale (radius of region) and its orientation (angle). With the descriptor value obtained as a numeral, the key point description is obtained. There is no relativity in the descriptor and the position of the point. Hence, irrespective of the position, the descriptor remains the same. At a different position, the descriptor value measured was equal valued. This is the result of image translation. The transformation of image is an unavoidable step in image processing. From an input image, 15 key point descriptors are extracted. Figure 5.7 depicts the 15 positions from where the key points are extracted from the input image. The forehead is the region of the face that shows expression as well as age details. The baby face in the age group of 0–5 has a forehead that is slightly bulged, showing the softness of the brain and the skin. The next age group, 6–12, shows flatness in the forehead region with minor texture variations. The forehead shows the signs of aging with a rounded shape that starts to flatten. The flat forehead then starts to become stiffer in an adult face. The flat forehead begins to show fine lines, which become wrinkles. The eyes are the region that show maximum changes in a face with respect to emotions and expressions as well as aging. A baby's eyes round and then lengthen a little, but the surrounding areas remain similar until the age of 10–12. The protruding of the eye region becomes normal in the teen age group from 13 years onward. The lower portion of the eye looks to be even after 19 years. The upper lid shows proper lines after 24 years. Aging is seen in the corners of the eyes from age 30. The outer eye corner shows fine dotted depressions. The prominence in this depression is seen after 34 years. The fine lines start after

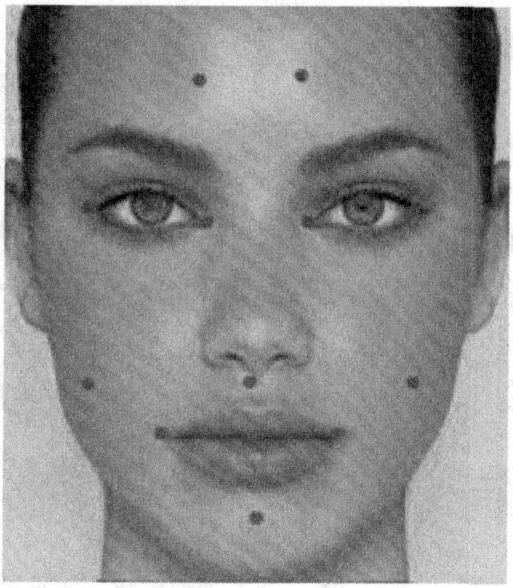

FIGURE 5.7 Key Points from an Input Image.

that. The amount of wrinkles at the corners of the eyes also increases. The inner eye corner starts showing signs of aging from 40 years. The lip corner palpably also shows the signs of aging, such as wrinkles.

5.3.2.4 Shape

The shape of the face provides basic attributes in the categorization of age. It is possible to identify the categorization of age from the shape features. The extracted feature is maintained to remain unaffected by scaling or any translation operations. The processing of the image is to retrieve essential features from the face. The feature considered with respect to shape should provide affine invariance. The linear mapping of the coordinate system should preserve the straightness and parallelism of lines. The extracted feature should not vary with affine transforms. The reduction of noise is done to maintain the shape feature. The features of the facial shape are retained such that part of the shape should not deviate in comparison to the actual shape. The statistical independence of the feature is required for the processing. The reliability of the shape is mandatory, which means the features remain the same.

Figure 5.8 shows the variation in the shape of the face as a person ages. The change in the shape is due to changes in craniofacial and muscular growth. The changes occur with a variation of the overall shape of the face, from circular in a child's face to a triangular shape and then to a square shape in adulthood due to the enlargement of the side muscles. The shape is a vital feature in the estimation of a person's age. The width of the face becomes broader in the upper portion of the face, and the lower portion of the face starts to widen in breadth at a young age due to the change in the facial bones. With the change in the facial bones, the muscles near the jaw also get loosened, and this difference becomes noticeable in the shape of the face. This variation in shape can also be noticed in images of faces that have facial

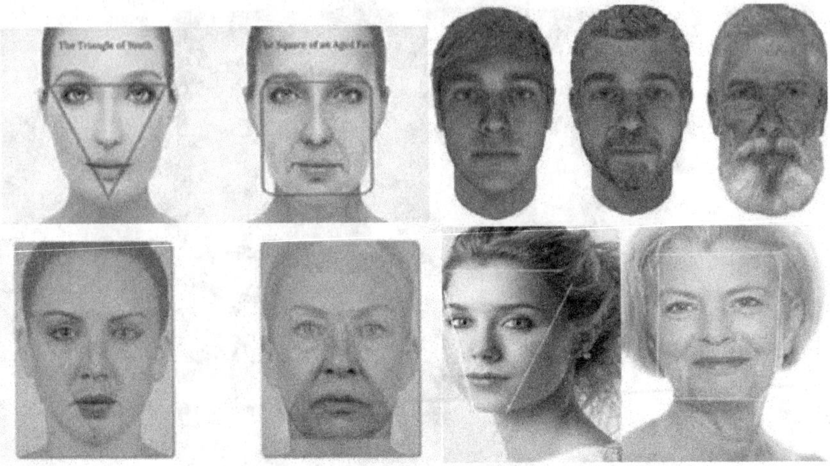

FIGURE 5.8 Variation in the Lower Jaw of the Face with Age Progression.

hair. A person with a beard also develops variation in the shape of the lower portion of the face due to the enlargement of the lower jaw. This difference has a great impact on the estimation of the person's age.

5.3.2.5 Color and Gradient

The color (Hsu et al., 2002) and gradient of the image provide unique information. The information that was obtained was found to be repeatable. The variations in photo and measures were accurately defined. The efficient features are substantially significant and robust in identifying the age group. The discrimination of the fractals is carried out easily by the color feature. The estimation process is eased with the delineation of the intensity variation. The directional change in the intensity of the image is provided by the gradient. The texture of the image is revealed with the extraction of the color and gradient values from the image. This provides smoothing effects to induce the difference in age. So, by concentrating both color value and texture values, the effect of aging is inferred. The gradient of the image retrieves information of the image. The gradient is achieved by filtering the image. By obtaining the gradient of the image, variation of intensity at several points in the image can be easily obtained as well.

5.3.3 SFTA and LDN

The 45 feature (Kumar et al., 2020) values were extracted based on the 5 criteria of orientation, color and gradient, texture, shape and the key points. These features were used for the classification of age. The features as extracted from the training images were stored in the database (Lin et al., 2012). When an input was given, the features were extracted, and the resultant values were matched with the database values and the classification was done. To make the estimation further defined, angular classification was done and was followed by matching the values with the training database. With the binary image, the border details were obtained. The border details were retrieved by the function HausDim(I). This function provides the detailing of the border of the

TABLE 5.2
Fractals Extracted from the Image.

Grid value	Index value in each domain	Rotation	Brightness	Contrast
11111100	1	6	111	0.111
11111200	1	7	223	–0.132
11111300	1	6	342	0.121
11111400	1	2	87	0.148
11112100	1	3	135	–0.091
11112200	1	4	–310	0.132
11112300	1	2	132	0.413
11112400	1	1	190	–0.156
11113100	1	3	99	0.234
11113210	7	4	–1.24	0.86
...	...	...	...	...

image I. The fractal dimensions were obtained by using this descriptor value (Liu et al., 2013), and the age was estimated. This function returned the fractal dimension D of an object represented by the binary image I. Nonzero pixels belonging to an object and 0 pixels constituted the background. The 45 fractal values obtained for each pixel position were stored in the database file for both the X and Y axis. The input facial image was segmented into 64 grids. The fractal features extracted from the image are shown in Table 5.2. The first column denotes the grid value. The second column is used to denote index value, which refers to the value of the position in the domain block. The third column is used to denote rotation value, which ranges from 0 to 7. The fourth column is used to denote the brightness, and the fifth column denotes the contrast value. The training of the system is done using these fractal feature codes. Both the fractal values and the local feature values play a vital role in the age estimation of a person from a facial image. The extraction of the fractal values is done with the modification of binary patterns in all eight directions. The edge detection process is made smoother with the segmentation, which achieves 64 grids after the process. An added advantage is the retained texture details. To obtain a standard contrast in the image, normalization is used. The histogram normalization is performed on each grid. This results in the achievement of the fractal values. The three-layered neural network trains the system in a fine manner. This makes the achievement of the exact age prediction of the person from the neural network. The pattern sets are created for every age group, and the training of the neural network system is done. The classification is done using the softmax classifier. The softmax function mapping is given by

$$f(x_i, W) = W_{x_i} \tag{5.2}$$

where x is the data and W is the weight matrix. The softmax function is given by

$$f_j(z) = \frac{e^{z_j}}{\sum_k e^{z_k}} \tag{5.3}$$

The softmax function compares the vectors of real values to vectors of values between zero and one. Using the softmax classifier, cross-entropy is minimized. It interprets the scores as log probabilities for each class and increases the normalized probability of the correct class. This interpretation of the score is given by

$$L_i = -\log\left(\frac{e^{z_j}}{\sum_k e^{z_k}}\right) \tag{5.4}$$

5.3.4 Classification

The input image after preprocessing renders 45 feature values. Classification (Malhotra et al., 2018) is the final step and provides the refined age of the person from the facial image. Neural networks are great techniques to handle many difficulties encountered in the field of image recognition, speech recognition and natural-language processing. The features that are aspired with respect to biological details are provided to the database, and the software observes the data and learns the features to adapt to avail classified age group. As per Wei-Lun Chao, deep neural network (DNN) based on the input data and the information of the data availed from the database provided a way of classification into an appropriate age group.

5.3.4.1 Angular Classification

The input image is analyzed to extract the features from it. Prior to extracting the features from the face, the face is analyzed for the first level of classification by applying the angular classification technique. In the angular classification technique (Deepa & Sasipraba, 2015), the dimensional measure of the face is retrieved as an angle value. This angle value is formed by a triangle between the two eye midpoints and the center of the lip. Since only facial information is required, the face must be perfectly fitted to the grid. This was achieved by using the average face size, which is the relationship between the face size and the facial feature distances. As the same features were to be extracted from all the images, the concentrated features remained common. To orient to the common features, all the images were required to be of a specific size. The images were hence resized. The initial values were 256 pixels in height and 192 pixels in width. The resizing was done, and the face portion of the image was obtained in the specified size. If the input image after cropping provided the image in accordance with the required size, the aspect ratio was checked and the input size was retained. Enlarging of the image or reducing the dimension of the image was maintained to obtain the image in the uniform size, which aggrandized the efficiency of the algorithm. Whether enlarging or reducing the size of the image, altering the number of pixels was done to have pixels in new positions and these values were then estimated. Bicubic interpolation was done to interpolate the values of the new pixels, and the output pixel value was found on the basis of the weighted average of pixels in the nearest four-by-four neighborhood. These distances were computed with the following equations. The coordinate position was taken as (x, y) and the measures were calculated. The midpoint of the eye positions was calculated by Equation (5.5).

$$MidpointEye = \left(LE(x) + RE(x)\right)/2 \tag{5.5}$$

where LE denotes left eye and RE denotes right eye. The distance between the eyes was given by Equation (5.6).

$$DistEyes = RE(x) - LE(x) \tag{5.6}$$

The distance between the eye and mouth was calculated by Equation (5.7).

$$DistEyeMouth = eyes(y) - mouth(y) \tag{5.7}$$

The triangle was obtained by finding the locations of the left eye, the right eye and the lip. The angle between right eyeball, mouth point, and left eyeball is called the face angle. In human age progression, the face angle changes over time. Hence, by calculating the face angle, the broad classification of age was obtained. Identify a rectangular face area from the given input face image.

1. From the face image, the eye region is identified.
2. The cropped image is then histogram equalized and is converted into a binary image.
3. The binary image is divided horizontally into two parts. The upper part, containing two eyes, is denoted by UP, and the lower part, containing the mouth, is denoted by LP.
4. The UP is divided vertically into two parts. The part that contains the right eye is denoted by RE, and the other part that contains the left eye is denoted by LE.
5. In row R1, minimum row sum of gray level in UP is found. In columns C1 and C2, minimum column sum of gray level in RE and LE are found. The (R1, C1) coordinate represents the middle point of the right eyeball, and (R1, C2) represents the middle point of the left eyeball.
6. In row R2, minimum row sum of gray level in LP is found. R2 represents the mouth.
7. The midpoint C3 of two eyeballs is calculated. C3 = (C1+ C2) /2 and the coordinate (R2, C3) represents the middle point of the mouth.
8. A triangle by joining three coordinate points—left eyeball (R1, C1), right eyeball (R1, C2) and mouth point (R2, C3)—is drawn.
9. The slope (m1) of the triangle side from mouth point (R2, C3) to right eyeball (R1, C1) and the slope (m2) of the triangle side from mouth point (R2, C3) to left eyeball (R1, C2) are calculated.
10. The face angle (A) is calculated using the formula given by Equation (5.8).

$$A = \tan^{-1}\frac{(m1 - m2)}{(1 + m1 * m2)} \tag{5.8}$$

11. The age group based on the face angle value (A) obtained is shown in Table 5.3.

When the angle value was obtained, the classification (Deepa 2015) into a broad category could be easily done. The angle formed from a baby's face was small compared to an older face. The angle value of less than 44 degrees was observed on the

TABLE 5.3
Age Group Classification.

Face angle in degrees (A)	Age group in years
< 44	< 19
44–48	20–24
49–54	25–34
55–60	35–44
> 60	45 and above

face of a person in the baby, child or young adult stage. When the angle value was between 44 and 48, the input image was of an adult between 20 and 24 years old. An angle value between 49 and 54 provided a classification of age in the range of 25 to 34 years. When the angle value was between 55 and 60, the age was found to be around 35–44 years. When the angle increases, the age becomes equal to or greater than 45 years and is classified as an elderly face. Compared of various face shapes, the angular classification has provided results with 81.7% accuracy. On considering images of 120 faces with various shapes, such as oval, long, round, square, heart and diamond, 98 images have provided exact age classification as per the angular classification. The other images provided classification in the adjacent groups.

5.3.4.2 Deep Neural Classification

DNN (Fard et al., 2013; Suo et al., 2010; Li et al., 2011) depicts large set of functions than shallow networks and aids easy decompositions. DNN provides various functionalities, like applying layering using pools for every feature. The vector values provided as input were connected to the entire neural network. The forward and backward propagation is possible to obtain the right age of every layer. As the duration of calculation was lengthy for calculating the entire gradient model, images were processed in batches, with each age group trained separately. The efficiency of DNN was enhanced by using a four-layered DNN to improve reliability and increase the robustness and by increasing the size of the data by collecting over 1,000 images, and the ages were also labeled. The extracted features were stored in the database. The DNN was used for classification. In DNN, the first layer grouped the pixels in the image facilitating the detection of edges, and the second layer grouped the edges together. In the first layer, matching of the angular value and the extracted feature value with the broad category of age group was done. In the second layer, XOR operation was applied to the extracted features. The third layer was used to store the class labels, which provided the age category. The fourth layer stored the details of the X and Y coordinate values, which were the threshold values extracted at various points. The function used to calculate the pass in the feed forward neural network is given by Equation (5.9).

$$nn = nnff\left(nn, x, y\right) \tag{5.9}$$

DNN requires a large size of training set to match with higher accuracy. The neural network was obtained with the formula given by Equation (5.10).

$$nn = nnsetup\left(\left[size\left(TrainX,\right)5006\right]\right) \tag{5.10}$$

Here, Train is the training image, and the coordinate position is in terms of the values of x and y. The prediction algorithm matched the values of the given input image with the values stored in the database. Classification of image provided an output thatwas likely to be in the predicted category based on the matching criteria with the data taken from the training set. As by Geng et al. (2007), DNN is a process where the principal of working is as that of a human brain. The DNNs are interconnected to achieve a refined output based on machine learning.

If the output obtained from the first level is found to be A, then the 45 feature values are matched with the class groups class 1, class 2 and class 3. Class 1 has the stored dataset of the age group 0–5. Class 2 has the stored dataset of the age group 6–12. Class 3 has the training dataset records of the age group 13–19. The angle value, when less than 44, will marshal within the three classes. This is because craniofacial growth is a more eminent feature of shape variation in these age groups. In case matching is found in this age group, the extracted 45 feature values of the input are matched firstly with class 1, then with class 2 and then with class 3. If no matching is found, the matching value is checked in the next angular classification group in class 4. Although the age group in class A is confined to 0–18 years of age, there are possibilities where the net computed result value can be in either class A or class B if the age is in the range of 18–19 years. This can happen in cases where the growth of the face is less than the expected size of the features. These variations are expected in human faces either due to internal factors (such as heredity) or external factors (such as pose variations). The result hence cannot be found from one age class only. The probability of matching will be possible only in the next age group. The same can occur when the age is 19, the angle matching if found in the class B. The angle is greater than 44, as the craniofacial growth results with the widening of the face.

The matching of the feature values is found in class 3. So, with every start node of the sub class labels, the matching is checked in the next adjacent layer, too. The propagation of the internodes was achieved by identifying the relevance propagation values, which were obtained by computing the rate of matching of the values with the stored database values. Class B denotes the age range to be from 19 to 25. This age group had an enlarged size of the face due to changes in the craniofacial growth and changes in the texture values and the fiducial values due to the effect of the growth of the facial bones and muscular changes. The broad classification of the 19–25 age group had the feature values matching with the several nodes in the second-layer class group. Class B had the angular classification values and the feature values found in several different categories. Class 4 provided the feature value classification in the age group 20–24. Class C had the age range from 26 to 35 years. The second level of classification to confined to the age group of either class 5 or class 6.

The class 5 level of classification was used to classify age in the range of 25 to 29 years, and class 6 to classify age in the 30–34 age group. Class D classified the age in the range of 36–45 years. The next level of the confinement of classification is to classify the age to either the group 35–39 years, as in class 7, or in the range of 40–44 years, as per class 8. The last class label, E, classified the age belonging to the group greater than 45 years of age. In this age group, the shape of the face changes a lot, and the refined classification was very challenging. The anomalies in this age group were due to the health factors and lifestyle of the person.

The relevance function was obtained from the probability score that was obtained from the input and hidden layers. The probability score was obtained when the MRP function was equal to the probability of the particular component in each layer. DNN here used the feed forwarding of the values obtained from one layer on to the next inner layers consecutively to obtain a unique classification.

Hence, to obtain a uniform standard value, a decomposition is obtained and given by the factor k. The value of k was taken as 0.01. This rotating factor was taken to maintain the standard value of the denominator. Every value was taken into consideration from each layer, and the probability comparison was done to provide a single value output. The rotation was obtained by aggregating the result of each layer with the next layer, and the final probability was obtained from the net values of the MRP function in each layer. Consider Figure 5.9, which explains the representation of the DNN layer in finding the matching class group. The input was the angle value calculated from the fiducial points on the image, such as the eye points, lip position and the 45 feature values that were extracted from the facial image processing through the SFTA method. The input was checked for matching of the angular classification. As per the angular classification, the resultant value was confined to a particular angular set. The output was obtained after processing from the next layer of classification. In the next layer of classification, there were 12 age groups: 0–5, 6–12, 13–19, 20–24, 25–29, 30–34, 35–39, 40–44, 45–49, 50–54, 55–59 and 60 and above. Class 1, class 2 and class 3 were the next layers from the angular set A. Class 4 is the next layer from angular set B. Class 5 and class 6 were the next layers of the angular set C. Angular set D had two confinement layers, class 7 and class 8. The final angular set, E, had four next layers in the refined age group: class 9, class 10, class 11 and class 12. Every class of the respective angular set was differentiated on the basis of the nodes of the particular set. There were links between the nodes in the second layer. This was used for interlinking between the nodes to obtain a unique result. Since there were possibilities for matching of angle value in one age group and refined age with the matching of the feature values in another class group, there was an interlinking between the nodes to provide the result based on the highest matched value. The outputs O1 to O12 provided the classified age in the range 0–5, 6–12, 13–19, 20–24, 25–29, 30–34, 35–39, 40–44, 45–49, 50–54, 55–59 and 60 and above.

5.4 OBTAINED RESULTS

The output was obtained in the form of a network structure that consisted of the updated activation layer, the error and the loss. In training the system, the functions associated were either of the direct training, which used only limited training set

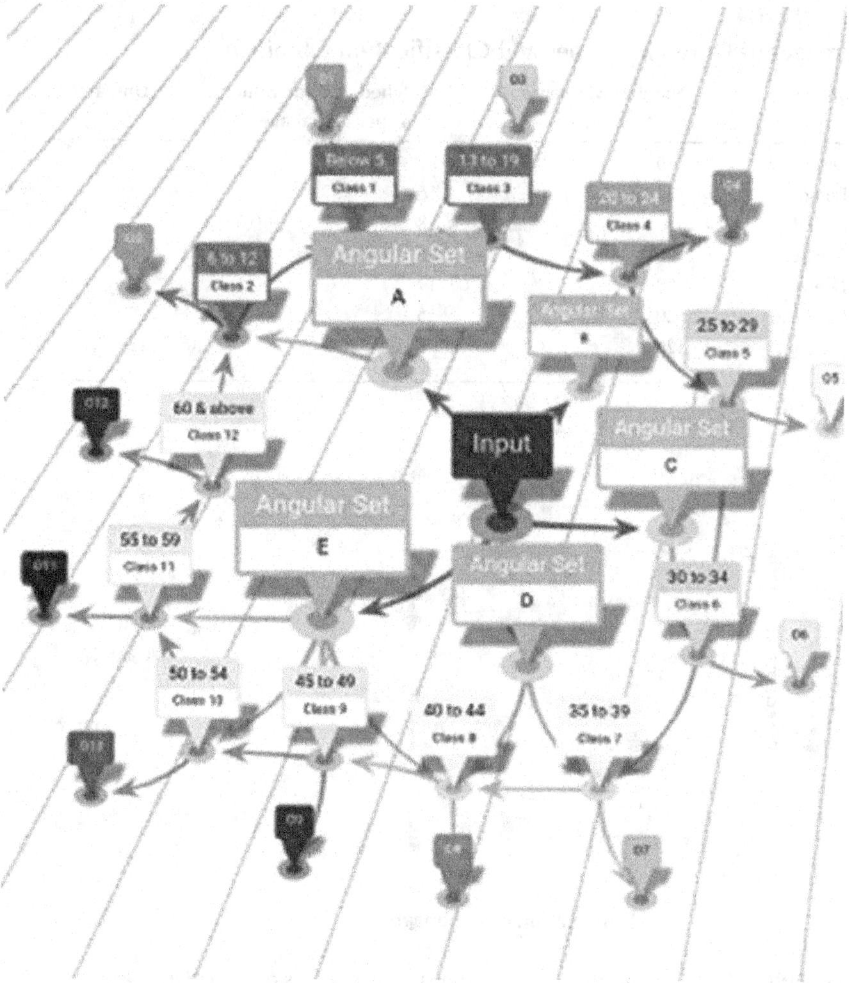

FIGURE 5.9 Deep Neural Classification of Age.

data, or of the training of the entire network. The direct training used approximation (Nguyen et al., 2014), which was calculated in terms of clustering. The clustering provided an approximated value. In using training of the entire network, DNN was used, so the resulted output rendered a better classification range.

Table 5.4 shows the classification of age obtained for different shapes of faces. This classification obtained, as per Figure 5.10, provides the details of images that show the classified age that matched the actual age group (Rein-Lien et al., 2002) of the face and also those images that did not match with the actual age group. The round shape of the face provided matching of age at a higher rate of accuracy compared to the other shapes. Similarly, the images of square face shapes provided less accuracy compared to the other face shapes. The angular classification provides almost nearing age group

TABLE 5.4
Images of Different Shapes and Classification Obtained.

Shape	Number of images	Matched angular value as per actual age	Unmatched images
Oval	20	17	3
Long	20	16	4
Round	20	18	2
Square	20	15	5
Heart	20	17	3
Diamond	20	15	5
Total	120	98	22

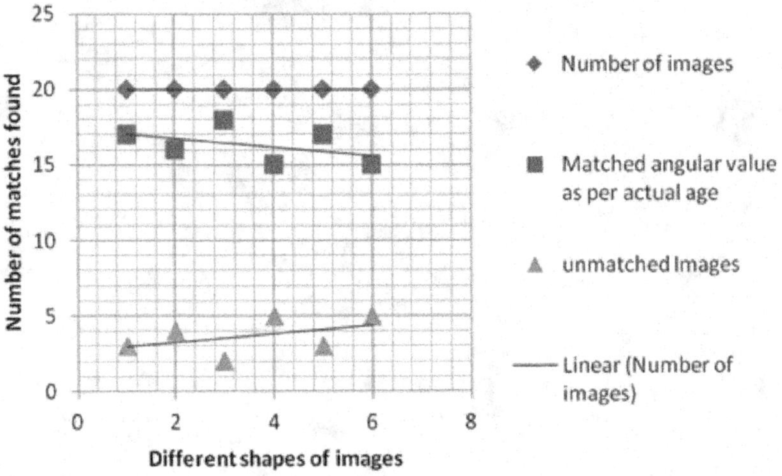

FIGURE 5.10 Classification of Age Compared with Faces of Different Shapes.

estimation. The difference was noticed in the images that belonged to the border ages. The refinement in the estimation could be obtained with the SFTA method, where the classification was based on the feature values extracted from the fiducial values (Shan, 2010). The angle value mismatch was also found in images that had pose variation and, when rotated, the image provided only a half portion of the face. The side pose images provided the least classification in estimation of the age. The angular classification provides the estimation of the angle value in the images with calculation of the angle value from the midpoint of the eyes to the middle position of the lips.

5.5 CONCLUSION

In the proposed method, various datasets of different age groups, pose and illumination were included. The inclusion of people from different geographic regions within various age groups and different poses was to check the accuracy of the algorithm to confirm whether the preprocessing was done in such away as to reach the standard size of the system and also whether the feature values that were extracted from the processed face render the values to match with the expected results. The softmax classifier was used to classify the results. The DNN provided better classification. The comparative analysis with the reference papers made the estimation classified in a better way. The classification was done on various input images and training sets to improve the efficiency of the algorithm. Since the estimation was obtained by considering varying criteria, better accuracy rate was obtained. The common drawbacks were phase variation, database set size, level of classification, light effects and poses. The classified age groups are also high with narrow age limits. The furtherance can be achieved with the inclusion of all these factors.

REFERENCES

Alley, R. (1998), *Social and Applied Aspects of Perceiving Faces* (Lawrence Erlbaum Associates, Inc, Hillsdale).

Al-Shannaq, S. and L. A. Elrefaei (2019), "Comprehensive analysis of the literature for age estimation from facial images", *IEEE Access*, Vol. 7, pp. 93229–93249.

Andreas, Lanitis, Chrisina Draganova, and Chris Christodoulou (2004), "Comparing different classifiers for automatic age estimation", *IEEE Transactions on Systems, Man, and Cybernetics—Part B: Cybernetics*, Vol. 34, No. 1.

Cheng Yaw, Low, Andrew Beng, Jin Teoh, and C. J. Ng (2019), "Multi-fold Gabor, PCA and ICA filter convolution descriptor for face recognition", *IEEE Transactions on Circuits and Systems for Video Technology*, Vol. 29, No. 1.

Chin-Teng, Lin, Dong-Lin Li, Jian-Hao Lai, Ming-Feng Han, and Jyh-Yeong Chang (2012, August), "Automatic Age Estimation System for Face Images", *International Journal of Advanced Robotic Systems*, Vol. 9.

Costa, A. F., G. E. Humpire-Mamani, and A. J. M. Traina (2012), "An efficient algorithm for fractal analysis of textures", in *SIBGRAPI 2012 (XXV Conference on Graphics, Patterns and Images)* (Ouro Preto, Brazil), pp. 39–46.

Cuixian, Chen, Yaw Chang, Karl Ricanek, and Yishi Wang (2010), "Face age estimation using model selection", *IEEE Computer Society Conference on Computer Vision and Pattern Recognition*, Vol. 1, pp. 93–99.

Cunjian, Chen, Antitza Dantcheva, and Arun Ross (2014), "Impact of facial cosmetics on automatic gender and age estimation algorithms", in *Proceedings of 9th International Conference on Computer Vision Theory and Applications (VISAPP)* (Lisbon, Portugal).

Deepa, A. and T. Sasipraba (2015, March), "Age estimation in facial images using angular classification technique", *Advances in Natural and Applied Science.*

Eason, G., B. Noble, and I. N. Sneddon (1955, April), "On certain integrals of Lipschitz-Hankel type involving products of Bessel functions", *Philosophical Transactions of the Royal Society. London*, Vol. A247, pp. 529–551. (references)

Fard, H. M., Khanmohammadi, S., Ghaemi, S., and Samadi, F. (2013), Human age-group estimation based on ANFIS using the HOG and LBP features. *Journal of Electrical and Electronics Engineering,* 2(1), 21–29.

Geng, X., Z. Zhau, and K. Smith-Miles (2007), "Automatic age estimation based on facial aging patterns", *IEEE Transactions on Pattern Analysis and Machine Intelligence*, Vol. 29, pp. 2234–2240.

Geng, Xin, Zhi-Hua Zhou, and Kate Smith-Miles (2007, December), "Automatic age estimation based on facial aging patterns", *IEEE Transactions on Pattern Analysis and Machine Intelligence*, Vol. 29, No. 12.

Gupta, S., K. Thakur, and M. Kumar (2020), "2D-human face recognition using SIFT and SURF descriptors of face's feature regions", *The Visual Computer*, pp. 1–10.

Han, Hu and Anil K. Jain (2014), "Age, gender and race estimation from unconstrained face images", *IEEE, MSU Technical Report CSE*, pp. 14–15.

Hlaing, Htake and Khaung Tin (2012a), "Subjective age prediction of face images using PCA", *International Journal of Information and Electronics Engineering*, Vol. 2, No. 3, pp. 296–299.

Hlaing, Htake and Khaung Tin (2012b, May), "Subjective age prediction of face images using PCA", *International Journal of Information and Electronics Engineering*, Vol. 2, No. 3.

Hsu, Rein-Lien, M. Abdel-Mottaleb, and A. K. Jain (2002), "Face detection in color images", *IEEE Transactions on Pattern Analysis and Machine Intelligence*, Vol. 24, No. 5, pp. 696–706.

Jianyi, L. A. N, Yao ma. *A. B*, Lixin Duan, Fangfang Wang and Yuehu Liu. (2013), Hybrid constraint SVR for facial age estimation, *Signal Processing, Pub Elsevier, 94, pp.* 576–582.

Kumar, M., S. Gupta, and N. Mohan (2020), "A computational approach for printed document forensics using SURF and ORB features", *Soft Computing*, pp. 1–12.

Li, Zhifeng, Unsang Park, and Anil K. Jain (2011, September), "A discriminative model for age invariant face recognition", *IEEE Transactions on Information Forensics and Security*, Vol. 6, No. 3.

Lin, C., Li, D., Lai, J., Han, M., and Chang, J. (2012), Automatic age estimation system for face images. *International Journal of Advanced Robotic Systems, 9*(5). https://doi.org/10.5772/52862

Ling, Haibin, Stefano Soatto, Narayanan Ramanathan, and David W. Jacobs (2010, March), "Face verification across age progression using discriminative methods", *IEEE Transactions on Information Forensics and Security*, Vol. 5, No. 1, pp. 82–91.

Liu, Jianyi, A. N. Yao Ma, A. B. Lixin Duan, Fangfang Wang, and Yuehu Liu (2013), "Hybrid constraint SVR for facial age estimation", *Signal Processing, Pub Elsevier*, Vol. 94, pp. 576–582.

Malhotra, Manisha, Harpal Singh, and Meenakshi Garg (2018), "Statistical feature based image classification and retrieval using trained neural classifiers", *International Journal of Applied Engineering Research*, Vol. 13, pp. 5766–5771.

Moghadam fard, Hamid, Sohrab Khanmohammadi, Sahraneh Ghaemi, and Farshad Samadi (2013), "Human age-group estimation based on ANFIS using the HOG and LBP features", *Electrical and Electronics Engineering*, Vol. 2, No. 1, pp. 21–29.

Nguyen, T., S. R. Cho, and K. R. Park (2014, Springer), "Human age estimation based on multi-level local binary pattern and regression method in future information technology", *Lecture Notes in Electrical Engineering*, Vol. 309, pp. 433–438.

Rein-Lien, H., and Abdel-Mottaleb. (2002), M and Jain, A. K. Face Detection in Color Images, in *IEEE Transactions on Pattern Analysis and Machine Intelligence*, *24*(5), 696–706.

Ramanathan, N. and R. Chellappa (2006), "Face verification across age progression", *IEEE Trans. Image Process*, Vol. 15, pp. 3349–3361, doi: 10.1109/TIP.2006.881993.

Shan, Caifeng (2010), "Learning local features for age estimation on real-life faces", ACM-MPVA'10: Proceedings of the 1st ACM International Workshop on Multimodal Pervasive Video Analysis, pp. 1878039–1878045.

Suo, Jinli, Song-Chun Zhu, Shiguang Shan, and Xilin Chen (2010), "A compositional and dynamic model for face aging", *IEEE Transactions on Pattern Analysis and Machine Intelligence*, Vol. 32, No. 3.

Wang, Peng-Hua, Bo-You Yu, and Po-Ning Chen (2013, May), "General expressions of derivative-constrained linear-phase type-I FIR filters," in *IEEE International Conference on Acoustics, Speech and Signal Process (ICASSP)*, pp. 5578–5582.

Zimbler, J. M. S., M. S. Kokosa, and J. R. Thomas (2001), "Anatomy and pathophysiology of facial aging", *Facial Plastic Surgery Clinics of North America*, Vol. 9, pp. 179–187.

6 The Potential of Machine Learning and Artificial Intelligence (AI) in the Health Care Sector

Rajni Goyal, Gittaly Dhingra, Vishal Goyal, Harpal Singh, and Priyanka Kaushal

CONTENTS

6.1 INTRODUCTION

6.1.1 ARTIFICIAL INTELLIGENCE

Artificial intelligence (AI) is a science about machines that can work smartly. It is the area of computer science in which machines perform those exercises that require intelligence. "Artificial" defines something that is not natural or is built by humanoids, and "intelligence" explains the capability to understand. AI is performed inside the system. The scope of AI is understood by distinctive analysts of researchers with their investigative viewpoints. The main approach behind planning and designing such machines that are as brilliant as humans was called the cognitive science

DOI: 10.1201/9781003184140-6

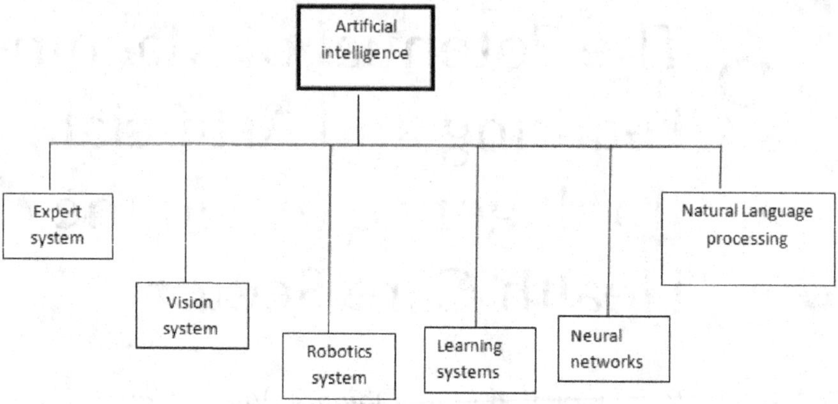

FIGURE 6.1 Basic Structure of the Artificial Intelligence System.

approach to AI (Alimadadi et al., 2020). According to this approach, human understanding and effort are involved to construct machines that can emulate the human brain's process. The second approach is a test developed by Alan Turing in 1940. According to this test, there was an interrogation held between a human and a computer that was programmed under some conditions. The examiner would not know which was the computer and which was the human. This interrogation was purely textual messages–based. On the occasion the examiner is unable to decide whether the reply was given by the computer or the human, the computer passes the test and is said to be brilliant. Figure 6.1 represents the basic framework of the AI system.

6.1.2 Classifications of Artificial Intelligence

Based on capabilities AI can be classified into three types:

- **Weak:** A weak AI system is defined as the formation of some computer-based tasks that are not able to solve explicit problems. This system is basically focused on narrow tasks. It is also known as the non-sentient or weak intelligence system.
- **Applied:** Its main aim is to provide commercially feasible smart schemes, such as security surveillance techniques that can identify and permit only certain people to the building. This system behaves smartly only for designated problems.
- **Strong:** The main aim is to build machines that can truly reason and solve problems. These machines should be self-aware, and their overall intellectual ability is indistinguishable from that of a human being. This system behaves as a super-intelligent system or recursive self-improvement system that behaves like a human mind.

6.1.3 Machine Learning

Machine learning is a subfield of AI in which earlier data or activities are used by the machine without being explicitly programmed by human beings (Amisha et al., 2019). With machine learning, various applications can modify themselves according to the data in real-time programming (A. McGregor et al., 2004). Data can be structured or semi-structured. Machine learning has wide use in Google Search algorithms, Facebook, auto-tagging, email spam filters, online recommended system, and more (W.G. Baxt, 1990). Figure 6.2 explains the framework of machine learning. Figure 6.3 illustrates the structure. A machine-learning system improves its performance from experience and enhance its performance.

6.1.4 Key Transformations Among Artificial Intelligence and Machine Learning

The main differences between AI and machine learning (W.G. Baxt, 1996) are described in Table 6.1.

6.1.5 Classifications of Machine Learning

Machine learning can be classified into three types as shown in Figure 6.4 and here:

6.1.5.1 Reinforcement Learning

Also called learning from errors, it works on the trial-and-error method. It features an algorithm that learns and improves from errors. Favorable results are encouraged and unfavorable are discouraged.

6.1.5.2 Supervised Learning

Supervised learning is task driven. The machines are trained on labeled data, and the data needs to be labeled accurately. It is very powerful when used in the right circumstances.

6.1.5.3 Unsupervised Learning

Unsupervised learning is data driven. In this type of learning, machines can work on unlabeled data. Human intervention is not required in this learning method.

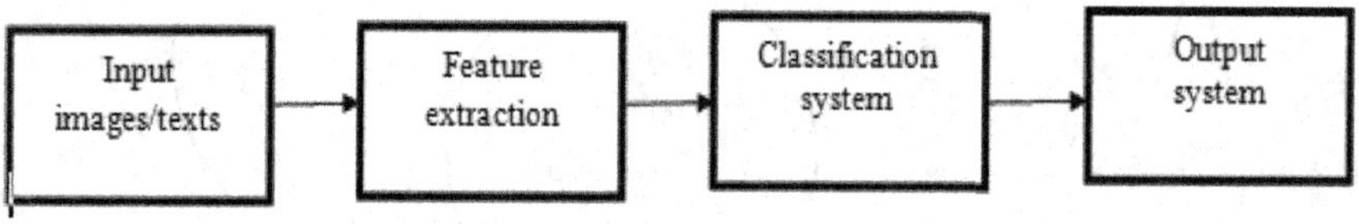

FIGURE 6.2 The Basic Framework of the Machine-learning System.

TABLE 6.1
The Basic Significant Modifications among the AI and Machine-Learning Models.

Artificial Intelligence (AI)	Machine Learning (ML)
1. AI is technology that allows a machine to behave like a human being.	1. ML is a subset of AI that acquires past data automatically without programming explicitly.
2. The goal of AI is to make computer systems smart so they can solve different problems.	2. The goal of ML is to read the data so that it can give the correct and desired output.
3. AI is concerned about increasing the possibility of success.	3. ML is concerned with pattern and accuracy only.
4. AI is used in an expert system, robots (Siri is an intelligent humanoid robot), customer support systems (using catboats), online game playing, etc.	4. Applications of ML include Facebook auto-tagging suggestions, the Google Search algorithm, online recommended systems, etc.
5. Deep learning and machine learning are two principal subsets of AI.	5. Deep learning is a principal subset of ML.
6. In AI, the smart system completes the task just like a human being would.	6. In ML, machines work on data to achieve a particular task and give exact results.
7. AI has a broad range of scope.	7. ML has limited scope.
8. AI makes the smart system that performs multiple tasks.	8. ML builds machines that can perform only the particular task for which they are prepared.
9. AI uses experience to acquire information.	9. ML learns from the trained data or information.
10. AI comprises reasoning, self-correction, modification, and learning.	10. ML includes self-correction and learning during the new data collection.
11. The AI framework is mainly concerned with the maximum chance of success ratio.	11. ML models are concerned with only patterns and accuracy.
12. AI can be divided into three categories: weak, applied, and strong.	12. ML can be sub divided into supervised learning, unsupervised learning, and reinforcement learning.
13. AI deals with structured, unstructured, and semi-structured data.	13. ML mainly works with structured and semi-structured data.

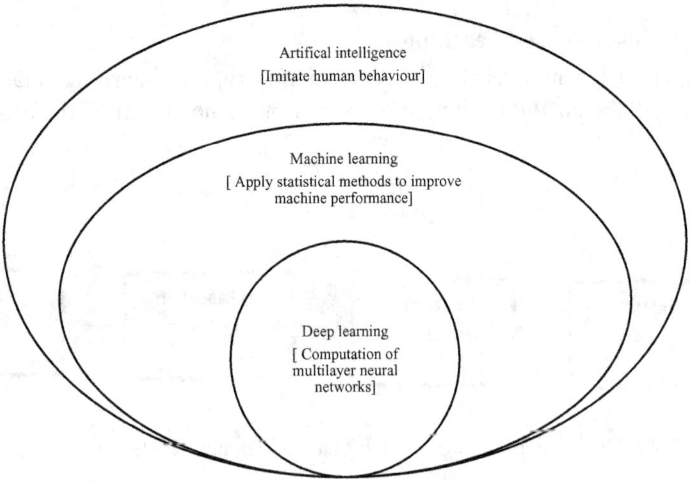

FIGURE 6.3 Structure of AI and Machine-Learning Models.

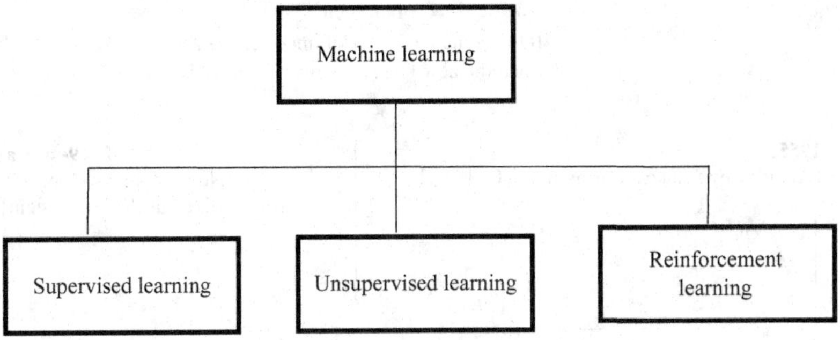

FIGURE 6.4 Classification of Machine-Learning Models.

This classification is done on the premise of the information of particular objects. This information is of two types:

Labeled data: Requires human intervention to label the data in the beginning, but the input and output parameters are a completely machine-readable pattern.

Unlabeled data: Has one or no parameters in machine-readable form, and it doesn't require much human intervention but does require more complex solutions.

6.2 ARTIFICIAL INTELLIGENCE IN HEALTH CARE

AI and machine learning play a crucial role in the judgment and handling of patients who need care (C. Kambhampati et al., 2004; Henson et al., 1997). The issues with the existing healthcare system are:

1. A lack of information
2. Unexpected passing due to human mistakes
3. Medications depend on the current study only

These main issues became the reason for upgrading the healthcare system with AI techniques (Shapiro, 1992). Figure 6.5 defines the timeline structure of AI in the healthcare system.

Apart from diagnosis and treatment, other background processes must take place to take care of a patient, such as (D. B. Henson et al., 1997):

- Information gathering from patients through test reports and interviews
- Analysis and processing results
- Gathering data from multiple sources to diagnose patients accurately
- Formulating and managing the selected treatment method
- Monitoring patients properly
- Aftercare and follow-ups

Technology and automation are already being used in the medical field, as medical histories are digital now (Steimann, 2001). Patient appointments can be arranged

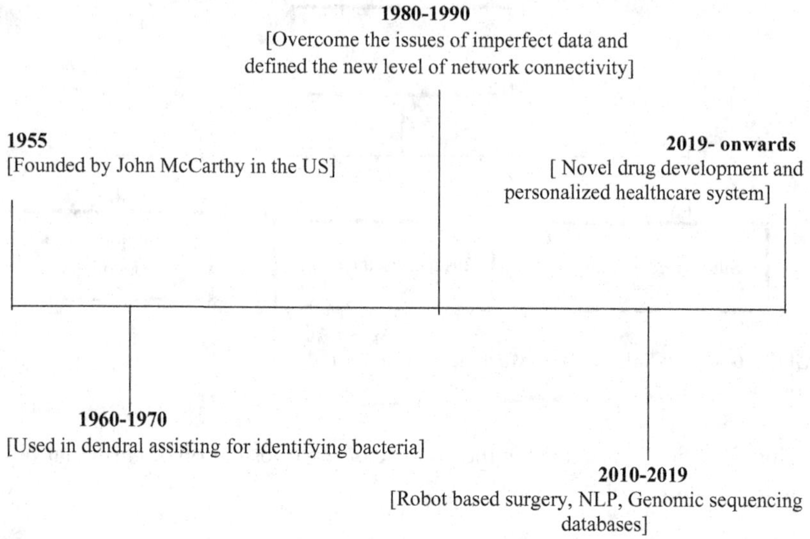

FIGURE 6.5 Timeline Structure of AI System.

online, and patients can access their test reports via patient portals. The basic structure of the AI system is graphically explained in Figure 6.6.

There are some fields of healthcare where AI is already being used.

1. **Decision support systems:** DXpalin is a clinical decision support system (CDSS) used by clinicians for generating stratified diagnoses based on input.
2. **Laboratory information system:** Germwatcher is an expert system used to detect and track infections in hospitalized patients.
3. **Robotics surgical systems:** The da Vinci surgical system is a robotics surgical system with robotic arms. It is used by various surgeons for precision surgery that is not possible manually. This robotic system is used in thyroid cancer removal, gastric bypass, hysterectomies, laparoscopic surgery, and a variety of other surgical procedures.
4. **Therapy:** AI therapy is an online course for people who are struggling with social anxiety.
5. **Online health advice:** Babylon is an online application where the patient can talk with a doctor or a therapist by video and can access health information in one place.
6. **Virtual health assistant:** AI has also presented virtual health supporters using speech, cognitive computing, body gesticulations, and augmented reality. Mainly, this technique can reduce the frequent hospital calls and burden from a medical professional.
7. **Patient service bots:** This application allows patients to ask questions regarding appointments, medical tests, bill payments, medical assistance, and medication refills, etc. This tool also decreases the burden of the medical assistants and receives positive feedback from the patients.

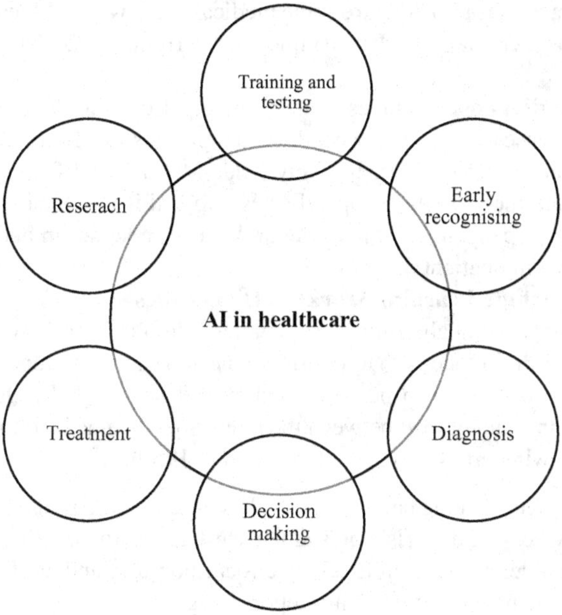

FIGURE 6.6 AI System Structure.

8. **Fraud detection:** Fraud occurs when medical institutions charge more to treat an injury or perform a surgery. AI-based tools help reduce such fraud attempts.
9. **Novel drug development:** AI healthcare tools help medical assistants develop new drugs to serve patients by scanning the already existing medicines and restructuring them more effectively. This application also reduces manual clinical times and cost.
10. **Doctor support system:** Doctors also benefit from AI system solutions because they help enhancing doctors' performance and experience.
11. **Treatment design:** AI tools are created to analyze patient reports, provide clinical expertise, and research personalized treatment tracks.
12. **Research on pandemics and diseases:** AI tools can compute and evaluate new technology and statistical methods to fight the COVID-19 pandemic or any other diseases.

Using AI in medicine is not only reducing the manual tasks and freeing up physicians' time, it is also increasing efficiency and productivity (R. Mayo and J. Leung, 2018).

6.3 APPLICATION AREAS OF ARTIFICIAL INTELLIGENCE AND MACHINE LEARNING IN THE MEDICAL CENTER

AI and machine learning have made great advancements in pharma and biotech efficiency (Y. Mintz and R. Brodie, 2019). From clinical research to doctors' work,

AI is playing a vital role. These are some medical fields where AI and machine learning are playing a crucial role (PavelHamet and J. Tremblay, 2017):

1. **Disease diagnosis:** It takes years of medical training for doctors to diagnose diseases correctly. Even then, diagnostics is often a strenuous and time-consuming process. In many fields, the demand for experts is much more than the available supply. For example, this has been seen during the COVID-19 pandemic, which has put doctors under strain and led to delays in life-saving patient diagnostics.
   How Machine Learning Works in Diagnostics
   Machine learning algorithms see a pattern similar to the way doctors see it. The only difference is that algorithms need a thousand concrete examples. These algorithms also need to be digitized neatly (J. R. Quinlan, 1986). A machine cannot read between the lines. So, machine learning is helpful in the following areas of diagnostics (Turing, 1950):

   i. Detecting lung cancer and strokes based on CT scans.
   ii. Accessing the risk of sudden cardiac death or other heart-related problems based on ECG (electrocardiogram) and cardiac MRI (magnetic resonance imaging) report images.
   iii. Classifying skin lesions from skin images.
   iv. Finding indicators of diabetic retinopathy in eye images.

   Machine learning algorithms are just as good at diagnostics as doctors due to the availability of data. These algorithms draw the results in fractions of a second. Combined data from multiple data resources, like CT scans, MRI, genomics, proteomics, patient data, and handwriting files, is helping in assuring the cause of disease and its progression (Walczak and Nowack, 2001).
   The whole structure is defined in Figure 6.7.
2. **Classifying diseases:** The opportunity of AI to examine images and distinguish patterns leads to the possibility of establishing algorithms to support doctors in identifying specific diseases quicker and more efficiently. Moreover, these processes can constantly learn, thus enhancing its subsequent value of predicting the right diagnosis. Deep learning has made huge advancements in automatically diagnosing diseases, making diagnostics very cheap and accessible.
3. **Improving the decision-making process:** Current solutions based on AI technology already help doctors to investigate obstacles, produce vast

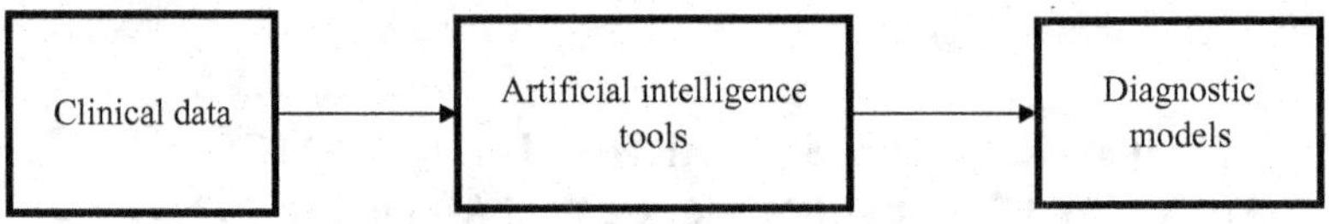

FIGURE 6.7 Functional Structure of AI System.

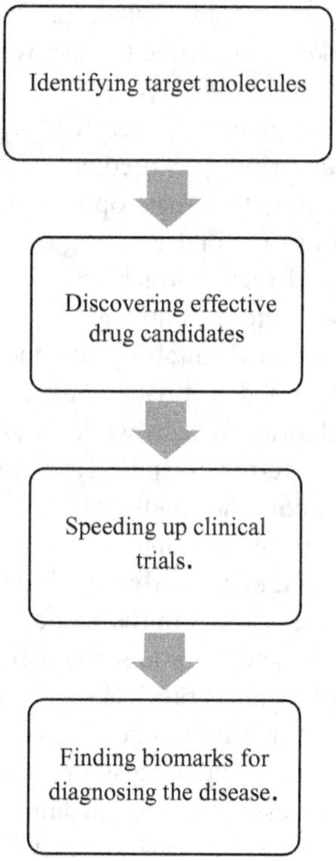

FIGURE 6.8 Four Stages of Drug Development.

quantities of health data fast, and ensure a holistic understanding of a patient's health.

4. **Develop drugs faster:** Developing a drug is a very long and expensive process. It takes many years, sometimes even decades, to develop a drug or vaccine for a particular disease. Machine learning plays a crucial role in this field too. There are four stages of drug development, as shown in Figure 6.8.

Machine learning works in all stages of drug development.

i. **Identifying target molecules:** The first step of drug development is to understand the biological origin of a disease and its resistance mechanism. After that, good targets are identified for treating the disease. Deep sequencing and short hairpin RNA (shRNA) screening are two high-throughput techniques that have made data available to identify good target proteins.

ii. **Discovering drug candidates:** After discovering the target molecules, compounds need to be find out, that can interact with the molecules identified in the first stage. This process requires thousands, or often millions, of potential compounds to find their effect on given molecules. Current software has produced bad and inaccurate results. So, it takes a long time to narrow it down to get the best drug candidates.

Machine learning algorithms learn from the structural fingerprints and molecular descriptors to predict the suitability of a molecule and then filter them to get the best options having minimal side effects. This saves a lot of time in drug design.

iii. **Speed up clinical trials:** Machine learning speeds up the work of clinical trials by automatically identifying suitable candidates. It's very difficult to find a suitable candidate. Machine learning differentiates between good and bad candidates by identifying the different patterns. Algorithms also work as an early warning system for researchers during clinical trials. These machine learning algorithms tell researchers about the candidates that will not produce the desired result, which saves in development time.

iv. **Identifying biomarkers for diagnosing the disease:** Biomarkers are molecules generally found in the blood, and these molecules provide absolute certainty about the disease a patient has or does not have. They make the diagnosis process secure and cheap. But finding suitable biomarkers for a particular disease is typically a hard, expensive, and time-consuming process. AI automates this manual task and speeds up the process. The machine language algorithms differentiate between the good and bad candidates, which helps the clinician focus on analyzing the best prospects.

v. **Personalize treatment:** Different patients have different reactions and schedules to react to the particular medicine. But it is very difficult to find out the factors that should affect the treatment choices. Machine learning automates statistical work and helps to discover those characteristics that indicate how a patient will respond to a particular treatment. This is done by cross-referencing similar patients and comparing their treatment and outcomes to help doctors design the right treatment plan.

vi. **Helping to fight the viral pandemic:** A viral pandemic is a serious threat. A large amount of information is needed to learn the internal structure of the virus. Machine learning plays a crucial role in identifying the following points:

- Identifying the people who are most at risk
- Diagnosing patients
- Finding new drugs
- Finding existing drugs that can help
- Predicting the spread of the disease
- Understanding the virus in a better way
- Mapping where the virus came from
- Predicting the next pandemic
- Providing suitable medicines

- Predicting the influence of the virus
- Spreading awareness
- Patient health management
- Detecting the cause of the virus
- Evaluating the side effects of the disease
- Predicting the severity of the virus

6.4 CONCLUSION

AI and machine learning counterfeit insights in the healthcare system and have inspected and looked into the foremost vital role in the system. Natural-language processing and machine learning are the subfields of the AI system. The AI system must have a machine learning system that can offer assistance for work with structured data, such as images, big data, and genetic data. Mainly, the advancements in the AI system can effectively and efficiently assist human thoughts, power, and resources. AI brings a revolution to the healthcare system.

6.5 FUTURE SCOPE

Machine learning and AI aid in data-driven results, key trend recognition, and analysis productivity. When it arises in the healthcare sector, machine-learning tools can be used in a variety of ways to improve disease detection, analysis, and treatment while also enhancing overall healthcare operations. Using open-source software and limiting customization reduces expense and complexity.

REFERENCES

Alimadadi, Ahmad, Sachin Aryal, Ishan Manandhar, Patricia B. Munroe, Bina Joe, and Xi Cheng. Artificial intelligence and machine learning to fight COVID-19. *AI and Machine Learning for Understanding Biological Processes Physiol Genomics* 2020; 52: 200–202; First published March 27, 2020; doi:10.1152/physiolgenomics.00029.2020.

Amisha, Paras Malik, Monika Pathania, and Vyas Kumar Rathaur. Overview of artificial intelligence in medicine. *Journal of Family Medicine and Primary Care* 2019 July; 8(7): 2328–2331. doi:10.4103/jfmpc.jfmpc_440_19

Anthony, McGregor, Mark Hall1, Perry Lorier, and James Brunskill. Flow clustering using machine learning techniques. Lecture Notes in Computer Science 3015. April 2004, pp. 205–214.

Baxt, W. G. Use of an artificial neural network for data analysis in clinical decision-making: The diagnosis of acute coronary occlusion. *Neural Computing* 1990; 2: 480–489.

Baxt, W. G. and J. Skora. Prospective validation of artificial neural network trained to identify acute myocardial infarction. *Lancet* 1996; 347: 12–5.

Hamet, Pavel, Johanne Tremblay. Artificial intelligence in medicine. *Metabolism* 2017 April; 69: S36–S40.

Henson, D. B, S. E. Spenceley, and D. R. Bull. Artificial neural network analysis of noisy visual field data in glaucoma. *Artificial Intelligence in Medicine* 1997; 10: 99–113.

Kambhampati, Chandra, John R. T. Monson, and Philip Drew. *Artificial intelligence in medicine* Article in of Annals of The Royal College Surgeons of England October 2004.

Mayo, R. C. and J. Leung. Artificial intelligence and deep learning-Radology's next frontier? *Clinical Imaging* 2018; 49: 87–88.

Mintz, Y. and R. Brodie. Introduction to artificial intelligence in medicine. *Minimally Invasive Therapy & Allied Technologies* 2019; 28: 73–81.

Pavel Hamet, Johanne Tremblay. Artificial intelligence in medicine. *Metabolism* 2017 April; 69: S36–S40.

Quinlan, J. R. Induction of decision trees. *Mach Learn* 1986; 1(1): 81–106.

7 Applications of Artificial Intelligence in Modern Health Care and Its Future Scope

Prince Sharma, Geetika Sharma, Mandeep Singh, Khushboo Sharma, Nimranjeet Kour, and Pooja Chadha

CONTENTS

7.1 INTRODUCTION

In recent years, there has been an immense number of discussions about the emergence of artificial intelligence (AI) as well as the consequences associated with its significance in numerous areas of life. But the classical application of AI can, however, be traced back not only to Aristotle's syllogisms in 300 BC (Ramesh et al., 2004; Warwick, 2013) but also to the theory of reasoning machine by Roman Llull's in 1300 AD. Nevertheless, it has been only since the 1950s that practical applications and a clearer definition have emerged (Simmons & Chappell, 1988; Kok et al., 2009).

The classical definition of AI can be summed up as the intelligence or reasoning ability of machines, rather than the intelligence of other living organisms or even

DOI: 10.1201/9781003184140-7

human beings (Minsky, 1961; Weng et al., 2001). In other words, AI is the analysis of 'intelligent agents', i.e., any firm or gadget that can perceive and understand the changes occurring in its environment and consequently take suitable action to increase the probability of reaching the set objective (Wooldridge & Jennings, 1995). AI refers to circumstances arising within the devices that can stimulate the ability to learn and analyze by the human brain and thus finds its usage in the problem-solving capability of the mind. Thus, this type of AI can also be described simply as machine learning (ML) (Huang et al., 2015).

If we consider the main aim of AI, it has been to illustrate the intelligence for the development of these devices that requires the use of automated computer programming. A formal description of AI holds it as 'a broader field of science that is primarily implicated with the calculation or computational understanding of intelligent behavior and creating intelligent agents that can exhibit such behavior' (Shapiro, 1992). A more simplified definition of AI can be described as 'machines capable of assuming human capabilities' and 'making modern computers that can do things that currently humans do,' yet a further more precise elucidation would be described as 'the science which is concerned with making intelligently efficient machines' (Huang et al., 2015; Weng et al., 2001; Szolovits & Pauker, 1978).

AI typically involves a system that is known to contain both the parts, i.e., software as well as hardware. From a software point of view, AI is specifically related to computation or algorithms (Hopfield, 1982). An ideal framework for implementing AI calculation is an artificial neural network (ANN). It can be considered an imitator of the functional human brain, thus representing an interdependent network of functional neurons connected through weighted communication channels between neurons (Watts & Strogatz, 1998). Multiple stimuli from a neighboring neuron can be perceived by one neuron; thus, whole interconnections can be stimulated, depending on the various inputs from the surroundings (Zucker & Regehr, 2002). Thus, the overall neural network (NN) has been known to generate outputs depending upon the stimuli from different natural changes occurring in the environment. Neural networks are thus predominantly piled complex structures of variously interconnected configurations. Researchers from time to time have come up with the following attributes that can aid neural networks:

1. Administered learning, in which the main task is to extrapolate a function that can map a source of input into another source of output relevant to various examples of input and output.
2. Unadministered learning, in which the major task is to grasp from a set of test data inputs that have never been classified or tabulated, to identify and classify the features that are common with in the data.
3. Reinforced learning, in which the main task assigned is to create a response according to the particular surrounding to boost the rewards and decrease the penalties concerned (Schmidhuber, 2015).

Due to the evolution of computational powers, neural networks have been understood 'deeper', i.e., further. More piles of neurons have to be incorporated within the set of the network to imitate the efficiency of a human brain and continue the process of learning. Additionally, more functions that can be inculcated into the neural network are integrating characteristic extraction as well as classifying the outcomes within a single network frame, thus justifying the technical term related to deep learning (LeCun et al., 2015; Arel et al., 2010).

Taking into account the hardware perspective, AI has been known to be engaged primarily with the application of neural network calculations or algorithms within a manual computational platform. The best approach is to enact the neural network algorithms into a central processing unit (CPU), within a multicore/multithread arrangement (Zucker & Regehr, 2002). Furthermore, good convolutional computation sources, such as graphical processing units (GPUs), are known to be better preferred over CPUs for the extensive production process of neural networks (Dinkelbach et al., 2012). Moreover, co-processing of CPU and GPU are better favored over only CPUs usage (Naveros et al., 2014, 2018). In recent times, some more customizable and programmable accelerator hardware platform, application-specific integrated circuits (ASICs), and field-programmable gate arrays (FPGAs) have been known to bring about the neural network toward more personalized applications that can work efficiently not only in terms of power efficiency but also in form factor (Nurvitadhi et al., 2016) and computation capability.

In comparison with CPU and GPU, these platforms have been so personalized that the applications can be customized according to the need of the system, thus making them more compact, compared to traditional CPU and GPU. For the use of AIs in edge devices, like mobile phones in wireless networks, further improvement in efficient power usage and form factor is the need of the hour.

Research all around the world has made efforts to administer AI algorithms by making use of memristors (Jeong & Shi, 2018; Prezioso et al., 2015; Wen et al., 2015), spintronics (Grollier et al., 2016), as well as analog integrated circuits (Kwon et al., 2018; Bartolozzi & Indiveri, 2007). Among these, a new platform, called memristor crossbar circuits (Zhang et al., 2018), can integrate algorithms with the memory of the device as one of the major problems, i.e., accessing the memory wall can be avoided, which is obligatory to update needed parameters. In recent times, researchers have improvised the efficiency of AI implementation by making a relative reduction in the use of several bits utilized for data description. The vital advantages associated with such an approach are faster computation, smaller form factor (Yoshida, 2018), and less power usage, yet the 'memory wall' limits cannot be concealed. For achievement of satisfactory performance of ANNs, the ratification of various training techniques (like using pre-training techniques [Erhan et al., 2010] or deep training [Marshall et al., 2001]), use of balanced datasets (Yuan et al., 2018), constant availability of datasets (Dudek-Dyduch et al., 2009), and sufficient amount of data (Rodríguez- Pe´rez & Bajorath, 2018) have to be taken into consideration.

Due to rapid development in the field of AI hardware and software technologies, AI has found its applications in various technical fields, like IoT (Internet of Things) (Chiang & Zhang, 2016), natural-language processing (Alshahrani & Kapetanios, 2016; Kim, 2010), machine vision (Guo et al., 2016), robotics (Schaal, 1999), and autonomous driving (Nguyen et al., 2018; Yang et al., 2018). With time, researchers have applied AI in the biomedical field to help improve not only analysis but also treatment aftermath, thus increasing the efficiency of health care (Yu et al., 2018; Mamoshina et al., 2016; Peng et al., 2010).

This chapter reviews the concept of AI and implementations of AI in the area of biomedicine, thus encompassing the mainstream area of biomedicine and the health industry. The prime goal for the application of AI for health care is to make it more participatory, preventive, and personal so that AI can efficiently customize vast benefactions in the biomedical stream. Thus, considering the progress made over time, it can be estimated that AI can develop into a major tool for biomedicine and healthcare.

7.2 VARIOUS TYPES OF AI THAT ARE RELEVANT TO HEALTH CARE

Predominantly, AI can be regarded as not just one technology; instead, it is a combination of a variety of several small technologies complementing one another. Most of the technologies that find their importance in AI do find their relevance in the health care field, but the task they perform and the specificity of the processes they control are quite wide. Major AI technologies having vast relevance to biomedicine and health care can be described as:

1. Learning more about machine language

Machine language has been described as a statistical technique that can be used to fit various models into the precise data and learn from such data. ML has undoubtedly been found to be the most ordinary form of AI. In 2018, 63% of companies surveyed were known to employ machine language for their business (Davenport & Kalakota, 2019).

In biomedical health care, the most common application found about the use of machine language has been depicted as the precision of medicine, i.e., prediction of the various protocols used for treatment that are most suitable to show maximum results on a patient, depending upon the specified attributes (Lee et al., 2018). A huge variety of device learning necessitated the use of datasets that are required from which outcome variables can be determined; this is called supervised learning.

A rather more complex form of machine learning has been determined to be a neural network, a technology used for categorization of various applications, such as determination of the aspects as the probability of a patient to acquire a particular disease (Sordo, 2002). It can also take into account the terms of input, output, and the weights of various variables that are associated with inputs and outputs.

The other aspect of ML comprises neural network models, commonly known as deep learning, which includes a variety of characters that can be used to predict various outcomes. One of the most common applications of deep learning in health care is the characterization of various cancerous lesions that are seen in radiological images (Fakoor et al., 2013). Deep learning can be well implemented in various fields, particularly radiometry. Both deep learning and radionics are useful in oncology-oriented image analysis, promising greater accuracy in diagnosis than computer-aided detection (CAD) (Vial et al., 2018).

Deep learning has also found its application for speech recognition and thus is an advanced area of natural-language processing (NLP). Contrasted to the age-old forms of statistical analysis of data, the characteristics implied in these deep-learning models have a simpler level of understanding for a human observer; thus, the explanation of the outcomes associated with the model can be very hard for interpretation.

2. Processing of natural language

The main goal concerning AI researchers since the 1950s has been to make sense of human language. Natural learning process (NLP) has been known to incorporate several applications, such as translations of the test material, total analysis of textual data, and several additional goals that can be coupled to language-learning analysis. Two fundamental approaches known to be associated with it are statistical NPL and semantic NPL. Statistical NLP is based on machine learning (deep learning neural networks in particular) and has contributed to a recent increase in accuracy of recognition. In health care, the prominent application of NPL involves understanding, classification, and creation of clinical documentation and published research. NPL system has been known to analyze unstructured clinical notes on patients, prepare reports, and also to direct colloquial AIs.

3. The heuristic rule approach system

In health care, an adept system centered on the collection of 'if-then' directives has been widely employed for purpose of clinical decision support that has been applied over the past couple of decades and is still in use in recent times (Vial et al., 2018). Presently, different electronic health recorders have put forth an ordain of norms for the systems in the present day. These expert systems instead of robots needed human assistance to design a particular set of norms within a described knowledge framework. However, when the rules are larger in number, the rules conflict within themselves, tend to break down, and are thus more time-consuming. With time, they are being replaced by more recent approaches based on ML computations.

4. Physical or materialized robots

Physical robots have been studied to carry out pre-determined assignments, such as welding, elevating, shifting, or assembling certain articles in relevant sites, such as industries, warehouses, garages, and factories. In recent times, robots are more collaborative with humans and can be easily trained. With the recent advancement in AI technologies, their brains are being fed by more intelligent operating systems. In the future, it seems likely that better improvements in implementing techniques to ameliorate intelligence, as seen in many areas of AI, can be integrated into physical robots that can help in their further improvement.

For example, surgical robots can provide superpowers to surgeons in the near future, thus upgrading their capability to generate, perceive, and minimize incisions and substantially more (Davenport & Glaser, 2002). Yet, it should never be overlooked that major important decisions will always be made by human surgeons; still, their assistance can improve the way surgeries are handled. Robotic surgery finds its applications in prostate surgery, gynecological surgeries, and more.

5. Automation of robotic processes

This particular automation has been known to perform various structural digital tasks. It is known to include information systems, in a way as if human users were following the manuscript or rules. In contrast to the other areas of AI, robotic process automation (RPA) is relatively inexpensive, transparent in its action, and easy to program. RPA doesn't involve robots but is only computer programs on the servers. In health care, they find their usage for monotonous work, like refurbishing patient records, prior authorization, or invoice process. But in addition to various other techniques in RPA, like image reasoning, it can be helpful in the extraction of data (Hussain et al., 2014). All these technologies described individually are more or less used in an integrated manner to improve the robots' AI-based brain. Maybe in the coming future all this technical knowledge can intermix so an amalgamated solution can be worked out.

7.3 APPLICATIONS OF AI IN HEALTH CARE

Recent advancements in AI techniques have been able to send huge waves across health care professionals regarding replacements of human physicians with AI doctors in the near future. However, it should be taken into account that the replacement of human physicians by machines is, of course, not possible; yet, AI can be a significant tool to assist physicians in making strong medical decisions or replacing the need for final human judgment in certain functional areas of biomedical health care (e.g., radiology).

7.3.1 Early Detection and Diagnosis

Diagnosis of a disease using AI has been a focus since at least 1970, around the time MYCIN was being developed for the diagnosis of bacterial infection. The classic system of the early rule-based system was utilized for accurate diagnosis and the treatment of disease. With the advancement of technology, various techniques are used that utilize AI and can help in various diagnostic purpose (Bush, 2018).

Presence of any disease is examined by variegated diagnostic tests. Before diagnosis, there are copious protocols that can determine whether we should move forward for diagnostic purposes or for halting diagnostic proceedings. Screening is one of the initial processes that examines the symptoms that are primarily visible in the patient. However, AI systems are also helpful in screening purposes. With the current scenario of COVID-19, there are artificial techniques that are repurposed for the betterment of the situation (Naudé, 2020). One of the most important devices during this time is a thermal scanner. For screening, there are various temperature scanners installed at diverse entry points, which are utilized to screen people for fever. However, with this technology, there is a need for frontline staff to operate the thermal scanner, but they're also at risk of coming into contact with the virus.

To solve this problem, repurposing of the thermal scanners is done by using AI to restrain the exposure of frontline personnel. AI-based cameras with multisensory imaging technology have been implemented by airports and medical centers. These cameras not only monitor the thermal temperature of the patient, they can also scan and recognize the patient. This camera allows authorities to keep an eye on the crowd and recognize people who have a higher body temperature. Tampa General Hospital in Florida, United States, was one of the first hospitals to utilize this AI-based multisensory camera technology that enables via thermal face scanning the screening of anyone entering the hospital (Chamola et al., 2020; Rysavy, 2013; Liu et al., 2018).

In recent years, a variety of voice-assisted AI techniques have been used. IBM Watson was one of them, and it gained a lot of media attention for its emphasis on precision medicine, especially for cancer detection and treatment, which is a combination of ML and NLP. Based on data-analysis programs, Watson works the application of programming interfaces, which includes various program-assisted techniques, like speech-language, vision, and ML (Ross & Swetlitz, 2017).

Applying AI in cancer proved to be very effective, especially in early-stage screening and diagnostic purposes. AI acts as a helping tool for the health care workers at different stages and helps them to diagnose and get better results with minimum error. A study by Nakamura et al. (2000), who developed an artificial neural networking–based computer, added a diagnosis scheme that proved to be effective in distinguishing between benign and malignant tumors. The ANN-based computer was trained and checked using 56 chest radiographs; 22 of these were benign nodules and 34 were malignant nodules. Several imaging objective features are included for the analysis and the outcome of the experimentation. It was observed that ANN with clinical parameters and objective features achieves a considerably higher performance than ANN with clinical parameters alone. Hence, the outcome of this experiment was that AI can help gain more accurate results (Dayhoff & DeLeo, 2001).

Matsuki et al. (2002) achieved similar findings using ANNs to differentiate between benign and malignant lung nodules. They looked at 155 lung nodules, 91 of which were malignant and 56 of which were benign, using clinical criteria and subjective radiological findings extracted by the attending radiologist.

The detection of a pattern of specific genomic derived mutation is also assessed by the availing of digital pathology slides by deep-learning algorithm. Further research is needed to compare the accuracy of an algorithm for classifying digital pathology slides identified as a micrometastases slide for breast cancer by pathologists with DNN (Coudray et al., 2018). However, the aforementioned ANN scheme was not automated because they use radiologist subjective rating as a feature of lung nodules (Ström et al., 2020; Summerton & Cansdale, 2019).

However, combining ANN with different imaging techniques also enhances the accuracy and result of the current scenario techniques, like PET and CT scan. A study by Nie et al. (2006) distinguishes various lung nodules in the patient for both thoracic CT and whole-body 18F-fluorodeoxyglucose (FDG) PET. They use clinical parameters together with PET features, CT features, and both CT and PET features. Legitimate increases in both CT and PET features were discerned rather than using them alone. Consequently, artificial intelligence could help in enhancing the ongoing research and future diagnostic perspective. Numerous techniques that include temporal subtraction, augmented by algorithm scan techniques, have been developed and successfully applied to various radiological approaches, such as plain radiography, scintigraphy, ultrasound, PET-CT, and MRI. They are useful in the identification of lessons as well as the diagnosis of various stages of cancer and pathological diagnosis. Diagnosis via AI is not limited to the surgical room, as it moves forward for scanning, screening, and diagnostic purposes (Lowe et al., 2009).

7.3.2 Therapeutic Aspects of Artificial Intelligence

Utilization of AI in different fields of science and technology and diverse aspects of daily life is a new common thing. More emphasis is given to those protocols that give precise results. This aim is achieved by AI that works more precisely, and it provides an unbiased result, while chances of error are mainly based on technical glitches (Hamet & Tremblay, 2017).

Applying AI in treatment should be done more precisely because the result can cost a patient's life. Using AI in treatment purpose involves many advancements in selecting the technique and is performed under guidance and supervision of the specialist who can handle both the aspect that is the AI system as well as knowledge of his specialized field of treatment. Disparate case studies reveal the usage of AI systems for treatment purposes that include trained AI systems with precise protocols and different algorithmic data analysis systems that can help in treating particular diseases with particular symptoms. ML systems can help in solving operational problems, such as diagnosis, treatment, and surgical purposes and also for predictive purposes (Davenport & Kalakota, 2019; Jordan & Mitchell, 2015).

Selection of the most appropriate treatment for a particular disease with a particular symptom is one of the important analytical features of using AI in treatment approaches. Diverse software systems are working for the analysis of samples, which can be in any form, such as slides, reports, imaging, etc. One of the software programs is TissUUmaps, which is an online tool for interactive examination of the sample that can provide different results by imaging-marker assessment and morphological assessment. These tools provide information by interpreting the gene marker and morphological markers for the particular tissue. This software can analyze and predict the better or relevant outcomes for the particular disease and can also predict the stage and time duration for further treatment and related outcomes of the treatment protocols (Nichols et al., 2019; Jiang et al., 2017).

When the system is combined with AI, treatment or assessment of the particular disease can give better results. Human mind power and AI work alongside each other and can provide defensible results and assess the nearby or other outcomes that can be skipped or ignored by human brain (Jamshidi et al., 2020).

Some of the challenges for improving health care conditions in different cases, such as cancer and dramatic arthritis, are locating the right individual at the right spot at the right time and also providing the right treatment and giving the right advice for tackling the situation. AI can also play role in assessing the right pathway for a given problem. Different software available online combined with AI as well as medical health care personal guidance can provide initial-level help (Romm & Tsigelny, 2020).

ML encompasses a wide variety of algorithms that make intelligent predictions based on data sets that are often large and contain millions of unique data points. ML has advanced to the point where it can appear to be at a human level of semantic comprehension and knowledge extraction and also has the ability to identify abstract patterns with greater precision than human experts. The use of ML can be applied for the prediction and analysis of the performance of stroke treatment in various cases. A study by Bentley et al. (2014) used an algorithm support vector machine (SVM), which is a type of ML data analysis, to predict whether a patient with intravenous thrombolysis treatment would develop symptomatic intracranial hemorrhage by analyzing CT scans. They accomplished this by using whole-brain imaging as an input for SVM results, which outperforms traditional radiological methods to enhance the intravenous thrombolysis diagnosis decision-making process. Love et al. (2013) suggested a stroke care model based on a Bayesian belief network review of practice recommendations, meta-analysis, and clinical trials. This model included 56 variables and 3 decisions for evaluating the diagnosis, treatment, and outcome prediction procedures. Compared to the conventional method, ML methods have the advantage in improving prediction performances by analyzing the data by ANN and SVM, obtaining a prediction accuracy above 70%. Rakhmetulayeva et al. (2018) use an ML method to classify factors that affect the outcome of endovascular embolization therapy for brain arteriovenous malformations.

Some of the tools involved in the health care system for treatment purposes are AESOP, Cyberknife, and Robodoc. AESOP (Automated Endoscopic System for Optimal Positioning) is a type of surgical robot for endoscopy approved by the FDA (US Food and Drug Administration) that works on voice-command, as few voice commands are sufficient for its fair motion (Camarillo et al., 2004). During the procedure, it is in direct contact with the tissue; still, its manipulations are non-invasive and it fairly controls its motion.

Another important tool for treatment purposes is Cyberknife, used to radiate a variety of tumors in various organs. The input required for the operative functions are preoperative CT scan images to a path-planning algorithm that generates a spatial path for linear accelerator carried on the robot. For the elimination of tumors, preoperative pathways are automatically analyzed that are corotated with the real-time radiographic images. Although Cyberknife works autonomously, before the procedure, the surgeon or radiotherapist must edit the computer-generated pathway to get more accurate results, as the radiation is sufficient for the destruction of cancerous cells along with the normal cells. Different generations of Cyberknife have been made, with advancements in every new generation. Similarly, another surgical robotics in health care is Robodoc, which is a bone-milling portion of total hip arthroplasty. For the operative purpose of these surgical robotic devices, various microdevices are required, such as sensors, actuators that are electrical, and technical components that are commonly referred to as micro-electrical mechanical systems (MEMS) (Azad et al., 2020; Dharmapalan, 2019).

For the operative purpose of robotics, AI is necessary, as it provides computer vision to navigate, sense, and also assess their reactions. Robotic task learning is provided by ML from humans and is part of computer programming and AI. As we have seen, AI in the health care system automates and predicts various processes (Banerjee et al., 2020).

7.3.3 Drug Development

Research and development of drugs are relatively complex, time-consuming, expensive, and have a high attrition rate (Waring et al., 2015). Drug attritions have been known to happen in clinical studies, including great loss of resources, and it can be evident from the fact that nine out of ten drug candidates are known to fail in the first phase of clinical trials and regulatory approval (Fleming, 2018). Furthermore, conventional drug development so far has five stages (Xue et al., 2018):

1. Development and discovery
2. Pre-clinical studies
3. Clinical trials
4. Review by FDA
5. FDA post-market safety development and monitoring

According to the recent reports by Eastern Research Group (ERG), it may take up to 10–15 years to develop a new molecular entity; still, the success rate only approaches 2.01% (Xue et al., 2018). In a recent estimate, certain pharmaceutical companies spent nearly US$2–6 billion in 2015, and up to $802 million in 2003, just for the development of a new chemical drug that was approved by the FDA (Zhou et al., 2020). With this data, it is evident that the pharmaceutical industry is constantly in a state of decline. But due to recent advancements in the field of AI, a rather accelerated and improved pharmaceutical R&D (research and development) has been possible. A subset of AI, particularly machine-language algorithms, have now been used with many achievements to decrease drug discovery time. It has now been evident that using AI can restore parts of the drug-development process in a much quicker, safer, and cheaper route. But at the same time, it should be kept in mind that all the stages of drug creation cannot be removed; AI can, rather, assist with the pace of these stages, such as the discovery of new chemical compounds that can act as possible drugs (Murali & Sivakumaran, 2018).

ML has until now proved its efficiency in the developmental process of drugs whenever there was a time of previous health emergency. For instance, Bayesian ML models were in subsequent use at the time of the Ebola epidemic to speed up the process of discovering the molecular inhibitors that can be used against the virus (Ekins et al., 2017). Similarly, Zhang et al. (2017) at the time of the frequent recurring influenza epidemic in China, adopted the use of machine language–assisted virtual screening against avian H7N9 virus's inhibitor development. At the time of the West Africa Ebola outbreak in 2014, a program featuring the use of AI as a tool to scan accessible medicines was used to redesign the fight against the virus. With success, two drugs were discovered within one day to reduce infectivity, while analysis of this kind of research usually takes months to years, a difference that was important to saving millions of lives (Agrawal, 2018). In the coming years, it can be possible that AI platforms can unite with memory-computing technology and thus can offer accelerated drug development and discovery; not only this, it can also help scientists find a new use for drugs (Murali & Sivakumaran, 2018). Various ML techniques have now been used to identify certain drug candidates by way of predicting the drug–target interactions (DTIs) that can exist between the existing drug and the viral proteins. Zhang et al. (2020), also developed a deep learning deeper-feature convolutional neural network (DFCNN) that has the ability to classify/identify the protein–ligand interactions with much more accuracy. Thus, in the modern world, the use of AI can help in suggesting the ideal candidates that can benefit from this technique and also analyze their effectiveness much more quickly.

All around the world, many companies have collaborated with AI companies for several in-house projects. Phenotypic drug discovery is one of the most important uses of AI platforms, where screening of compounds is done on cell and animal models for those compounds that have the ability to cause a desirable change, without any previous knowledge of the biological target that is to be hit (Lamberti et al., 2019). A recent AI platform known as AiCure has developed a mobile application that can measure medication adherence within phase two of the trials of patients who are known to suffer from schizophrenia. It has recently been reported that AiCure has increased the adherence to 25%, compared to the traditional 'directly modified observed therapy' (Bain et al., 2017). The selection of the patients for the clinical trial is the most crucial step. This helps in evaluating the relationship that exists between various human-relevant biomarkers with *in vitro* phenotypes that can help in providing a more precise quantifiable and relatively predictable assessment of the uncertainty that exists within the therapeutic response in a specific patient. The evolution of AI has helped in the identification and prediction of various human-relevant biomarkers of certain diseases that allow the enrollment of a particular patient population that is in the second and third phases of clinical trials. It has been estimated that AI-predictive modeling for the selection of patient population would lead to an increased success rate in the clinical trials (Deliberato et al., 2017; Perez-Gracia et al., 2017).

Thus, the new advancements in AI have led the foundation to future rational drug development and optimization, which can help remarkably to optimize the impact on drug development procedures, which eventually can improve human health.

7.3.4 Drug Repurposing

In addition to the most important assistance of AI in drug development, the other most relevant aspect of its importance is to help in the identification of existing drugs that can be repurposed (Chamola et al., 2020). The four most important steps in drug repurposing are:

1. Identification of compounds
2. Compound acquisition
3. Clinical trials
4. FDA post-market monitoring and development

A repositioned drug can omit the initial steps and can directly go to preclinical testing along with the clinical trials, thus lowering the risk and the costs. The basic idea of drug repurposing is that a number of diseases may have a common molecular pathway. It is also accountable for a large amount of precise information that can be made accessible on the toxicity, formulation, dosage, pharmacology, and clinical trial data, shelved, approved, or even the discontinuation of the drug (Ngo et al., 2016). Knowing the substantial costs, high attrition rates, and comparative low pace of drug development, exploitation of known drugs can help to further enhance the efficiency of the drug while minimizing the side effects arising from its clinical trials. Sir James Black, a Nobel laureate, said, 'The most fruitful way for the discovery of a new drug is to start with an old one' (Chong & Sullivan, 2007). Drug repurposing/drug re-tasking/re-profiling/repositioning is a strategy to find several new applications for approved and drugs that are undergoing clinical trials or failed in clinical trials. It provides an advantage over the conventional method, as the safety issues of using the drug have been revealed from the clinical trial results for other applications, thus, re-profiling of the known drug can actually speed up the process of medication to the patients at a relatively lower cost than the development of a new drug. For decades, academic institutions and science funders have championed the idea that screening libraries of existing drugs with various tests could uncover new applications and have made observations that have led to medicines designed for one disease finding uses in another. Some best examples are thalidomide for myeloma (Laubach et al., 2009), sildenafil citrate for erectile dysfunction (Ghofrani et al., 2006), and remdesivir for COVID-19 (Beigel et al., 2020).

Various *in silico* methods have been designed for predicting the pharmacological properties of drug and drug repurposing with the help of transcriptomic data of several biological systems that have been conditioned through DL applications. Deep neural networks (DNNs), which are essentially multilayer systems known to contain connected and interconnected artificial neurons with the ability to perform various data transformations, provide high-level representations of data for these described methods (Lozano-Diez et al., 2017). Aliper et al. (2016) demonstrated therapeutic-based categorization of drugs, such as efficacy, functional class, toxicity, and therapeutic use with the help of DNNs that can classify complex drug action mechanisms on the pathway level.

7.3.5 Drug Toxicity and Safety

The major challenge in bringing a novel drug to market is its safety. The most unexpected toxicity is the major cause of attrition at the time of clinical trials and also at the time of post-marketing safety concerns, which is the leading cause of unnecessary mortality and morbidity. An unexpected effect that can occur after a dose of the drug includes adverse effects (AEs) or adverse drug reaction (ADRs). In 2008–2017, FDA has validated 321 new drugs. During the same tenure, FDA Adverse Event Reporting System (FAERS) have reported some 10 million AE that are known to include 5.8 million serious reports and 1.1 million AE-related deaths (Basile et al., 2019).

The use of AI techniques in preclinical drug safety has been known for its important role in premarketing drug safety, specifically for toxicity evaluation. The toxicity evaluation in drug design is a crucial step that involves the identification of the adverse effect of the chemicals on animals, plants, human beings, and even the environment (Raies & Bajic, 2016). Having a preclinical evaluation is obligatory for the prevention of toxic drugs to reach clinical trials. Despite the various efforts, increased toxicity has still been the most common issue leading to drug failure, thus accounting for drug withdrawals of nearly two-thirds of post-marketing drugs (Onakpoya et al., 2016).

For toxicity assessment of a drug, animal studies have been the most traditional approach (Segall & Barber, 2014). But the main reason accounting for the constrained approach is time cost and the ethical considerations associated with this aspect. Various *in silico* approaches have been demonstrated to estimate the toxicity of the drug candidates. The main approach used by this technique is quantitative structural-activity relationships (QSARs) and target-based predictions (Basile et al., 2019).

7.3.6 Virus Modeling and Analysis

Viral infections are one of the most common infections to human health and still pose a major threat to humans. For the development of successful treatment, one needs to understand the virus itself. According to WHO, AIDS was the cause of death for nearly one million people in 2016 (Amarasinghe et al., 2011). All around the world, dengue fever cases have increased in the last decade. Amarasinghe et al. (2011) pointed out that there are nearly 50 million annual cases and that it was the possible cause of 25,000 deaths (Guzman et al., 2010; Yang et al., 2020). A virus infects the host's cell using a lock-and-key mechanism for binding to the host's cell receptors. Thus, investigations of the human–virus interactions are of prime importance, which, in turn, is the initial step for the immense efforts to know the possible ways viruses hijack and utilize host cell functions so as to carry their own life process. Inhibitors are the main roadblocks that can obstruct the overall mechanism by hindering the interaction of the virus to the host cell receptor. Hence, to design an effective viral inhibitor, scientists must model the binding mechanism of the virus and host cell. Likewise, protein–protein interaction (PPI) plays a vital role for host immune responses to viral infection and also serves as the foundation of cell communication between human and viruses in complex human–virus interaction systems (Dyer et al., 2008; Yang et al., 2019). Thus, human–virus PPI is important for a complete understanding of the pathogenicity of the virus, which leads to the foundation of better effective prevention strategies that can be assigned to combat the disease.

Various biochemical, biophysical, and biological methods can be used to individually determinePPIs, but recently mass spectroscopy (MS) and yeast two-hybrid (Y2H) has been used for large-scale determination of PPIs (Ito et al., 2001; Puig et al., 2001; Shoemaker & Panchenko, 2007) and to understand the corresponding biological processes. However, all these experiments have been applied to identify only intraspecies PPIs (Ho et al., 2002; Ito et al., 2000; Rual et al., 2005); on the other hand, interspecies interactions have been relatively understudied. Moreover, it has come to light that experimental determinations of PPIs are comparatively more time consuming, and their complete interactomes are relatively hard to obtain. Thus, good computational methods for PPI predictions, along with the experimental techniques, can provide experimentally testable hypotheses, and limiting the range of PPI candidates can exclude protein pairs that show low interaction probability (Yang et al., 2020).

ML can be calculated as one of the most useful techniques for building new and improved models. Previously, ML models skilled with protein data have shown success in predicting PPIs between human body cells and the H1N1 virus, thus eliminating the need for any model to find the entire virus–host interactome (Shapira et al., 2009). Moreover, through ML, the model for the protein-folding mechanisms used by a virus to survive can also be predicted. For instance, Senior et al. (2020) made use of deep learning (DL) algorithms and analyzed its amino acid sequence so as to predict the three-dimensional structure of the protein. In the current COVID-19 pandemic, the AlphaFold system of DeepMind, Google's AI company, has been used to predict SARS-CoV-2-associated protein structures. The overall structure of the virus can be better understood by these predictions and thus help to develop medicine/a cure for the treatment of the COVID-19 virus (Chamola et al., 2020).

7.3.7 Disease Surveillance and Risk Prediction

Human history has been full of infectious diseases like the Black Death, the 1918 influenza pandemic, the Ebola outbreak, and the current COVID-19 pandemic, that result in loss of life as well as livelihoods (Lederberg, 1997; Jones et al., 2008; Swerdlow and Finelli, 2020; Holmdahl and Buckee, 2020).

AI-based disease surveillance systems are in work for the help of health care experts for better control and curbing of disease. This system has an advantage over traditional monitoring systems, as it allows mass monitoring of the population as well as a record of analyzed data in an unbiased manner. These pros make it a technique worth using, as it is timing-saving, cost-effective, and can help in saving lives. In the current scenario of COVID-19, AI-based surveillance systems prove to be one of the biggest boons of AI, as it helps in both disease surveillance as well as in predicting the risk of infection in different situations and tackling infectious diseases above and beyond COVID-19 (Holmdahl and Buckee, 2020; Kuziemski and Misuraca, 2020; Simsek and Kantarci, 2020).

Some of the early systems for disease surveillance and risk prediction are ProMED-mail, which is used for checking widespread notifications on the outbreak of any disease (Madoff & Woodall, 2005). For feedback, each of these tools uses different types of online interactions. Some of them enable people to actively search out health information online, thus implicitly providing useful public health surveillance data. People can, on the other hand, choose to exchange health-related information on the internet, either unconsciously or consciously, for a multitude of reasons. Forecasting seeks to foresee the future, while now casting tries to map the current state of occurrences in near real-time. Digital surveillance was based on Google search trends, page views, social media postings, and participatory surveillance efforts, which rely on different online interactions as input, whereas some require active users who seek health care information online (Salathé, 2018). However, for the AI health care search engine, specific keywords, symptoms related to the disease, must be mentioned to get significant accuracy of surveillance predictions. In most digital surveillance systems, supervised learning is used to equate data patterns to a user-specified result. Supervised ML has shown promise in solving prediction problems, especially in image analysis, including classification of skin lesions (Esteva et al., 2017), identification of early breast cancer indicators (Wang et al., 2016) and detection of diabetic retinopathy (Gulshan et al., 2016). For the geographical spread of infectious disease, digital data, such as Twitter geolocations, user's social media posts, etc., are assessed, such as in the case of the chikungunya virus tracking in Europe (Rocklöv et al., 2019). When combined with mechanistic flu models, additional data can be used to predict peak time and duration, as well as epicentral locations (Zhang et al., 2017). Hence, for the betterment of health care and risk, prediction regarding any disease outbreak is passed through an AI-based surveillance system (Aiello et al., 2020).

Prediction of risk of getting infected, risk of developing severe symptoms, risk of coming in contact with a contaminant, as well as after results of infection that are choosing a specific line of treatment are AI assimilated risk predictions (Naudé, 2020). Age, sex, hygienic behaviors, health problems, travel history, family medical history, and other factors all influence the likelihood of being infected. We can't apply direct mathematical formulas for assessment, hence comprehensive analysis of these factors integrated with an advanced AI system can provide relevant prevision of the risk profile for an individual. DeCaprio et al. (2020) describe the vulnerability index for individuals susceptible to COVID-19 via stratagem based on ML that can also predict survival rate and need for ICU treatment if that patient contracted COVID-19. Bayesian health is a similar kind of startup that predicts the early warning system for tracing acute respiratory distress syndrome (ARDS), which predicts the chances of occurrence of COVID-19 (Strickland, 2020); the accuracy of this model has shown to be about 70–80% in predicting severe cases of COVID-19. Additionally, AI techniques, especially ML, are used to correlate drug utility on a patient via analyzing the patient's data parameter. Hence, these predictive analyses can help doctors and health care professionals achieve better results and prepare for further aspects of the disease (Chamola et al., 2020).

7.3.8 Clinical Care

Using AI techniques, especially deep learning, seems to be a very fruitful technique for reducing the workload of clinicians' specialty doctors to paramedics. DNN (deep neural networking) can interpret medical scans, pathology slides, skin lesions, retinal images, electrocardiograms, endoscopy, faces, and vital signs. Using these AI techniques, there is marked improvement in assessment and a reduction in misinterpretations (Lindsey et al., 2018). DNNs have been applied across various medical scans, including bone fractures, the aging process (Gale et al., 2017), computed tomography (CT) scans for lung nodules (Shadmi et al., 2018), different types of cancers (Liu et al., 2018; Shadmi et al., 2018), MRI (Lieman-Sifry, et al., 2017), echocardiograms (Madani et al., 2018), and mammographies (Lehman et al., 2019).

Utilizing electronic medical record data, machine- and deep-learning algorithms can predict various suitable clinical parameters for the patient according to situation, conditions, and data regarding the patient's health condition, which can benefit the patient as well as health care personnel. Many firms, such as Careskore, which provides health care systems with an estimated risk of readmission and mortality based on electronic health record data, are already marketing such algorithms (Rayan, 2019). AI-assessed smartwatches are quite useful, as photoplethysmography and accelerometer sensors on the watch learn the user's heart rates at rest and with physical activity, and in case of any variation in the normal pattern, it can warn the user. These smart gadgets can also benefit patients with kidney disease, as deep learning of the ECG pattern on a smartwatch can reliably detect elevated potassium levels in the blood. Also, there are various AI systems with access to a smartwatch that can monitor the different patterns of a user's metabolism, along with keeping a record of the data it sends it to the linked users, which can be doctors, physicians, medical care personnel, or family members; hence, it can help in the betterment of health care services and help the user in case of an emergency (Beam & Kohane, 2016; Thiébaut & Thiessard, 2018).

7.3.9 Disease on the Focus of AI

Despite increasingly vast AI literature in biomedical fields, the scope of medical research mainly focuses on a few disease types, particularly nervous system diseases, cancer, and cardiovascular diseases.

1. Cancer: In 2017, it came into account that IBM Watson for Oncology could be a legitimate AI system that could be used for evaluation of cancer by making use of double-blind affirmation studies (Somashekhar et al., 2017). Similarly, clinical images can be used to identify skin cancer using an AI system (Esteva et al., 2017).
2. Neurology: Bouton et al. (2016) proposed an AI device that can help quadriplegic patients regain control of their movements. Farina et al. (2017) validated the capacity of an offline man–machine interface for controlling upper-limb prostheses using different discharge timings of spinal motor neurons.
3. Cardiology: Dilsizian and Siegel (2014) showed how an AI system would aid in the diagnosis of heart diseases by analyzing cardiac images. The FDA recently granted Arterys approval to market Arterys Cardio DL applications, which are using AI to provide the automated and adjustable ventricle segmentations that are usually based on traditional cardiac MRI images (Marr, 2017).

According to various reports, these three diseases are one of the major causes of death; thus, early diagnosis is critical in preventing patients' health from deteriorating. Despite having applications in the diagnosis of these major diseases, AI has recently been applied in various other diseases as well. For example, Long et al. (2017) analyzed the data of ocular images to diagnose congenital cataract disease, and Gulshan et al. (2016) analyzed referable diabetic retinopathy by way of retinal fundus photography.

7.4 CONCLUSION AND FUTURE SCOPE

AI breakthrough in the life sciences and other fields has been primarily expeditious, with a centralized path toward authentication and increased inclination in the scientific sphere for execution. The introduction of AI at several stages, including language searches for the biomedical literature, formulating new molecules, imagining off-target effects, and toxicity, is changing the way we think about drug development in the future (Smalley, 2017; Lowe, 2018). There is a newfangled approach, that preclinical animal trials can be mitigated via machine-erudition assessment of toxicity (Luechtefeld et al., 2018). Amalgamation of the robustness of human clinicians with the potency of AI deep learning systems truncated the mistakes in diagnostics that are deeply rooted in the current system (Patel et al., 2009). The anticipating capabilities of AI made it an integral part of medical technology; presently, AI tools are availed in different fields, including cancer, cardiology, neurology, and more (Johnson et al., 2018; Thompson et al., 2018).

AI will perpetuate a more progressive role in diverse fields of science and technologies. Future vision involves the manufacture of machines that are supremely more intelligent than humans. In today's scenario, this work is in the embryonic stage, and its fate relies on ongoing research. If the researcher disentangles the enigma of the human brain, AI may procure attributes of the human brain. Human consciousness unified with machines is a subject yet to be known (Hussain, 2018).

Agricultural and industrial food production implements computer vision systems and AI. It can be used subsequently in grading fruits and other materials. This methodology dispenses comprehensible objective analysis and instigates precise descriptive data. Diversifying the practice of robotics, autonomous vehicle, with drones and agriculture machines are the harbingers of future work (Patricío & Rieder, 2018). There is an ongoing presumption that in the future AI will personate a significant role in health care. Although preliminary attempts at stipulating detection and treatment have evinced laborious outcomes with time and fidelity, AI will subjugate that section also. The substantial questions to AI in the health care field are not about the utilitarian values of this technology but, rather, certifying their espousal in day-to-day clinical practice. For the profitable expansion of AI, regulatory validation by AI systems, proper standardization of products so they operate comparably, proper inculcation in clinicians, remuneration by public or private sectors, and modernization over time in different domains is needed (Davenport & Kalakota, 2019). A lucrative AI system must acquire the ML component for maneuvering structured data and natural-language processing for computing unstructured texts (Jiang et al., 2017).

AI innovations will emerge in the plausible future. Refinement in speech, video conferencing, and face and voice identification will be the area of future outlook in AI (Duan et al., 2016). Augmentation of AI in the field of robotics in the environment can help in agricultural fields and other areas, too; additionally, in self-navigated cars, delivery robots will be more productive, thus minimizing domestic chores (Zhu et al., 2017; Cruz et al., 2015; Vinciarelli et al., 2015). AI technology in the upcoming future will cover emanating and conventional technologies. AI-assisted robotics have diverged future objectives, like better surveillance, defense, and ambush, without jeopardizing human life in a combat zone. In addition, space exploration, military applications, and industrial processes are works in progress (Dilsizian & Siegel, 2014).

Disaster management and unmanned drones with image identification and refining processes will assist in estimating the destruction of infrastructure and impart prognoses aimed at eluding traffic obstructions. Furthermore, the precedence of AI can be seen in imminent times in the colonizing world, even in the absence of human beings (Shabbir & Anwer, 2018).

REFERENCES

Agrawal, P. (2018). Artificial intelligence in drug discovery and development. *Journal of Pharmacovigilance*, *6*(2). https://doi.org/10.4172/2329-6887.1000e173

Aiello, A. E., Renson, A., & Zivich, P. N. (2020). Social media—and internet-based disease surveillance for public health. *Annual Review of Public Health*, *41*, 101–118.

Aliper, A., Plis, S., Artemov, A., Ulloa, A., Mamoshina, P., & Zhavoronkov, A. (2016). Deep learning applications for predicting pharmacological properties of drugs and drug repurposing using transcriptomic data. *Molecular Pharmaceutics*, *13*(7), 2524–2530. https://doi.org/10.1021/acs.molpharmaceut.6b00248

Alshahrani, S., & Kapetanios, E. (2016). Are deep learning approaches suitable for natural language processing? In *International Conference on Applications of Natural Language to Information Systems* (pp. 343–349). Springer, Cham.

Amarasinghe, A., Kuritsk, J. N., Letson, G. W., & Margolis, H. S. (2011). Dengue virus infection in Africa. *Emerging Infectious Diseases*, *17*(8), 1349–1354. https://doi.org/10.3201/eid1708.101515

Arel, I., Rose, D. C., & Karnowski, T. P. (2010). Deep machine learning-a new frontier in artificial intelligence research [research frontier]. *IEEE Computational Intelligence Magazine*, *5*(4), 13–18.

Azad, A., Sharma, A. K., Wadhawan, S., & Mittal, S. K. (2020). Computerized health care systems: Optimistic digitalization in radiotherapy. *Available at SSRN 3734096*.

Bain, E. E., Shafner, L., Walling, D. P., Othman, A. A., Chuang-Stein, C., Hinkle, J., & Hanina, A. (2017). Use of a novel artificial intelligence platform on mobile devices to assess dosing compliance in a phase 2 clinical trial in subjects with schizophrenia. *JMIR MHealth and UHealth*, *5*(2), e18. https://doi.org/10.2196/mhealth.7030

Banerjee, I., Robinson, J., Kashyap, A., Mohabeer, P., & Sathian, B. (2020). COVID-19 and artificial intelligence: The pandemic pacifier. *Nepal Journal of Epidemiology*, *10*(4), 919.

Bartolozzi, C., & Indiveri, G. (2007). Synaptic dynamics in analog VLSI. *Neural Computation*, *19*(10), 2581–2603.

Basile, A. O., Yahi, A., & Tatonetti, N. P. (2019). Artificial intelligence for drug toxicity and safety. In *Trends in Pharmacological Sciences* (Vol. 40, Issue 9, pp. 624–635). Elsevier Ltd. https://doi.org/10.1016/j.tips.2019.07.005

Beam, A. L., & Kohane, I. S. (2016). Translating artificial intelligence into clinical care. *Jama*, *316*(22), 2368–2369.

Beigel, J. H., Tomashek, K. M., Dodd, L. E., Mehta, A. K., Zingman, B. S., Kalil, A. C., Hohmann, E., Chu, H. Y., Luetkemeyer, A., Kline, S., Lopez de Castilla, D., Finberg, R. W., Dierberg, K., Tapson, V., Hsieh, L., Patterson, T. F., Paredes, R., Sweeney, D. A., Short, W. R., . . . Lane, H. C. (2020). Remdesivir for the treatment of covid-19 — Final report. *New England Journal of Medicine*, *383*(19), 1813–1826. https://doi.org/10.1056/NEJMoa2007764

Bentley, P., Ganesalingam, J., Jones, A. L. C., Mahady, K., Epton, S., Rinne, P., Sharma, P., Halse, O., Mehta, A., & Rueckert, D. (2014). Prediction of stroke thrombolysis outcome using CT brain machine learning. *NeuroImage: Clinical*, *4*, 635–640.

Bouton, C. E., Shaikhouni, A., Annetta, N. V., Bockbrader, M. A., Friedenberg, D. A., Nielson, D. M., Sharma, G., Sederberg, P. B., Glenn, B. C., Mysiw, W. J., Morgan, A. G., Deogaonkar, M., & Rezai, A. R. (2016). Restoring cortical control of functional movement in a human with quadriplegia. *Nature*, *533*(7602), 247–250.

Bush, J. (2018). How AI is taking the scut work out of health care. *Harvard Business Review*, *5*.

Camarillo, D. B., Krummel, T. M., & Salisbury Jr, J. K. (2004). Robotic technology in surgery: Past, present, and future. *The American Journal of Surgery*, *188*(4), 2–15.

Chamola, V., Hassija, V., Gupta, V., & Guizani, M. (2020). A comprehensive review of the COVID-19 pandemic and the role of IoT, drones, AI, blockchain, and 5G in managing its impact. *IEEE Access*, *8*, 90225–90265. https://doi.org/10.1109/ACCESS.2020.2992341

Chiang, M., & Zhang, T. (2016). Fog and IoT: An overview of research opportunities. *IEEE Internet of Things Journal*, *3*(6), 854–864.

Chong, C. R., & Sullivan, D. J. (2007). New uses for old drugs. *Nature*, *448*(7154), 645–646. https://doi.org/10.1038/448645a

Coudray, N., Ocampo, P. S., Sakellaropoulos, T., Narula, N., Snuderl, M., Fenyö, D., . . . Tsirigos, A. (2018). Classification and mutation prediction from non—small cell lung cancer histopathology images using deep learning. *Nature medicine*, *24*(10), 1559–1567.

Cruz, F., Twiefel, J., Magg, S., Weber, C., & Wermter, S. (2015, July). Interactive reinforcement learning through speech guidance in a domestic scenario. In *2015 International Joint Conference on Neural Networks (IJCNN)* (pp. 1–8). IEEE, Killarney, Ireland.

Davenport, T. H., & Glaser, J. (2002). Just-in-time delivery comes to knowledge management. *Harvard Business Review*, *80*(7), 107–111.

Davenport, T. H., & Kalakota, R. (2019). The potential for artificial intelligence in healthcare. *Future Healthcare Journal*, *6*(2), 94.

Dayhoff, J. E., & DeLeo, J. M. (2001). Artificial neural networks: Opening the black box. *Cancer: Interdisciplinary International Journal of the American Cancer Society*, *91*(S8), 1615–1635.

DeCaprio, D., Gartner, J., Burgess, T., Kothari, S., & Sayed, S. (2020). Building a COVID-19 vulnerability index. *arXiv preprint arXiv:2003.07347*.

Deliberato, R. O., Celi, L. A., & Stone, D. J. (2017). Clinical note creation, binning, and artificial intelligence. *JMIR Medical Informatics*, *5*(3), e24. https://doi.org/10.2196/medinform.7627

Dharmapalan, B. (2019). From AI to Robotics to Precision Medicine—Revolutionising the Healthcare Sector. In *Science Reporter* (pp. 14–19). NISCAIR-CSIR, New Delhi, India.

Dilsizian, S. E., & Siegel, E. L. (2014). Artificial intelligence in medicine and cardiac imaging: Harnessing big data and advanced computing to provide personalized medical diagnosis and treatment. *Current cardiology reports*, *16*(1), 441.

Dinkelbach, H. Ü., Vitay, J., Beuth, F., & Hamker, F. H. (2012). Comparison of GPU-and CPU- implementations of mean-firing rate neural networks on parallel hardware. *Network: Computation in Neural Systems*, *23*(4), 212–236.

Duan, Y., Chen, X., Houthooft, R., Schulman, J., & Abbeel, P. (2016, June). Benchmarking deep reinforcement learning for continuous control. In *International Conference on Machine Learning* (pp. 1329–1338). JMLR Workshop and Conference Proceedings, New York, USA.

Dudek-Dyduch, E., Tadeusiewicz, R., & Horzyk, A. (2009). Neural network adaptation process effectiveness dependent of constant training data availability. *Neurocomputing*, *72*(13–15), 3138–3149.

Dyer, M. D., Murali, T. M., & Sobral, B. W. (2008). The landscape of human proteins interacting with viruses and other pathogens. *PLoS Pathogens*, *4*(2), e32. https://doi.org/10.1371/journal.ppat.0040032

Ekins, S., Freundlich, J. S., Clark, A. M., Anantpadma, M., Davey, R. A., & Madrid, P. (2017). Machine learning models identify molecules active against the Ebola virus in vitro. *F1000Research*, *4*, 1091. https://doi.org/10.12688/f1000research.7217.3

Erhan, D., Courville, A., Bengio, Y., & Vincent, P. (2010, March). Why does unsupervised pre-training help deep learning? In *Proceedings of the Thirteenth International Conference on Artificial Intelligence and Statistics* (pp. 201–208). JMLR Workshop and Conference Proceedings, New York, USA.

Esteva, A., Kuprel, B., Novoa, R. A., Ko, J., Swetter, S. M., Blau, H. M., & Thrun, S. (2017). Dermatologist-level classification of skin cancer with deep neural networks. *Nature*, *542*(7639), 115–118.

Fakoor, R., Ladhak, F., Nazi, A., & Huber, M. (2013). Using deep learning to enhance cancer diagnosis and classification. In *Proceedings of the International Conference on Machine Learning* (Vol. 28). ACM, New York.

Farina, D., Vujaklija, I., Sartori, M., Kapelner, T., Negro, F., Jiang, N., Bergmeister, K., Andalib, A., Principe, J., & Aszmann, O. C. (2017). Man/machine interface based on the discharge timings of spinal motor neurons after targeted muscle reinnervation. *Nature Biomedical Engineering*, *1*(2), 1–12.

Fleming, N. (2018). How artificial intelligence is changing drug discovery. *Nature*, *557*, S55–S57.

Gale, W., Oakden-Rayner, L., Carneiro, G., Bradley, A. P., & Palmer, L. J. (2017). Detecting hip fractures with radiologist-level performance using deep neural networks. *arXiv preprint arXiv:1711.06504*.

Ghofrani, H. A., Osterloh, I. H., & Grimminger, F. (2006). Sildenafil: From angina to erectile dysfunction to pulmonary hypertension and beyond. *Nature Reviews Drug Discovery*, *5*(8), 689–702. https://doi.org/10.1038/nrd2030.

Grollier, J., Querlioz, D., & Stiles, M. D. (2016). Spintronic nanodevices for bioinspired computing. *Proceedings of the IEEE*, *104*(10), 2024–2039.

Gulshan, V., Peng, L., Coram, M., Stumpe, M. C., Wu, D., Narayanaswamy, A., Venugopalan, S., Widner, K., Madams, T., Caudros, J., Kim, R., Raman, R., Nelson, P. C., Mega, J., & Webster, D. R. (2016). Development and validation of a deep learning algorithm for detection of diabetic retinopathy in retinal fundus photographs. *JAMA*, *316*(22), 2402–2410.

Guo, Y., Liu, Y., Oerlemans, A., Lao, S., Wu, S., & Lew, M. S. (2016). Deep learning for visual understanding: A review. *Neurocomputing*, *187*, 27–48.

Guzman, M. G., Halstead, S. B., Artsob, H., Buchy, P., Farrar, J., Gubler, D. J., Hunsperger, E., Kroeger, A., Margolis, H. S., Martí-nez, E., Nathan, M. B., Pelegrino, J. L., Simmons, C., Yoksan, S., & Peeling, R. W. (2010). Dengue: A continuing global threat. *Nature Reviews Microbiology*, *8*(12), S7–S16. https://doi.org/10.1038/nrmicro2460

Hamet, P., & Tremblay, J. (2017). Artificial intelligence in medicine. *Metabolism*, *69*, S36–S40.

Ho, Y., Gruhler, A., Heilbut, A., Bader, G. D., Moore, L., Adams, S.-L., Millar, A., Taylor, P., Bennett, K., Boutilier, K., Yang, L., Wolting, C., Donaldson, I., Schandorff, S., Shewnarane, J., Vo, M., Taggart, J., Goudreault, M., Muskat, B., . . . Tyers, M. (2002). Systematic identification of protein complexes in Saccharomyces cerevisiae by mass spectrometry. *Nature*, *415*(6868), 180–183. https://doi.org/10.1038/415180a

Holmdahl, I., & Buckee, C. (2020). Wrong but useful—what covid-19 epidemiologic models can and cannot tell us. *New England Journal of Medicine*, *383*(4), 303–305.

Hopfield, J. J. (1982). Neural networks and physical systems with emergent collective computational abilities. *Proceedings of the National Academy of Sciences*, *79*(8), 2554–2558.

Huang, G., Huang, G. B., Song, S., & You, K. (2015). Trends in extreme learning machines: A review. *Neural Networks*, *61*, 32–48.

Hussain, A., Malik, A., Halim, M. U., & Ali, A. M. (2014). The use of robotics in surgery: A review. *International Journal of Clinical Practice*, *68*(11), 1376–1382.

Hussain, K. (2018). Artificial Intelligence and its Applications goal. *Artificial Intelligence*, *5*(01).

Ito, T., Chiba, T., Ozawa, R., Yoshida, M., Hattori, M., & Sakaki, Y. (2001). A comprehensive two-hybrid analysis to explore the yeast protein interactome. *Proceedings of the National Academy of Sciences of the United States of America*, *98*(8), 4569–4574. https://doi.org/10.1073/pnas.061034498

Ito, T., Tashiro, K., Muta, S., Ozawa, R., Chiba, T., Nishizawa, M., Yamamoto, K., Kuhara, S., & Sakaki, Y. (2000). Toward a protein-protein interaction map of the budding yeast: A comprehensive system to examine two-hybrid interactions in all possible combinations between the yeast proteins. *Proceedings of the National Academy of Sciences of the United States of America*, *97*(3), 1143–1147. https://doi.org/10.1073/pnas.97.3.1143

Jamshidi, M., Lalbakhsh, A., Talla, J., Peroutka, Z., Hadjilooei, F., Lalbakhsh, P., . . . Sabet, A. (2020). Artificial intelligence and COVID-19: Deep learning approaches for diagnosis and treatment. *IEEE Access*, *8*, 109581–109595.

Jeong, H., & Shi, L. (2018). Memristor devices for neural networks. *Journal of Physics D: Applied Physics*, *52*(2), 023003.

Jiang, F., Jiang, Y., Zhi, H., Dong, Y., Li, H., Ma, S., Wang, Y., Dong, Q., Shen, H., & Wang, Y. (2017). Artificial intelligence in healthcare: Past, present and future. *Stroke and Vascular Neurology*, *2*(4), 230–243.

Johnson, K. W., Soto, J. T., Glicksberg, B. S., Shameer, K., Miotto, R., Ali, M., & Dudley, J. T. (2018). Artificial intelligence in cardiology. *Journal of the American College of Cardiology*, *71*(23), 2668–2679.

Jones, K. E., Patel, N. G., Levy, M. A., Storeygard, A., Balk, D., Gittleman, J. L., & Daszak, P. (2008). Global trends in emerging infectious diseases. *Nature*, *451*(7181), 990–993.

Jordan, M. I., & Mitchell, T. M. (2015). Machine learning: Trends, perspectives, and prospects. *Science*, *349*(6245), 255–260.

Kim, T. H. (2010, June). Emerging approach of natural language processing in opinion mining: A review. In *International Conference on Ubiquitous Computing and Multimedia Applications* (pp. 121–128). Springer, Berlin, Heidelberg.

Kok, J. N., Boers, E. J., Kosters, W. A., Van der Putten, P., & Poel, M. (2009). Artificial intelligence: Definition, trends, techniques, and cases. *Artificial Intelligence*, *1*, 1–20.

Kuziemski, M., & Misuraca, G. (2020). AI governance in the public sector: Three tales from the frontiers of automated decision-making in democratic settings. *Telecommunications Policy*, 101976.

Kwon, M. W., Baek, M. H., Hwang, S., Park, K., Jang, T., Kim, T., Lee, J., Cho, S., & Park, B. G. (2018). Integrate-and-fire neuron circuit using positive feedback field effect transistor for low power operation. *Journal of Applied Physics*, *124*(15), 152107.

Lamberti, M. J., Wilkinson, M., Donzanti, B. A., Wohlhieter, G. E., Parikh, S., Wilkins, R. G., & Getz, K. (2019). A study on the application and use of artificial intelligence to support drug development. *Clinical Therapeutics*, *41*(8), 1414–1426. https://doi.org/10.1016/j.clinthera.2019.05.018

Laubach, J. P., Richardson, P. G., & Anderson, K. C. (2009). Thalidomide maintenance in multiple myeloma. *Nature Reviews Clinical Oncology*, *6*(10), 565–566. https://doi.org/10.1038/nrclinonc.2009.131

LeCun, Y., Bengio, Y., & Hinton, G. (2015). Deep learning. *Nature*, *521*(7553), 436–444.

Lederberg, J. (1997). Infectious disease as an evolutionary paradigm. *Emerging Infectious Diseases*, *3*(4), 417.

Lee, S. I., Celik, S., Logsdon, B. A., Lundberg, S. M., Martins, T. J., Oehler, V. G., Estey, E. H., Miller, C. P., Chien, S., Dai, J., Saxena, A., Blau, A., & Becker, P. S. (2018). A machine learning approach to integrate big data for precision medicine in acute myeloid leukemia. *Nature Communications*, *9*(1), 1–13.

Lehman, C. D., Yala, A., Schuster, T., Dontchos, B., Bahl, M., Swanson, K., & Barzilay, R. (2019). Mammographic breast density assessment using deep learning: Clinical implementation. *Radiology*, *290*(1), 52–58.

Lieman-Sifry, J., Le, M., Lau, F., Sall, S., & Golden, D. (2017, June). FastVentricle: Cardiac segmentation with ENet. In *International Conference on Functional Imaging and Modeling of the Heart* (pp. 127–138). Springer, Cham.

Lindsey, R., Daluiski, A., Chopra, S., Lachapelle, A., Mozer, M., Sicular, S., . . . Potter, H. (2018). Deep neural network improves fracture detection by clinicians. *Proceedings of the National Academy of Sciences*, *115*(45), 11591–11596.

Liu, X., Chen, K., Wu, T., Weidman, D., Lure, F., & Li, J. (2018). Use of multimodality imaging and artificial intelligence for diagnosis and prognosis of early stages of Alzheimer's disease. *Translational Research*, *194*, 56–67.

Long, E., Lin, H., Liu, Z., Wu, X., Wang, L., Jiang, J., An, Y., Lin, Z., Li, X., Chen, J., Li, J., Cao, Q., Wang, D., Liu, X., Che, W., & Li, J. (2017). An artificial intelligence platform for the multihospital collaborative management of congenital cataracts. *Nature Biomedical Engineering*, *1*(2), 1–8.

Love, A., Arnold, C. W., El-Saden, S., Liebeskind, D. S., Andrada, L., Saver, J., & Bui, A. A. (2013). Unifying acute stroke treatment guidelines for a Bayesian belief network. *Studies Health Technology and Informatics*, *192*, 1012.

Lowe, D. (2018). AI designs organic syntheses. *Nature*, *555*, 592–593

Lowe, V. J., Kemp, B. J., Jack, C. R., Senjem, M., Weigand, S., Shiung, M., . . . Petersen, R. C. (2009). Comparison of 18F-FDG and PiB PET in cognitive impairment. *Journal of Nuclear Medicine*, *50*(6), 878–886.

Lozano-Diez, A., Zazo, R., Toledano, D. T., & Gonzalez-Rodriguez, J. (2017). An analysis of the influence of deep neural network (DNN) topology in bottleneck feature based language recognition. *PLoS One*, *12*(8), e0182580. https://doi.org/10.1371/journal.pone.0182580

Luechtefeld, T., Marsh, D., Rowlands, C., & Hartung, T. (2018). Machine learning of toxicological big data enables read-across structure activity relationships (RASAR) outperforming animal test reproducibility. *Toxicological Sciences*, *165*(1), 198–212.

Madani, A., Arnaout, R., Mofrad, M., & Arnaout, R. (2018). Fast and accurate view classification of echocardiograms using deep learning. *NPJ Digital Medicine*, *1*(1), 1–8.

Madoff, L. C., & Woodall, J. P. (2005). The internet and the global monitoring of emerging diseases: Lessons from the first 10 years of ProMED-mail. *Archives of Medical Research*, *36*(6), 724–730.

Mamoshina, P., Vieira, A., Putin, E., & Zhavoronkov, A. (2016). Applications of deep learning in biomedicine. *Molecular Pharmaceutics*, *13*(5), 1445–1454.

Marr, B. (2017). First FDA approval for clinical cloud-based deep learning in healthcare. *Forbes*. First FDA Approval for Clinical Cloud-Based Deep Learning in Healthcare (forbes.com) (accessed 10 December 2020).

Marshall, R. C., Freed, D. B., & Karow, C. M. (2001). Learning of subordinate category names by aphasic subjects: A comparison of deep and surface-level training methods. *Aphasiology*, *15*(6), 585–598.

Matsuki, Y., Nakamura, K., Watanabe, H., Aoki, T., Nakata, H., Katsuragawa, S., & Doi, K. (2002). Usefulness of an artificial neural network for differentiating benign from malignant pulmonary nodules on high-resolution CT: Evaluation with receiver operating characteristic analysis. *American Journal of Roentgenology*, *178*(3), 657–663.

Minsky, M. (1961). Steps toward artificial intelligence. *Proceedings of the IRE*, *49*(1), 8–30.

Murali, N., & Sivakumaran, N. (2018). Artificial intelligence in healthcare-a review. *International Journal of Modern Computation, Information and Communication Technology*, *1*(6), 103–110. http://ijmcict.gjpublications.com

Nakamura, K., Yoshida, H., Engelmann, R., MacMahon, H., Katsuragawa, S., Ishida, T., . . . Doi, K. (2000). Computerized analysis of the likelihood of malignancy in solitary pulmonary nodules with use of artificial neural networks. *Radiology*, *214*(3), 823–830.

Naudé, W. (2020). *Artificial Intelligence against COVID-19: An Early Review* (IZA Discussion Paper). IZA Institute of Labor Economics, Bonn.

Naveros, F., Garrido, J. A., Carrillo, R. R., Ros, E., & Luque, N. R. (2018). Corrigendum: Event- and time-driven techniques using parallel CPU-GPU Co-processing for spiking neural networks. *Frontiers in Neuroinformatics*, *12*, 24.

Naveros, F., Luque, N. R., Garrido, J. A., Carrillo, R. R., Anguita, M., & Ros, E. (2014). A spiking neural simulator integrating event-driven and time-driven computation schemes using parallel CPU-GPU co-processing: A case study. *IEEE Transactions on Neural Networks and Learning Systems*, *26*(7), 1567–1574.

Ngo, H. X., Garneau-Tsodikova, S., & Green, K. D. (2016). A complex game of hide and seek: The search for new antifungals. *MedChemComm*, *7*(7), 1285–1306. https://doi.org/10.1039/c6md00222f

Nguyen, H., Kieu, L. M., Wen, T., & Cai, C. (2018). Deep learning methods in transportation domain: A review. *IET Intelligent Transport Systems*, *12*(9), 998–1004.

Nichols, J. A., Chan, H. W. H., & Baker, M. A. (2019). Machine learning: Applications of artificial intelligence to imaging and diagnosis. *Biophysical Reviews*, *11*(1), 111–118.

Nie, Y., Li, Q., Li, F., et al. (2006). Integrating PET and CT information to improve diagnostic accuracy for lung nodules: A semiautomatic computer-aided method. *Journal of Nuclear Medicine*, *47*, 1075–1080.

Nurvitadhi, E., Sheffield, D., Sim, J., Mishra, A., Venkatesh, G., & Marr, D. (2016, December). Accelerating binarized neural networks: Comparison of FPGA, CPU, GPU, and ASIC. In *2016 International Conference on Field-Programmable Technology (FPT)* (pp. 77–84). IEEE.

Onakpoya, I. J., Heneghan, C. J., & Aronson, J. K. (2016). Worldwide withdrawal of medicinal products because of adverse drug reactions: A systematic review and analysis. *Critical Reviews in Toxicology*, *46*(6), 477–489. https://doi.org/10.3109/10408444.2016.1149452

Patel, V. L., Shortliffe, E. H., Stefanelli, M., Szolovits, P., Berthold, M. R., Bellazzi, R., & Abu-Hanna, A. (2009). The coming of age of artificial intelligence in medicine. *Artificial Intelligence in Medicine*, *46*(1), 5–17.

Patnaik, A. K., Patra, R., & Bhuyan, P. K. (2020). Comparison of artificial intelligence based roundabout entry capacity models. *International Journal of Intelligent Transportation Systems Research*, *18*(2), 288–296.

Patrício, D. I., & Rieder, R. (2018). Computer vision and artificial intelligence in precision agriculture for grain crops: A systematic review. *Computers and Electronics in Agriculture*, *153*, 69–81.

Peng, Y., Zhang, Y., & Wang, L. (2010). Artificial intelligence in biomedical engineering and informatics: An introduction and review. *Artificial Intelligence in Medicine*, *48*(2–3), 71–73.

Perez-Gracia, J. L., Sanmamed, M. F., Bosch, A., Patiño-Garcia, A., Schalper, K. A., Segura, V., Bellmunt, J., Tabernero, J., Sweeney, C. J., Choueiri, T. K., Martín, M., Fusco, J. P., Rodriguez- Ruiz, M. E., Calvo, A., Prior, C., Paz-Ares, L., Pio, R., Gonzalez-Billalabeitia, E., Gonzalez Hernandez, A., . . . Melero, I. (2017). Strategies to design clinical studies to identify predictive biomarkers in cancer research. In *Cancer Treatment Reviews* (Vol. 53, pp. 79–97). W.B. Saunders Ltd. https://doi.org/10.1016/j.ctrv.2016.12.005

Prezioso, M., Merrikh-Bayat, F., Hoskins, B. D., Adam, G. C., Likharev, K. K., & Strukov, D. B. (2015). Training and operation of an integrated neuromorphic network based on metal-oxide memristors. *Nature*, *521*(7550), 61–64.

Puig, O., Caspary, F., Rigaut, G., Rutz, B., Bouveret, E., Bragado-Nilsson, E., Wilm, M., & Séraphin, B. (2001). The tandem affinity purification (TAP) method: A general procedure of protein complex purification. *Methods*, *24*, 218–229. https://doi.org/10.1006/meth.2001.1183

Raies, A. B., & Bajic, V. B. (2016). In silico toxicology: Computational methods for the prediction of chemical toxicity. *Wiley Interdisciplinary Reviews: Computational Molecular Science*, *6*(2), 147–172. https://doi.org/10.1002/wcms.1240

Rakhmetulayeva, S. B., Duisebekova, K. S., Mamyrbekov, A. M., Kozhamzharova, D. K., Astaubayeva, G. N., & Stamkulova, K. (2018). Application of classification algorithm based on SVM for determining the effectiveness of treatment of tuberculosis. *Procedia Computer Science*, *130*, 231–238.

Ramesh, A. N., Kambhampati, C., Monson, J. R., & Drew, P. J. (2004). Artificial intelligence in medicine. *Annals of The Royal College of Surgeons of England*, *86*(5), 334.

Rayan, R. (2019). Precision medicine in the context of artificial intelligence. *SIGMOD Record, Forthcoming, 1–8.*

Rocklöv, J., Tozan, Y., Ramadona, A., Sewe, M. O., Sudre, B., Garrido, J., . . . Semenza, J. C. (2019). Using big data to monitor the introduction and spread of Chikungunya, Europe, 2017. *Emerging Infectious Diseases*, *25*(6), 1041.

Rodríguez-Pe´rez, R., & Bajorath, J. (2018). Prediction of compound profiling matrices, part II: Relative performance of multitask deep learning and random forest classification on the basis of varying amounts of training data. *ACS Omega*, *3*(9), 12033–12040.

Romm, E. L., & Tsigelny, I. F. (2020). Artificial intelligence in drug treatment. *Annual Review of Pharmacology and Toxicology*, *60*, 353–369.

Ross, C., & Swetlitz, I. (2017). IBM pitched its Watson supercomputer as a revolution in cancer care. It's nowhere close. *Stat.*

Rual, J. F., Venkatesan, K., Hao, T., Hirozane-Kishikawa, T., Dricot, A., Li, N., Berriz, G. F., Gibbons, F. D., Dreze, M., Ayivi-Guedehoussou, N., Klitgord, N., Simon, C., Boxem, M., Milstein, S., Rosenberg, J., Goldberg, D. S., Zhang, L. V., Wong, S. L., Franklin, G., . . . Vidal, M. (2005). Towards a proteome-scale map of the human protein-protein interaction network. *Nature*, *437*(7062), 1173–1178. https://doi.org/10.1038/nature04209

Rysavy, M. (2013). Evidence-based medicine: A science of uncertainty and an art of probability. *AMA Journal of Ethics*, *15*(1), 4–8.

Salathé, M. (2018). Digital epidemiology: What is it, and where is it going? *Life Sciences, Society and Policy*, *14*(1), 1.

Schaal, S. (1999). Is imitation learning the route to humanoid robots? *Trends in Cognitive Sciences*, *3*(6), 233–242.

Schmidhuber, J. (2015). Deep learning in neural networks: An overview. *Neural Networks*, *61*, 85–117.

Segall, M. D., & Barber, C. (2014). Addressing toxicity risk when designing and selecting compounds in early drug discovery. In *Drug Discovery Today* (Vol. 19, Issue 5, pp. 688–693). Elsevier Ltd. https://doi.org/10.1016/j.drudis.2014.01.006

Senior, A. W., Evans, R., Jumper, J., Kirkpatrick, J., Sifre, L., Green, T., Qin, C., Žídek, A., Nelson, A. W. R., Bridgland, A., Penedones, H., Petersen, S., Simonyan, K., Crossan, S., Kohli, P., Jones, D. T., Silver, D., Kavukcuoglu, K., & Hassabis, D. (2020). Improved protein structure prediction using potentials from deep learning. *Nature*, *577*(7792), 706–710. https://doi.org/10.1038/s41586-019-1923-7

Shabbir, J., & Anwer, T. (2018). Artificial intelligence and its role in near future. *arXiv preprint arXiv:1804.01396.*

Shadmi, R., Mazo, V., Bregman-Amitai, O., & Elnekave, E. (2018, April). Fully-convolutional deep-learning based system for coronary calcium score prediction from non-contrast chest CT. In *2018 IEEE 15th International Symposium on Biomedical Imaging (ISBI 2018)* (pp. 24–28). IEEE, Washington, DC, USA.

Shapira, S. D., Gat-Viks, I., Shum, B. O. V., Dricot, A., de Grace, M. M., Wu, L., Gupta, P. B., Hao, T., Silver, S. J., Root, D. E., Hill, D. E., Regev, A., & Hacohen, N. (2009). A physical and regulatory map of host-influenza interactions reveals pathways in H1N1 infection. *Cell*, *139*(7), 1255–1267. https://doi.org/10.1016/j.cell.2009.12.018

Shapiro, S. C. (1992). *Encyclopedia of Artificial Intelligence* (2nd ed.). John Wiley & Sons, Inc., New York, USA.

Shoemaker, B. A., & Panchenko, A. R. (2007). Deciphering Protein—Protein Interactions. Part I. Experimental Techniques and Databases. *PLoS Computational Biology*, *3*(3), e42. https://doi.org/10.1371/journal.pcbi.0030042

Simmons, A. B., & Chappell, S. G. (1988). Artificial intelligence-definition and practice. *IEEE Journal of Oceanic Engineering*, *13*(2), 14–42.

Simsek, M., & Kantarci, B. (2020). Artificial intelligence-empowered mobilization of assessments in COVID-19-like pandemics: A case study for early flattening of the curve. *International Journal of Environmental Research and Public Health*, *17*(10), 3437.

Smalley, E. (2017). AI-powered drug discovery captures pharma interest. *Nature Biotechnology*, *35*(7), 604–606.

Somashekhar, S. P., Kumarc, R., Rauthan, A., Arun, K. R., Patil, P., & Ramya, Y. E. (2017). Abstract S6–07: Double blinded validation study to assess performance of IBM artificial intelligence platform, Watson for oncology in comparison with Manipal multidisciplinary tumour board—First study of 638 breast cancer cases.

Sordo, M. (2002). Introduction to neural networks in healthcare. *Open Clinical Knowledge Management for Medical Care*, 1–17.

Ström, P., Kartasalo, K., Olsson, H., Solorzano, L., Delahunt, B., Berney, D. M., . . . Iczkowski, K. A. (2020). Artificial intelligence for diagnosis and grading of prostate cancer in biopsies: A population-based, diagnostic study. *The Lancet Oncology*, *21*(2), 222–232.

Summerton, N., & Cansdale, M. (2019). Artificial intelligence and diagnosis in general practice. *British Journal of General Practice*, *69*(684), 324–325.

Swerdlow, D. L., & Finelli, L. (2020). Preparation for possible sustained transmission of 2019 novel coronavirus: Lessons from previous epidemics. *JAMA*, *323*(12), 1129–1130.

Szolovits, P., & Pauker, S. G. (1978). Categorical and probabilistic reasoning in medical diagnosis. *Artificial Intelligence*, *11*(1–2), 115–144.

Thiébaut, R., & Thiessard, F. (2018). Artificial intelligence in public health and epidemiology. *Yearbook of Medical Informatics*, *27*(1), 207.

Thompson, R. F., Valdes, G., Fuller, C. D., Carpenter, C. M., Morin, O., Aneja, S., & Rosenthal, S. A. (2018). Artificial intelligence in radiation oncology: A specialty-wide disruptive transformation? *Radiotherapy and Oncology*, *129*(3), 421–426.

Vial, A., Stirling, D., Field, M., Ros, M., Ritz, C., Carolan, M., Holloway, L., & Miller, A. A. (2018). The role of deep learning and radiomic feature extraction in cancer-specific predictive modelling: A review. *Translational Cancer Research*, *7*(3), 803–816.

Vinciarelli, A., Esposito, A., André, E., Bonin, F., Chetouani, M., Cohn, J. F., & Heylen, D. (2015). Open challenges in modelling, analysis and synthesis of human behaviour in human—human and human—machine interactions. *Cognitive Computation*, *7*(4), 397–413.

Wang, J., Yang, X., Cai, H., Tan, W., Jin, C., & Li, L. (2016). Discrimination of breast cancer with microcalcifications on mammography by deep learning. *Scientific Reports*, *6*(1), 1–9.

Waring, M. J., Arrowsmith, J., Leach, A. R., Leeson, P. D., Mandrell, S., Owen, R. M., Pairaudeau, G., Pennie, W. D., Pickett, S. D., Wang, J., Wallace, O., & Weir, A. (2015). An analysis of the attrition of drug candidates from four major pharmaceutical companies. *Nature Reviews Drug Discovery*, *14*(7), 475–486. https://doi.org/10.1038/nrd4609

Warwick, K. (2013). *Artificial Intelligence: The Basics*. Routledge, London and New York.

Watts, D. J., & Strogatz, S. H. (1998). Collective dynamics of 'small-world' networks. *Nature*, *393*(6684), 440–442.

Wen, S., Huang, T., Zeng, Z., Chen, Y., & Li, P. (2015). Circuit design and exponential stabilization of memristive neural networks. *Neural Networks*, *63*, 48–56.

Weng, J., McClelland, J., Pentland, A., Sporns, O., Stockman, I., Sur, M., & Thelen, E. (2001). Autonomous mental development by robots and animals. *Science*, *291*(5504), 599–600.

Wooldridge, M. J., & Jennings, N. R. (1995). Intelligent agents: Theory and practice. *The Knowledge Engineering Review*, *10*(2), 115–152.

Xue, H., Li, J., Xie, H., & Wang, Y. (2018). Review of drug repositioning approaches and resources. *International Journal of Biological Sciences*, *14*(10), 1232–1244. https://doi.org/10.7150/ijbs.24612

Yang, D., Jiang, K., Zhao, D., Yu, C., Cao, Z., Xie, S., Xiao, Z. Y., Jiao, X. Y., Wang, S., & Zhang, K. (2018). Intelligent and connected vehicles: Current status and future perspectives. *Science China Technological Sciences*, *61*(10), 1446–1471.

Yang, S., Fu, C., Lian, X., Dong, X., & Zhang, Z. (2019). Understanding human-virus protein-protein interactions using a human protein complex-based analysis framework. *MSystems*, *4*(2), 303–321. https://doi.org/10.1128/msystems.00303-18

Yang, X., Yang, S., Li, Q., Wuchty, S., & Zhang, Z. (2020). Prediction of human-virus protein- protein interactions through a sequence embedding-based machine learning method. *Computational and Structural Biotechnology Journal*, *18*, 153–161. https://doi.org/10.1016/j.csbj.2019.12.005

Yoshida, J. (2018). *IBM Guns for 8-Bit AI Breakthroughs [Internet]*. Cambridge: EE Times. Available from www.eetimes.com/document.asp?doc_id=1334029&utm_source=eetimes&utm_ medium=networksearch (accessed 10 December 2020).

Yu, K. H., Beam, A. L., & Kohane, I. S. (2018). Artificial intelligence in healthcare. *Nature Biomedical Engineering*, *2*, 719–731.

Yuan, X., Xie, L., & Abouelenien, M. (2018). A regularized ensemble framework of deep learning for cancer detection from multi-class, imbalanced training data. *Pattern Recognition*, *77*, 160–172.

Zhang, H., Saravanan, K. M., Yang, Y., Hossain, M. T., Li, J., Ren, X., Pan, Y., & Wei, Y. (2020). Deep learning based drug screening for novel coronavirus 2019-nCov. *Interdisciplinary Sciences: Computational Life Sciences*, *12*(3), 368–376. https://doi.org/10.1007/s12539-020-00376-6

Zhang, L., Ai, H.-X., Li, S.-M., Qi, M.-Y., Zhao, J., Zhao, Q., & Liu, H.-S. (2017). Virtual screening approach to identifying influenza virus neuraminidase inhibitors using molecular docking combined with machine-learning-based scoring function. *Oncotarget*, *8*(47), 83142–83154. https://doi.org/10.18632/oncotarget.20915

Zhang, Q., Perra, N., Perrotta, D., Tizzoni, M., Paolotti, D., & Vespignani, A. (2017). Forecasting seasonal influenza fusing digital indicators and a mechanistic disease model. In *Proceedings of the 26th International Conference on World Wide Web* (pp. 311–319). International World Wide Web Conferences Steering Committee, Republic and Canton of Geneva, CHE. https://doi.org/10.1145/3038912.3052678

Zhang, X., Huang, A., Hu, Q., Xiao, Z., & Chu, P. K. (2018). Neuromorphic computing with memristor crossbar. *Physica Status Solidi (a)*, *215*(13), 1700875.

Zhou, Y., Wang, F., Tang, J., Nussinov, R., & Cheng, F. (2020). Artificial intelligence in COVID-19 drug repurposing. In *The Lancet Digital Health* (Vol. 2, Issue 12, pp. e667–e676). Elsevier Ltd. https://doi.org/10.1016/S2589-7500(20)30192-8

Zhu, Y., Mottaghi, R., Kolve, E., Lim, J. J., Gupta, A., Fei-Fei, L., & Farhadi, A. (2017, May). Target-driven visual navigation in indoor scenes using deep reinforcement learning. In *2017 IEEE International Conference on Robotics and Automation (ICRA)* (pp. 3357–3364). IEEE, Singapore.

Zucker, R. S., & Regehr, W. G. (2002). Short-term synaptic plasticity. *Annual Review of Physiology*, *64*(1), 355–405.

8 *En*-Fuzzy-ClaF *A Machine Learning–Based Stack-Ensembled Fuzzy Classification Framework for Diagnosing Coronavirus*

Sourabh Shastri, Sachin Kumar, Kuljeet Singh, and Vibhakar Mansotra

CONTENTS

DOI: 10.1201/9781003184140-8

8.1 INTRODUCTION

The world is healing from the catastrophic phase of the coronavirus (COVID-19) pandemic. Last year was unprecedented for the whole world, as almost every achievement of the mankind has failed to combat the novel coronavirus (Shi et al., 2015). In December 2019, COVID-19 came into existence in the city of Wuhan in Hubei province, China, and killed thousands of people within days, spreading over the globe and killing millions more with its abnormal behavior (Tomar & Gupta, 2020). The rapid spread of this disease led the World Health Organization (WHO) to declare it a global pandemic. Within a year, WHO reported more than 111,434,637 confirmed cases, 2,467,383 deaths and 86,283,245 recovered cases (*WHO Coronavirus Disease [COVID-19] Dashboard*, n.d.). The virus spreads though human contact or through air in overcrowded areas, which led many governments around the world to enforce lockdown measures (Aristovnik et al., 2020). Policy- and decision-makers are facing many challenges in trying to stop the spread and transmission of COVID-19. People are forced to abide the government's rules and laws to social distance and wear face masks so that the exponential growth of COVID-19 can be controlled and the mortality rate can decrease. The modern era is the age of data science and artificial intelligence (AI) (Chaturvedi et al., 2021). Machine learning (ML) and deep learning (DL) algorithms are able to extract the required important features out of huge data to further classify the dataset into different classes. These techniques also assist in diagnosing COVID-19, classification, drug discovery, and vaccination procurement. But the most important thing is to work with AI techniques and to train models by collecting huge amounts of data so that feature extraction may hold its results with more validity (Gautam, 2021)(Islam et al., 2020). Real-time polymerase chain reaction (RT-PCR) is widely used for the detection of COVID-19, but the kits of RT-PCR are expensive and the confirmation reports take around seven hours. Also, the RT-PCR technique provides false-negative results, which leads to its low sensitivity (Kumar et al., 2018). To overcome this issue, the clinical imaging–based radiological imaging, such as computed tomography (CT) is used to detect COVID-19. Normal classification methods are lacking due to the data insufficiency and its nature as heterogeneous and incomplete (Shastri, Singh, Kumar et al., 2021). In this paper, we have taken the features of COVID-19 as a text dataset to carry out the experimental work. Due to the pandemic situation, goals like quality education, no poverty, and zero hunger are at greater risk. As the catastrophic phase of COVID-19 has stunned the world, the more responsibility policymakers, decision-makers, and researchers have to find the solution for this pandemic and suggest some ways by which the world can resume its regular functionalities. To mitigate the COVID-19 effect, policymakers are applying the following two strategies that are not sufficient for densely populated nations like India:

1. Moving testing facilities to the doorstep of individuals
2. Building different COVID-19 care centers where individuals have to go for testing

The first step is hardly feasible and costly to implement in real time. The second step is feasible but a bit risky and hectic for those who are not infected by COVID-19. There must exist some experimental study to predict the future effect, diagnose and classify COVID-19 for most affected nations so that policymakers can accordingly subjugate the COVID-19 effect down to the end (Ribeiro et al., 2020). One such study is proposed in this research article.

The rest of the paper is organized as follows: Section 8.2 represents the literature survey of some scholarly articles from the same domain. Section 8.3 describes the research methodology with data description, data preprocessing and methods used in this study. Section 8.4 contains the proposed experimental setup and results. Section 8.5 represents the comparative analysis of the proposed model. Section 8.6 contains the conclusion and future work of the study.

8.2 LITERATURE SURVEY

The twofold obstacle of data heterogeneity and its incompleteness overcomes in Hamed et al. (2020). Here, authors proposed an algorithm based on K-Nearest Neighbor variant to classify COVID-19 patients. A deep neural network (DNN)–based model is designed using two stages to differentiate the COVID-19 confirmed cases from the healthy persons with a five-fold cross-validation technique and training validation techniques (Jain et al., 2020). A hybrid model is proposed for face mask detection using deep and classical ML. Resnet50 is used for feature extraction and ensemble algorithms, decision trees, and a support vector machine (SVM) algorithm is used for the classification of the face mask detection (Loey et al., 2021). Various stacking–voting techniques are used for the classification of many diseases (Shastri, Kour et al., 2020). The Xception model is used to design a deep transfer learning–based automated system to detect COVID-19 using chest X-ray images. The automated system can reduce the time for analysis of COVID-19 and train the weights of the large network and fine-tune them on small datasets (Narayan Das et al., 2020). The combination of different classifiers using majority voting of deep transfer learning outputs the ensemble models to classify the COVID-19 chest X-ray images. Various pre-trained models based on convolutional neural networks (CNNs), such as DenseNet121, ResNet-50, COVID19XrayNet, SeResnet 50, Xception, DenseNet, ResNext50, NasNetMobile, CCSHNet, Inception_resnet_v2, and EfficientNets(B0-B5), are used to learn features and provide automated diagnostic techniques and tools (Ahuja et al., 2021; Gifani et al., 2021; Pathak et al., 2020; Wang et al., 2021; R. Zhang et al., 2020; Y.-D. Zhang et al., 2021). Two DL techniques have been proposed to automatically detect COVID-19 using artificial neural networks (ANNs) and recurrent neural network (RNN)–based bi-directional long short-term memory (LSTM). Both the techniques use AlexNet architecture and take advantage of transfer learning (Aslan et al., 2021). RNN-based convolutional LSTM, bi-directional LSTM and stacked LSTM are used to forecast the COVID-19 confirmed and death cases of India and the USA (Shastri, Singh et al., 2020). Deep transfer learning is adopted to recede the abnormalities within the dataset and authors have used two datasets of COVID-19 from the publicly available medical X-ray imaging datasets (Apostolopoulos & Mpesiana, 2020). Two deep transfer learning–based ensemble models are used for the diagnosis of COVID-19 chest X-ray images. The pre-trained models are used in both the ensemble models for better performance. The models are able to differentiate between the COVID-19, pneumonia and bacterial infection (Gianchandani et al., 2020).

8.3 RESEARCH METHODOLOGY

8.3.1 Dataset Description

The dataset used for the experiment was taken from Hamed (n.d.), which is composed of two different datasets. The first dataset is of 68 COVID-19 cases from the SIRM database (*Covid-19 Database | Sirm*, 2020), and the second is from 62 flu (non-COVID-19) cases from the influenza research database (IRD) (*Influenza Research Database—Influenza Genome Database with Visualization and Analysis Tools*, n.d.). The two decision labels of this dataset are: {COVID-19, non-COVID-19}.

8.3.1.1 Data Preprocessing

The data-mining techniques are used for the data-preprocessing phase. The original dataset (D_O) was not cleaned and preprocessed. D_O has many missing values and unwanted features. So, we first apply data-imputation methods to replace all the missing values for numeric attributes in a dataset with the modes/means of the attributes categorical with an asterisk (*). The imputed dataset (say D_{DI}) so obtained is then used for feature selection, where we select only important features using a wrapper feature selection technique. A total of 13 features are picked out by the wrapper method from imputed dataset (D_{DI}) having 17 features. The dataset obtained after selecting important features (D_{FS}) is then used for building several models. The dataset of selected features with description is described in Table 8.1.

8.3.2 Methods

To carry out the experimental setup, we have used various algorithms ("Data Mining with Open Source Machine Learning Software in Java," 2011) which are

TABLE 8.1
Dataset Description.

Feature	Type	Value
Age	Numerical	[4–90]
Gender	Categorical	{Male, Female}
Fever	Categorical	{Yes, No}
Dyspnea	Categorical	{Yes, No}
Nasal	Categorical	{Yes, No}
Cough	Categorical	{Yes, No}
Partial pressure of oxygen (PO2)	Numerical	[32–292]
Asthenia	Categorical	{Yes, No}
Exposure to COVID-19 patients	Categorical	{Yes, No}
From high-risk zone	Categorical	{Yes, No}
Temperature	Numerical	[35.7–40]
Medical history	Categorical	{Cancer, Croonic, Asthma, COPD, Chronic, DM}
Decision label	Categorical	{COVID-19, Non-COVID-19}

also discussed in the subsequent sections. The experiment is carried out with and without stacking (Shastri, Kour et al., 2021) and is explained in the following sections.

8.3.2.1 Hoeffding Tree

Thr Hoeffding tree is also called the anytime decision tree induction algorithm and is incremental in nature. Thr Hoeffding tree is capable of learning from the enormous data streams, and it assumes that the distribution remains constant over time. To find the best splitting attribute, the small sample cannot be used according to Hoeffding trees.

8.3.2.2 BayesNet

The conditional probabilities of different variables can be utilized using the Bayesian network (BayesNet). It is the process of using probability to predict the likelihood of certain events occurring in the future. The conditional dependencies of the group of variables can be represented by BayesNet with the help of a directed acyclic graph (DAG).

8.3.2.3 Naïve Bayes

The high input dimensionality can be handled using the Naïve Bayes classifier technique. It can outperform other sophisticated classification models. Naïve Bayes predicts on the basis of probability of an object, and hence it is called as a probabilistic classifier. To build rapid ML models, Naïve Bayes is the most suited classification technique and makes for quick forecasting.

8.3.2.4 Deep Neural Network (DNN)

DNN is a type of ML based on ANNs, where numerous layers of nodes are used to derive high-level functions from input information.

8.3.2.5 Support Vector Machine (SVM)

SVM is a supervised learning algorithm that can be used both for regression and classification problems. This algorithm uses the extreme vectors that help in the creation of a hyperplane. These vectors or points are known as support vectors. Hence, the algorithm is termed a support vector machine. It can handle both multiple continuous and categorical variables.

8.3.2.6 Multilayer Perceptron (MLP)

The multilayer perceptron is based on feedforward neural networks and consists of three layers: input, hidden, and output. The nonlinear activation function is used by hidden and output layers and not by input nodes. It uses the backpropagation supervised learning method to train the models. The networks in MLP can be built by simple heuristic function.

8.3.2.7 Stochastic Gradient Descent (SGD)

Learning from linear models is implemented through stochastic gradient descent (SGD). The linear models are like squared loss, binary class SVM, binary class logistic regression, and Huber loss. In general, SGD transforms the nominal attributes and replaces all the missing values to get the binary attributes.

8.3.2.8 Fuzzy Classifier

Fuzzy classification is the method of grouping elements into a set of fuzzy whose membership function can be described by the truth values of a fuzzy propositional function. Its simplest form is rule-based fuzzy classifier (or if-then fuzzy system). Let's take an example of two classes whose fuzzy classifier can be constructed by identifying the if-then rule:

IF *fever* is normal AND cough is *normal* THEN class is non-COVID-19.
IF *fever* is moderate AND cough is *severe* THEN class is COVID-19.
IF *fever* is severe AND cough is *moderate* THEN class is COVID-19.

8.3.3 Stacking Ensemble

Stacking or stacked generalization is an ensemble model (strong learner) that uses a meta model (meta classifiers) to learn how to best combine the predictions from two or more base models (weak learners) to achieve better performance. Two or more base models are also called level-0 models, whereas the meta model is called a level-1 model.

- Base models (level-0 models): models that fit on training data and whose predictions are compiled
- Meta model (level-1 model): model that learns how to best combine the predictions of level-0 models

The stacking ensemble algorithm is shown in Algorithm 8.1.

Algorithm 8.1: *Stacking Ensemble*

Input: Training data, $D = \{p_i, q_i\}$, where $i = 1$ to n
Output: Stacked Ensemble Classifier (E)

Step 1. Base Classifiers Learning
 for $c = 1$ to C ***do***
  learn e_c based upon D
 end for
Step 2. Construction of new dataset of predictions
 for $i = 1$ to n ***do***
  $De = \{p_i^1, q_i\}$, where $p_i^1 = \{e_1(p_i), \ldots, e_C(p_i)\}$
 end for
Step 3. Meta Classifier Learning
 learn E based on D_e
 return E

8.4 EXPERIMENTAL SETUP AND RESULTS

8.4.1 Experiment Environment

The open-source machine learning application WEKA 3.8.3 was used for constructing *En*-Fuzzy-ClaF and different classifiers on a system based on Intel (R) Core (TM) i5–5200U CPU @ 2.20 Gigahertz and 8.00 Gigabytes RAM under the Windows 10 operating system.

8.4.2 Proposed *En*-Fuzzy-ClaF Framework

This research introduces a new stacking ensemble–based fuzzy classification framework named *En*-Fuzzy-ClaF as a tool to identify COVID-19. The framework follows ML methodology as its backend, which starts with preprocessing state and ends with a trained model. The first step is data preprocessing, where we clean and preprocess the raw data into a usable form. The preprocessing phase was discussed in Section 8.3.1.1. After the preprocessing step, either we split the data or do cross-validation to train and build a model or framework. Here, our proposed framework has been validated using K=5 and K=10 fold cross-validation. Numerous models have been designed where only one best model is hand-picked for the detection of COVID-19. *En*-Fuzzy-ClaF stacking ensemble model is one of them, as shown in Figure 8.1.

We first build many single-classification models (model without ensemble) using preprocessed dataset at five and ten-fold cross-validation then note down the classification accuracy, f-measure, ROC, and kappa statistics as its evaluation measures. The single-classification model that gives the best prediction results (P1) is selected for comparison with stacking ensemble models. While building models without ensemble, the topmost classification algorithms in terms of accuracy is also selected for stacking ensemble to use as base and meta classifiers. We constructed six such

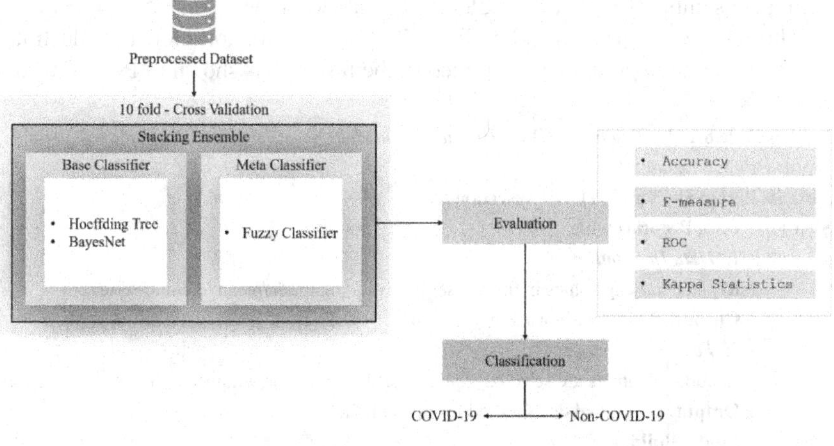

FIGURE 8.1 Proposed *En*-Fuzzy-ClaF Framework.

TABLE 8.2
Stacking Ensemble Descriptions.

Ensemble Number	Stacking	
	BASE Classifiers	META Classifier
E-1	Hoeffding ree and BayesNet	Naïve Bayes
E-2		DNN
E-3		SVM
E-4		MLP
E-5		SGD
E-6		Fuzzy

stacking ensemble models numbered from E-1 to E-6. As discussed in Section 8.3.3, stacking ensemble is a strong learner that is a combination of two models (or classifiers): base and meta. Each stacking ensemble gives better performance when we used two classification algorithms as the base and one as the meta classifier in the current dataset. The two classification algorithms acting as base classifiers are Hoeffding tree and BayesNet, which remain static throughout the experiment, but there is a variation of the classification algorithm in meta classifiers and these variants of classification algorithms includes: Naïve Bayes, DNN, SVM, MLP, SGD, and fuzzy classifier. The full description of stacking ensemble with ensemble number is shown in Table 8.2.

Among all, the stacking ensemble with ensemble number E-6 outperforms best in terms of classification accuracy, f-measure, ROC, and kappa statistics. In the E-6 stacking ensemble, the base classifier is the same as the one that has been used for other stacking ensembles, that is Hoeffding tree and BayesNet, whereas fuzzy classifier is used as meta classifier. The predicted results (P2) attained from stacking ensemble model E-6 is then compared with the prediction results of the best model without ensemble (P1). These results are discussed in the next section.

The schematic view of finding best COVID-19 detection model is illustrated in Figure 8.2 and the pseudo code applied in the research is shown in Pseudo Code 8.1.

Pseudo Code 8.1: *Finding best COVID-19 detection model*
Input: Original Dataset (D_O)
Output: Best model for COVID-19 detection
Step 1. **Data Preprocessing**

1. Data Imputation

Replace missing values in the dataset D_O by taking mode/mean (column-wise)

Output: D_{DI} is dataset used in next step

2. Feature Selection

Important features are selected from dataset D_{DI} using the wrapper feature selection technique

Output: D_{FS} is final dataset used for model building

Step 2. **Model Building**

1. Without Ensemble

a. Select classifier one by one from classifier list

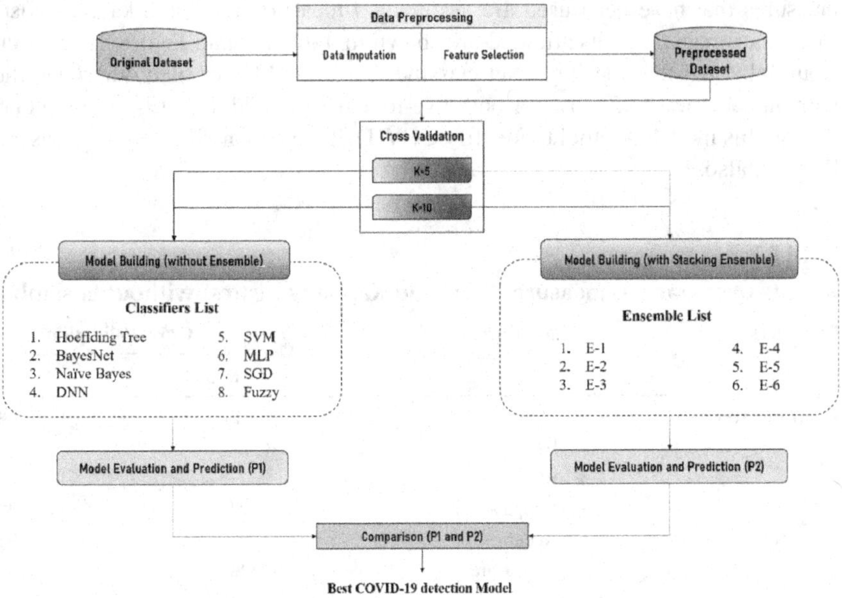

FIGURE 8.2 Flow Diagram for Finding Best COVID-19 Detection Model.

b. Apply classifiers on dataset D_{FS} by taking cross-validation five-fold and ten-fold
c. Model evaluation and prediction (P1)

2. ***With Stacking Ensemble***
a. Select stacking ensemble one by one from ensemble list
b. Apply stacking ensemble on dataset D_{FS} by taking cross-validation five-fold and ten-fold
c. Model evaluation and prediction (P2)

Step 3. **Model Comparison**
Compare models P1 and P2 by taking measures: accuracy, f-measure, ROC and kappa statistics
Output: Best model for COVID-19 detection

The original raw dataset has been preprocessed at the priliminary step to get the data that is preprocessed and is fit to use for experimentation. A cross-validation technique is used for K = 5 and K = 10 to build the model with different classifiers and their ensembles. We have compared all model combinations and the classifiers with ensemble and without ensemble to extraxt the best possible model for feature extraction and learning.

8.4.3 Results

8.4.3.1 Results without Ensemble

In this section, the experiment focuses on evaluating the prediction performance of each single classifier with cross-validation five-fold and ten-fold. The four evaluation measures that have been used are: accuracy, f-measure, ROC and kappa statistics. The experimental results attained are shown in Table 8.3 and Figure 8.3. It is clear from Table 8.3 that the BayesNet classifier performed better than other classifiers, with an accuracy of 93.85% for both five-fold and ten-fold. But it is not too good to choose this model as a tool to identify COVID-19 detection. So, we go for ensemble learning also.

TABLE 8.3
Results (Accuracy, F-measure, ROC and Kappa Statistics) without Ensemble.

Classifiers	Measures (%)	Cross-Validation	
		K=5	K=10
Hoeffding tree	Accuracy	68.46	68.46
	F-measure	65.60	65.60
	ROC	95.60	95.50
	Kappa statistics	38.58	38.58
BayesNet	Accuracy	93.85	93.85
	F-measure	93.80	93.80
	ROC	97.70	98.20
	Kappa statistics	87.70	87.70
Naïve Bayes	Accuracy	81.54	79.23
	F-measure	81.40	79.00
	ROC	92.80	90.70
	Kappa statistics	63.36	58.87
DNN	Accuracy	79.23	80.77
	F-measure	79.20	80.80
	ROC	83.50	84.70
	Kappa statistics	58.52	61.54
SVM	Accuracy	73.08	73.08
	F-measure	73.10	73.00
	ROC	73.30	73.30
	Kappa statistics	46.31	46.38
MLP	Accuracy	76.92	72.31
	F-measure	76.90	72.30
	ROC	83.20	80.70
	Kappa statistics	53.68	44.50
SGD	Accuracy	76.92	75.38
	F-measure	76.90	75.40
	ROC	76.80	75.30
	Kappa statistics	53.68	50.66
Fuzzy	Accuracy	73.85	73.08
	F-measure	73.90	73.00
	ROC	73.90	72.80
	Kappa statistics	47.66	45.85

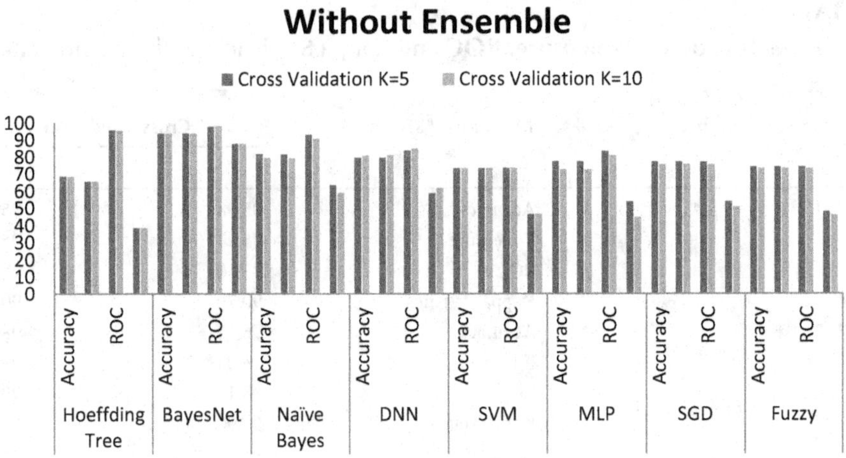

FIGURE 8.3 Performance without Ensemble

8.4.3.2 Results with Stacking Ensemble

As discussed earlier, we made six stacking ensembles (E-1 to E-6). These six stacked ensembles, having the same base classifiers and different meta classifiers, are called E-1 (Stacking {*Base*: Hoeffding tree and BayesNet, *Meta*: Naïve Bayes}), E-2 (Stacking {*Base*: Hoeffding tree and BayesNet, *Meta*: DNN}), E-3 (Stacking {*Base*: Hoeffding tree and BayesNet, *Meta*: SVM}), E-4 (Stacking {*Base*: Hoeffding tree and BayesNet, *Meta*: MLP}), E-5 (Stacking {*Base*: Hoeffding tree and BayesNet, *Meta*: SGD}) and E-6 (Stacking {*Base*: Hoeffding tree and BayesNet, *Meta*: Fuzzy}). The prediction performance of each ensemble is evaluated using the same measures used for single classifiers and validated on five-fold and ten-fold. The experimental results obtained are presented in Table 8.4 and Figure 8.4. It can be seen from Table 8.4 that stacking ensemble E-6, which uses fuzzy as its meta classifier, had a better performance than other stacking ensembles. The highest accuracy achieved by E-6 is 99.23% when cross-validation ten-fold (K=10) is used. Whereas f-measure, ROC and kappa statistics are 99.51%, 99.51% and 97.75%, respectively, when K=10. Two other models that give satisfactory results are E-6 at K=5, with an accuracy of 97.6%, and E-2 when K=10, with an accuracy of 98.46%. But to implement the model in real-time detection of COVID-19, it is better to go with model having apex performance. So, stacking ensemble E-6 (Stacking {*Base*: Hoeffding tree and BayesNet, *Meta*: Fuzzy}) when K=10, which we named "*En*-Fuzzy-ClaF," can be used as a tool to detect COVID-19.

8.5 COMPARATIVE ANALYSIS

A comparative analysis of the proposed *En*-Fuzzy-ClaF model with the already published scholarly articles within the same domain is shown in Table 8.5.

TABLE 8.4
Results (Accuracy, F-measure, ROC and Kappa Statistics) with Stacking Ensemble.

Ensemble Number	Measures (%)	Cross-Validation	
		K=5	K=10
E-1	Accuracy	95.38	95.38
	F-measure	95.40	95.40
	ROC	97.50	97.70
	Kappa statistics	90.76	90.78
E-2	Accuracy	96.92	98.46
	F-measure	98.04	99.10
	ROC	98.10	99.10
	Kappa statistics	90.90	95.56
E-3	Accuracy	95.38	95.38
	F-measure	95.40	95.40
	ROC	95.40	95.50
	Kappa statistics	90.76	90.78
E-4	Accuracy	95.38	94.61
	F-measure	95.40	94.60
	ROC	97.20	97.60
	Kappa statistics	90.76	89.23
E-5	Accuracy	95.38	94.61
	F-measure	95.40	94.60
	ROC	95.40	94.70
	Kappa statistics	90.76	89.23
E-6	Accuracy	97.69	99.23
	F-measure	98.51	99.51
	ROC	98.50	99.51
	Kappa statistics	93.42	97.75

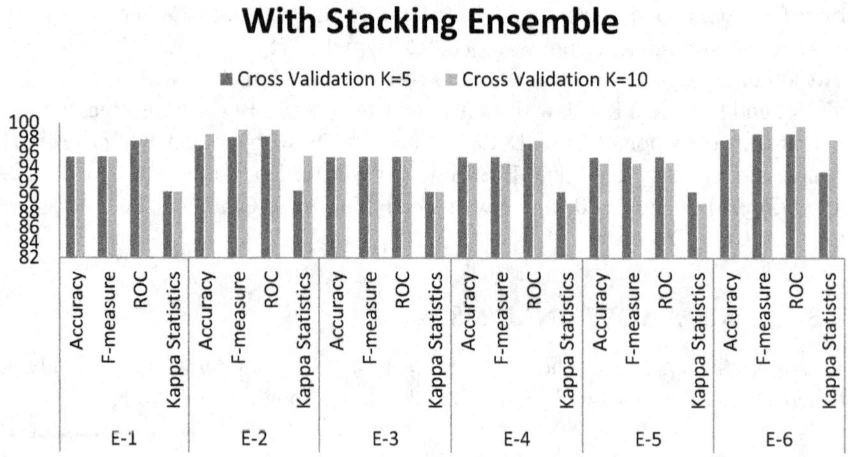

FIGURE 8.4 Performance with Stacking Ensemble.

TABLE 8.5
Comparative Analysis of the Proposed *En*-Fuzzy-ClaF Model with Previous Benchmark Studies.

Related Study	Method	Accuracy
Gupta, A. (Gupta et al., 2021)	InstaCovNet-19	99.08
Nour, M. (Nour et al., 2020)	Deep machine learning model	98.97
Afshar, P. et al. (Afshar et al., 2020)	Capsule network	95.70
Khan, A. (Khan et al., 2020)	Transfer learning	95.00
Narin, A. (Narin et al., 2020)	Transfer learning	98.00
Asif, S. (Asif et al., 2020)	Inception V3 transfer learning	98.00
Rahimzadeh, M. (Rahimzadeh & Attar, 2020)	Xception + ResNet50V2	91.40
Apostolopoulos, I. (Apostolopoulos & Mpesiana, 2020)	Transfer learning	93.48
Farooq, M. (Farooq & Hafeez, 2020)	ResNet-50 transfer learning	96.23
Shastri, S. (Shastri, Singh, Deswal et al., 2021)	CoBiD-Net Ensemble	99.13
Proposed En-Fuzzy-ClaF Model	***Stacking Ensemble***	**99.23**

8.6 CONCLUSION

The catastrophic phase of COVID-19 has led to many changes off which humans can't thrive. The technological achievements of mankind can't defend the havoc that has been created by COVID-19. Millions of people died and many more were infected, which led the scientific community to work tirelessly to combat this virus using technological tools and techniques. One such practice is carried out in the proposed research work of this study. In this research article, multiple ML-DL-based models with and without ensemble techniques to diagnose COVID-19 using classification were designed, and to frame the ensemble network, the stacking ensemble technique is used. Out of the total six ensemble models developed, the fuzzy classification ensemble model outperformed the single classifiers and other ensemble models in terms of accuracy, f-measure, ROC curve and kappa statistics. Our proposed fuzzy classification ensemble model achieved 99.23% accuracy with f-measure, 99.51% accuracy with ROC and 97.75% accuracy with kappa statistics. To sum up, this is the first fuzzy classifier–based COVID-19 nested ensemble study for the classification of COVID-19 confirmed and normal cases. In the future, we can deal with the economic loss and retrieval-based economic case studies of various countries. Also, we will analyze vaccination development, trials, distribution, effectiveness and their impact on the human body.

ACKNOWLEDGEMENTS

Funding information: This research did not receive any specific grant from funding agencies in the public, commercial, or not-for-profit sectors.

Conflicts of interest: The authors declare that they have no conflicts of interest.

Ethical approval: This article does not contain any studies with human participants or animals performed by any of the authors.

Informed consent: Informed consent was not required, as no humans or animals were involved.

REFERENCES

Afshar, P., Heidarian, S., Naderkhani, F., Oikonomou, A., Plataniotis, K. N., & Mohammadi, A. (2020). COVID-CAPS: A capsule network-based framework for identification of COVID-19 cases from X-ray images. In *Pattern Recognition Letters* (Vol. 138, pp. 638–643). https://doi.org/10.1016/j.patrec.2020.09.010

Ahuja, S., Panigrahi, B. K., Dey, N., Rajinikanth, V., & Gandhi, T. K. (2021). Deep transfer learning-based automated detection of COVID-19 from lung CT scan slices. *Applied Intelligence*, *51*(1), 571–585. https://doi.org/10.1007/s10489-020-01826-w

Apostolopoulos, I. D., & Mpesiana, T. A. (2020). Covid-19: Automatic detection from X-ray images utilizing transfer learning with convolutional neural networks. *Physical and Engineering Sciences in Medicine*, *43*(2), 635–640. https://doi.org/10.1007/s13246-020-00865-4

Aristovnik, A., Keržič, D., Ravšelj, D., Tomaževič, N., & Umek, L. (2020). Impacts of the COVID-19 pandemic on life of higher education students: A global perspective. *Sustainability (Switzerland)*, *12*(20), 1–34. https://doi.org/10.3390/su12208438

Asif, S., Wenhui, Y., Jin, H., Tao, Y., & Jinhai, S. (2020). Classification of COVID-19 from Chest X-ray images using deep convolutional neural networks. In *medRxiv*. https://doi.org/10.1101/2020.05.01.20088211

Aslan, M. F., Unlersen, M. F., Sabanci, K., & Durdu, A. (2021). CNN-based transfer learning—BiLSTM network: A novel approach for COVID-19 infection detection. *Applied Soft Computing*, *98*, 106912. https://doi.org/10.1016/j.asoc.2020.106912

Chaturvedi, K., Vishwakarma, D. K., & Singh, N. (2021). COVID-19 and its impact on education, social life and mental health of students : A survey. *Children and Youth Services Review*, *121*(July 2020), 105866. https://doi.org/10.1016/j.childyouth.2020.105866

Covid-19 Database | Sirm. (2020). www.sirm.org/en/category/articles/covid-19-database/

Data Mining with Open Source Machine Learning Software in Java. (2011). *The university of waikato*. www.cs.waikato.ac.nz/ml/weka/

Farooq, M., & Hafeez, A. (2020). COVID-ResNet: A deep learning framework for screening of COVID19 from radiographs. In *arXiv*. https://github.com/lindawangg/COVID-Net.

Gautam, Y. (2021). Transfer learning for COVID-19 cases and deaths forecast using LSTM network. *ISA Transactions*, *1*. https://doi.org/10.1016/j.isatra.2020.12.057

Gianchandani, N., Jaiswal, A., Singh, D., Kumar, V., & Kaur, M. (2020). Rapid COVID-19 diagnosis using ensemble deep transfer learning models from chest radiographic images. *Journal of Ambient Intelligence and Humanized Computing*, *0123456789*. https://doi.org/10.1007/s12652-020-02669-6

Gifani, P., Shalbaf, A., & Vafaeezadeh, M. (2021). Automated detection of COVID-19 using ensemble of transfer learning with deep convolutional neural network based on CT scans. *International Journal of Computer Assisted Radiology and Surgery*, *16*(1), 115–123. https://doi.org/10.1007/s11548-020-02286-w

Gupta, A., Anjum, Gupta, S., & Katarya, R. (2021). InstaCovNet-19: A deep learning classification model for the detection of COVID-19 patients using Chest X-ray. *Applied Soft Computing*, *99*, 106859. https://doi.org/10.1016/j.asoc.2020.106859

Hamed, A. (n.d.). *COVID-19 Dataset* (V1 ed.). Harvard Dataverse. https://doi.org/doi:10.7910/DVN/LQDFSE

Hamed, A., Sobhy, A., & Nassar, H. (2020). Accurate classification of COVID-19 based on incomplete heterogeneous data using a KNN variant algorithm. *Research Square*. https://doi.org/10.21203/rs.3.rs-27186/v1

Influenza Research Database. (n.d.). *Influenza genome database with visualization and analysis tools*. Retrieved February 21, 2021, from www.fludb.org/brc/home.spg?decorator=influenza

Islam, Z., Islam, M., & Asraf, A. (2020). A combined deep CNN-LSTM network for the detection of novel coronavirus (COVID-19) using X-ray images. *Informatics in Medicine Unlocked*, *20*, 100412. https://doi.org/10.1016/j.imu.2020.100412

Jain, G., Mittal, D., Thakur, D., & Mittal, M. K. (2020). A deep learning approach to detect Covid-19 coronavirus with X-Ray images. *Biocybernetics and Biomedical Engineering*, *40*(4), 1391–1405. https://doi.org/10.1016/j.bbe.2020.08.008

Khan, A. I., Shah, J. L., & Bhat, M. M. (2020). CoroNet: A deep neural network for detection and diagnosis of COVID-19 from chest x-ray images. *Computer Methods and Programs in Biomedicine*, *196*, 105581. https://doi.org/10.1016/j.cmpb.2020.105581

Kumar, J., Goomer, R., & Singh, A. K. (2018). Long short term memory recurrent neural network (LSTM-RNN) based workload forecasting model for cloud datacenters. *Procedia Computer Science*, *125*, 676–682. https://doi.org/10.1016/j.procs.2017.12.087

Loey, M., Manogaran, G., Taha, M. H. N., & Khalifa, N. E. M. (2021). A hybrid deep transfer learning model with machine learning methods for face mask detection in the era of the COVID-19 pandemic. *Measurement: Journal of the International Measurement Confederation*, *167*, 108288. https://doi.org/10.1016/j.measurement.2020.108288

Narayan Das, N., Kumar, N., Kaur, M., Kumar, V., & Singh, D. (2020). Automated deep transfer learning-based approach for detection of COVID-19 Infection in Chest X-rays. *Irbm*. https://doi.org/10.1016/j.irbm.2020.07.001

Narin, A., Kaya, C., & Pamuk, Z. (2020). Automatic detection of coronavirus disease (COVID-19) using X-ray images and deep convolutional neural networks. In *arXiv*. arXiv. http://arxiv.org/abs/2003.10849

Nour, M., Cömert, Z., & Polat, K. (2020). A novel medical diagnosis model for COVID-19 infection detection based on deep features and bayesian optimization. *Applied Soft Computing*, *97*, 106580. https://doi.org/10.1016/j.asoc.2020.106580

Pathak, Y., Shukla, P. K., Tiwari, A., Stalin, S., & Singh, S. (2020). Deep transfer learning based classification model for COVID-19 disease. *Irbm*. https://doi.org/10.1016/j.irbm.2020.05.003

Rahimzadeh, M., & Attar, A. (2020). A modified deep convolutional neural network for detecting COVID-19 and pneumonia from chest X-ray images based on the concatenation of Xception and ResNet50V2. *Informatics in Medicine Unlocked*, *19*, 100360. https://doi.org/10.1016/j.imu.2020.100360

Ribeiro, M. H. D. M., da Silva, R. G., Mariani, V. C., & Coelho, L. dos S. (2020). Short-term forecasting COVID-19 cumulative confirmed cases: Perspectives for Brazil. *Chaos, Solitons and Fractals*, *135*. https://doi.org/10.1016/j.chaos.2020.109853

Shastri, S., Kour, P., Kumar, S., Singh, K., & Mansotra, V. (2021). GBoost: A novel grading-AdaBoost ensemble approach for automatic identification of erythemato-squamous disease. *International Journal of Information Technology (Singapore)*. https://doi.org/10.1007/s41870-020-00589-4

Shastri, S., Kour, P., Kumar, S., Singh, K., Sharma, A., & Mansotra, V. (2020). A nested stacking ensemble model for predicting districts with high and low maternal mortality ratio (MMR) in India. *International Journal of Information Technology (Singapore)*. https://doi.org/10.1007/s41870-020-00560-3

Shastri, S., Singh, K., Deswal, M., Kumar, S., & Mansotra, V. (2021). CoBiD-net: A tailored deep learning ensemble model for time series forecasting of covid-19. *Spatial Information Research*, 1–14. https://doi.org/10.1007/s41324-021-00408-3

Shastri, S., Singh, K., Kumar, S., Kour, P., & Mansotra, V. (2020). Time series forecasting of Covid-19 using deep learning models: India-USA comparative case study. *Chaos, Solitons & Fractals*, *140*, 110227. https://doi.org/10.1016/j.chaos.2020.110227

Shastri, S., Singh, K., Kumar, S., Kour, P., & Mansotra, V. (2021). Deep-LSTM ensemble framework to forecast Covid-19: An insight to the global pandemic. *International Journal of Information Technology (Singapore)*. https://doi.org/10.1007/s41870-020-00571-0

Shi, X., Chen, Z., & Wang, H. (2015). Convolutional LSTM network : A machine learning approach for precipitation nowcasting. *NIPS'15: Proceedings of the 28th International Conference on Neural Information Processing Systems*, *1*, 802–810.

Tomar, A., & Gupta, N. (2020). Prediction for the spread of COVID-19 in India and effectiveness of preventive measures. *Science of the Total Environment*, *728*, 138762. https://doi.org/10.1016/j.scitotenv.2020.138762

Wang, S. H., Nayak, D. R., Guttery, D. S., Zhang, X., & Zhang, Y. D. (2021). COVID-19 classification by CCSHNet with deep fusion using transfer learning and discriminant correlation analysis. *Information Fusion*, *68*(October 2020), 131–148. https://doi.org/10.1016/j.inffus.2020.11.005

WHO Coronavirus Disease (COVID-19) Dashboard. (n.d.). Retrieved February 21, 2021, from https://covid19.who.int/

Zhang, R., Guo, Z., Sun, Y., Lu, Q., Xu, Z., Yao, Z., Duan, M., & Liu, S. (2020). COVID19XrayNet : A Two-Step transfer learning model for the COVID-19 detecting problem based on a limited number of chest X-ray images. *Interdisciplinary Sciences: Computational Life Sciences*, *512*. https://doi.org/10.1007/s12539-020-00393-5

Zhang, Y.-D., Satapathy, S. C., Zhang, X., & Wang, S.-H. (2021). COVID-19 diagnosis via DenseNet and optimization of transfer learning setting. *Cognitive Computation*, 0123456789. https://doi.org/10.1007/s12559-020-09776-8

9 Efficient Approach for Lung Cancer Detection Using Artificial Intelligence

Er. Vinod Kumar and Dr. Brijesh Bakariya

CONTENTS

DOI: 10.1201/9781003184140-9

9.1 INTRODUCTION

Rapid technical advances make it possible to identify and diagnose cancer by assessing artificial intelligence (AI). Eventually, a rational choice is strategic thinking. Several fields, like cardiology, pulmonology, intensive care, and mental wellbeing, have also been encouraging to integrate AI. For a variety of reasons, cancer is now an area where the quick impact of AI is predicted to be beneficial, and the efficacy of the AI-based primary care approaches is improved in several cases (Asan et al. 2020). First of all, cancer is one common option that places a critical amount of stress, harm, emotional trauma, and continuing economic burdens on the patients and community. For all cancer types, therefore, it is important, with no exception, to increase the survival rate of cancer.

The key to improving the rate of survival by five years is early diagnosis and rehabilitation. Indian cancers would rise 12 percent in the coming five years; by 2025, 1.5 million people may remain inactive and 1.39 million by 2020, based on Indian Medical Research Council's (ICMR) modern medical trends (Patel et al. 2020). Besides, by 2020, it is predicted that new advanced lung cancer cases will reach 67,000 per year. Imaging techniques have shown to be an exceptionally efficient instrument for early detection of diseases that can minimize mortality by up to 20 percent, though the technique has proven effective. CT imaging has indeed been demonstrated as an extremely promising early detection method that can reduce the number of deaths by 20 percent; thus, innovation has proven itself to be reliable, a thorough study has been performed on other forthcoming technologies, such as AI.

This technology might further improve nodule detection, identification, and scale, thereby decreasing falsified positive effects (Asan et al. 2020).

Cancer is a field in which results of AI within a brief period are likely to be substantial and disease is a responsive option to boost the credibility of AI global health strategies in many aspects. Secondly, due to various unique illnesses, societal dissatisfaction, and financial consequences, cancer is frequently reproduced and puts substantial burdens on victims and society overall. In all types of cancer, it is therefore important to increase the cancer-related results. Furthermore, the need for successful disease diagnosis, prognostics, and strategies of therapeutics is high at the conscious level; cancer is a major cause of morbidity and mortality, and this is alarming when considering cancer as a lethal condition since no other aspects exist. Then there are huge decision problems with tumorigenesis, particularly from the therapist's viewpoint because of actual data strain. The conventional cancer specialist has, frighteningly, been studying for more than 20 hours per working day to keeping up with medical advances. In recent decades, the techniques of AI have been successful, as many analytical, purposive approaches or optimization and parameters are used to better previous instances and to classify noisy or multidimensional data sources in endogenous patterns, which are useful clinically. It summarizes current and emerging trends in IA approaches at all stages of the experience of cancer survivors (Kumar and Bakariya 2021).

It is possible to describe the object as two-dimensional f (x, y), whereas x, y is just the co-ordinate, and (f) is the magnitude of any co-ordination pair (x, y). The magnitude of such a location is also considered an amplification of the gray area or image (f). This pixel value consists of a group of x and y things, both equally important. The elements of these images are often called pixels. The visualizations later are to be described by the three-dimensional method (x, y, z), and the voxels are often called independent elements. The performance of the imaging is an important radiological factor, and the regional term range ensures that its image can identify two adjacent objects (Wang et al. 2019). A further essential resolution attribute is the transition level that shows image quality and performance. DICOM is a standard policy for managing and revealing medical image data and is used in radiology, cardiology, oncology, obstetrics, and dentistry. DICOM archives can move systems and images into applications in DICOM format to obtain medical information and images. The image quality of a pixel contains position, patient identity, image, and image attributes. Also, the DICOM files (Murillo 2018) enable distinguishing between the patient and the image data challenging. It is associated with the patient as well. There is no concept where fundamental image analysis ends and where the further study on image and machine learning starts. Computer vision is often described as a measure consisting of one image both the operation's input and output.

9.2 BACKGROUND AND RELATED WORK

Sets of data are an essential part of any artificially intelligent framework. The validity of the information is conducive to analytic development, preparedness, and improvement so that the data gathered will be useful, and machine learning expertise can explain and mark it. This part provides AI data on early work to detect lung cancer (Johora et al. 2018; Kim and Jeter 2021). Lung cancer has become one of the toughest diseases in the world to fight. There are various ways in which researchers have tried to raise lung cancer awareness. CT scans some errors by neural networks and lung granulation tissue recognition. A deeper neural genotype structure (Bhatia et al. 2019) details the images extracted. The author's outcome indicates that as the nodule broadens, the greater the likelihood of tumor progression. The testing set contains features inside a linear support vector machine (SVM) that increases the performance of its tests compared to the classification task, also referred to as the feature classification. CT images are produced by compression ROI (Kumar et al. 2015) processing. Each image of ROI is split into a better approach for discrete view switch, as well as certain classification ranges for SVM-observable GLCM (Gupta et al. 2020), as versatility and precision boost a ranking. One approach also uses a much smaller nodule size (Kumar et al. 2015), which demonstrates excellent performance for the identification of types of cancer than all this ROI isolation. The physical dimensional level is known as its feature extraction (Taher et al. 2016), along with the Otsu partition thresholds and the GLCM. From this viewpoint, the influence of these characteristics detects cancer nodules, so it detects the first stages of cancer prevention. The median filter is a sample preparation system for noise removal processing by salt and pepper (Pradhan and Chawla 2020), as well as the value for the entire pixel circle, is calculated in numeric form (Kumar and Bakariya 2021; Altaf et al. 2019). A few of the interaction processes take place at the Otsu level. A review of these definitions shows that their GLCM or SVM helps to improve lung cancer classifications (Mayan et al. 2014). The lucrative contract is only distinct from many other meanings when using these images within a CAO system until the system is based on patterns such as NC factor, circularity, etc., and sets the cell's threshold for estimations that coincide with rule-based labeling. AI, machine learning (ML), and deep training are displayed as a diagnosis of lung cancer in Figure 9.1,

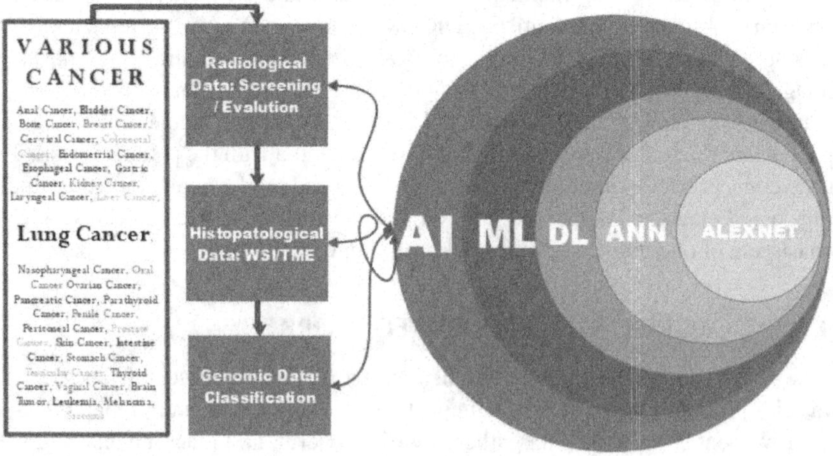

FIGURE 9.1 Cancer Diagnosis Using Artificial Intelligence, Machine, and Deep Learning.

including radiological, histopathological, and genomic data for the classification of cancers.

While the extracting approach is even more egalitarian than other strategies, there isn't only the required performance in the definition classification, like SVM and GLCM. In the earliest stages of carcinogenesis growth, many tactics for stalking approaches have been used to try to rectify the image by using Gabor filters (Kumar and Garg 2012), comprising the harmonic form and logarithmic factor to isolate the features by watersheds such as the field, perimeter, and eccentricities. That approach gives a rather automatic feature segmentation duly focused on CT slides that produces an original argument beyond human interaction (Layman and Greenspan 2020). The min-max normalization process was already viewed as a strategy of phase radionics. Enhancements may be analyzed for predictive accuracy using decision trees that include several standards for the identification of a level as malignant or benign (Deep Prakash et al. 2017) that is focused mostly on SCM and that will help distinguish the classification technique as malignant or benign. The strategy of GLCM is envisioned by numerous nodules. To aid in the classification of malignancies, the SCM filtering is set up with a Laplace and Gaussian distribution of nodules in metastases (Altaf et al. 2019).

This strategy improves the classification of all aspects that accurately combine classification with outcomes for the individuals, including decision-making, ML, and SVM. Some research is centered on the concept of automotive technology (Pelc and Wang 2020). The assessment of lung cancer is based only on SVM and even certain ML strategies (Deep Prakash et al. 2017) if the implementation of classification algorithms in contrast to individuals is extremely accurate. The author also looks at major predictions of lung cancer and identifies some shortcomings and abilities. The deep-learning (DL) methodology is good for non-invasive lung cancer (Wu and Qian 2019). To examine the fermented aspects of quantitative images, use DL, biological imaging concepts, including patient stratification. It will just need a 2D-UNet web service to have comprehensive tools for cancer detection (Altaf et al. 2019). 2D cancer masks are derived from specialized datasets with unique technologies and improved specifications and have built up a good neural net to recognize respiratory cancer in images of histopathology (Wu and Qian 2019). Findings indicate that ML could be successful mostly in the difficult process of image classification but can support lung cancer oncologists. In this context, the researcher utilizes a specific seed positioning and diagnosis validator (Li et al. 2018) to connect the neural network with a recurrent neural network (RNN), including medications and treatment for NSCLC (non-small cell lung cancer). As X-ray enhancements are made to the obtained results through their concerned images (Liu et al. 2018), pro isotopic characteristics derived from medical imaging techniques have been assessed. The technique for diagnosing lung cancer has also increased performance by using the correlation coefficient with different features. It improves cancerous tumor cell responses through a large amount of data via the pre-and post-radiation therapies and Bayesian networks (Wu et al. 2018). By the Markov method and wrapper-based exposure, local control interpretation is enhanced. To forecast respiratory cancer, the article proposes a cancer framework via an earlier estimate for radiology (Bhatia et al. 2019). In fairly limited forms, this experimental radioactive method is being used to overcome certain characteristics of the analytical procedures along with confidence and theoretical research. Besides, the closest classification functions for forecasting therapy success are proposed. Another goal is to assess lung cancer with such a range of high dimensionality specializations to select these features for more treatment, based on the selected sub-predictive characteristics (Echegaray et al. 2016; Kim et al. 2016). The neural networks indicated that a person with a localized tumor could predict its survival rate after radioactive treatments (Luo et al. 2018). Also, increasing predictive performance requires temporal correlation only with sequential pattern conditions. To respond to this 1D coverage, the overall and full 2D distributions became linked centrally and fully. Although the large-scale model improves performance relative to the fully optimized multilayered architecture (Wang et al. 2018), it can be enhanced by use of an image analytical technique for the diagnosis and prediction of lung cancer, which demonstrates its usefulness of SVM.

Work in progress—The whole article concerned with increasing efficiency, reducing diagnostic delays, recognizing tumors, and classifying features and functions. So, many trials have been performed to classify the most dangerous and uncommon kinds of lung cancer. Such lung nodules will then be analyzed, which will improve their sensitivity, accuracy, and specificity. The main aim of the research is to assess and develop diagnostic techniques for lung cancer.

Motivation—The major priorities are performance improvement, diagnostic period minimization, and the location of nodules and features. Several tests have even been carried out to determine that lung cancer is now the most dangerous and damaging form of cancer. As a result, each lung node is analyzed along with its specificity, precision, and accuracy. Lung cancer diagnoses will have to be improved and enhanced.

9.3 PROPOSED MODEL

Figure 9.2 shows five usual phases that describe the current layout. The first step seems to be the smooth analysis and processing of image data with different filters. The next step is to specify the image that extracts the correct portion. A nodule is made, with multiple features to be removed later. That function is selected after so many features are retrieved. A broad variety of specific classifications are also used for machines and deep training. The final phase then classifies half of the works as high quality, works where most of the pixels cost more than 250 and the other bits are excluded (Manikandan and Bharathi 2016). The third approach relies on the AlexNet proposal. The last process relies on the suggested reduction and labeling of features by AlexNet and SVM. A clear majority decision throughout the following phase demonstrates the type of the labels and indicates the label classification.

9.3.1 DATASET (LIDC-IDRI)

Databases are an essential part of any AI approach. The consistency of data is beneficial to analytical development, readiness, and progression. The data gathered should be validated and masked by ML health care professionals for this advancement to be useful. This section incorporates AI knowledge from early experiments on lung

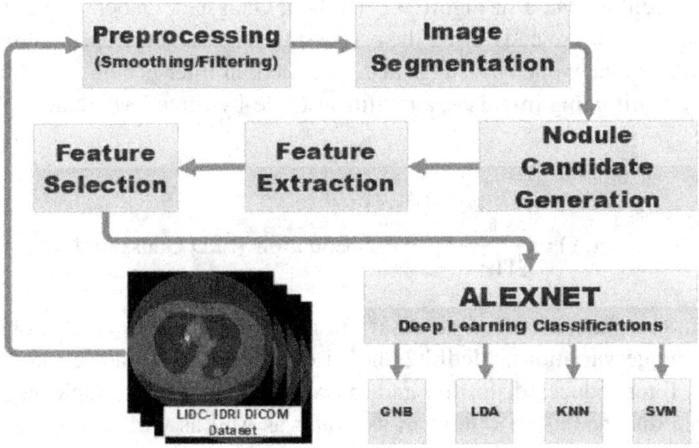

FIGURE 9.2 Diagram of the Proposed Architecture of the System.

cancer identification (Tabish et al. 2017; Manikandan and Bharathi 2016). There are 1,018 cases in the sample across 7 research organizations and 8 diagnostic imaging companies. For any event, an XML document is provided with CT-scan annotations (Chen et al. 2020). In three such processes, four trained thoracic radiologists conduct such annotations. The findings are split into two categories ("nodule supply 3 mm," "nodule < 3 mm," and "nodule supply 3 mm") (Razzak et al. 2018) for each radiography specialist. In the second stage, every radiologist then anonymously analyzes their classification and some other radiologists' classifications. Each nodule is examined by four radiologists independently. Averages above 3 are marked as malignant and below 3 as benign lesions. One of four radiologists confirmed a total of three radiologists, often with anomalies and identity, and removed several nodules from its study. The DCM files, seen by three or five radiologists in Figure 9.3(a), with a benign or malignant stage of the disease, are easier to understand by scanning with the pylidc tool (Mathur et al. 2020). Figure 9.3(b) shows a nodule in the LIDC-IDRI-0084 patient: slice size: 1,250 mm; pixel geometry: 0,703; nodule-like structure: 1 of 3 annotations almost slice 173, and details about annotations (Composite-4, Internal Structure-1, Sphericity-3, Margin-4, Lobulation-2, Speculation-5), Texture-5, Malignancy-5 (AI) (Kim and Jeter 2021).

9.3.2 Image Enhancement

The point of implementation is divided into three parts, as shown in Figure 9.2. The very first goal has been to get image data from established LIDC-IDRI CT images. The second phase distinguishes the pulmonary nodes from the expertise (Kavitha et al. 2019). The different ML and DL classifiers can also be evaluated, and efficiency gains can be performed. Image smoothing removes distortions in the entire image or even other minor irregularities, creating flickering of the outlines and recursively minimizing noise. It selects the desired images and examines them with certain filters. As demonstrated in Figure 9.2, median, Gaussian, Gabor, and contemporary techniques are introduced during the processing stage, and the optimal response was modified and a new one was suggested. The median filter is being used to boost the image by minimizing impulses, a nonlinear three-by-three level feature (Dartmouth 2019).

$$g(x,y) = \frac{1}{2\Pi\sigma^2} e^{\frac{-(x^2+y^2)}{2\sigma^2}} \quad \text{Equation 1: 2D Gaussian filter}$$

The Gaussian filter is indeed necessary to remove higher-frequency components to prevent image variations. Medfilt2 should only be used for noise removal. Such a low-pass filter reduces distortion and improves the surface's suppleness and precision. That distance of an axle from its center as well as y of both the axis from its center of the axis is a standard deviation of a Gaussian distribution (Kumar and Garg 2012). Processed features are extracted with a watershed segmentation during preprocessing. The images show the identified cancer lesions.

The Gabor functionality is a linear filter for image assessment and is still effective. Gaussian and harmonic features would be used, so it enhances tightness near objects and nodules. For equation expression of the 2D Gabor filter (Riti et al. 2016), β is its width; μ is its norm of parallel rotation thin strips; μ is the offset process; β is the normal differences; and μ is its proportion of spatial aspect. More features, such as a surface area, eccentricity, and perimeter, among other features, like the centroid, diameter, or key intensity pixels, are formed throughout medical diagnostics. Mostly during the development phase of medical imaging, various specifications are also established, particularly surface areas, perimeters, and eccentricities, as well as many other features, such as centroid, diameter, and pixels for principal strength. Since the malignant node has in fact been found, the features of the node can be accurately identified and calculated.

However, it is not clear whether it is benign or malignant.

$$g(x,y) = \exp\left(\frac{x'^2 + \gamma^2 y'^2}{2\sigma^2}\right)\cos\left(2\pi\frac{x'}{\lambda} + \phi\right)$$

$$x' = x\cos\theta + y\sin\theta \qquad \text{Equation: 2D Gabor filter}$$

$$y' = -x\cos\theta + y\sin\theta$$

Its next level would be to define specialized nodes, for instance, AlexNet, GoogLeNet, as well as other classifiers, like ML, linear discriminant analysis (LDA), SVM, Gaussian Naive Bayes (GNB): Naive Bayes Gaussian is an implementation of Naive Bay variations managed and supported by Gaussian standard distribution. A naive Bayes algorithm consists of a collection of supervised algorithms that operate based on the Bayes rule. It has a high performance even though it is a basic grading approach, or k-nearest neighbors (KNN). For creating a training set, the training features are being used. Its next phase is a qualified forecast model that classifies the unidentified nodes of cancer.

9.4 IMAGE SEGMENTATION

A dividing threshold segmentation (i.e., global or regional) is the best strategy for the isolation of the grayscale details through target gray data. The Otsu algorithm has been the most frequently used. A first seed value is applied that combines just the same pixels from the outside badge. The value would be that while noise or disruption is often caused, it typically distinguishes between related factors and consistently provides better results. The advantage there is that the features typically are isolated from the related fields to provide better segmentation performance data, and the downside is the more expensive processing facilities that lead to extreme noise or frequent finding. Pre-filtered images can be converted into various formats, partly eliminating the ambiguity with the front pixels also on the edges. The lung-separation procedure comes with various steps, while lung segmentation strategies are implemented in many steps, for example, preprocessing, median filter deployment, and Gaussian filter seeding, then new procedures are initiated to distinguish the lung in its area, as shown in Figures 9.4 (c), and 9.4 (d).

9.4.1 Watershed Segmentation

This analysis evaluates the subject or background in a certain place in the image. The color image was then turned into a gray image. The differential segmentation mechanism was established. The object was somewhat lower in gradient, whereas the highest gradient being was found to have been the object's boundary. After that, the first item was labeled. Blobs of the pixels of the object should all be connected to the foreground purpose. Within the creation and removal of dark spots and trimmed marks around each object, morphological techniques are used (Kuan et al. 2017). It showed a black background pixel and a framework to decide the impact of the foreground. Which seems to be the reason. The explanation seems to be that the background markers are too close to the object's segmentation. Figures 9.4 (a), 9.4 (b), 9.4 (c), and 9.4 (d) show a distinction in approaching sections that using a watershed segmentation in just such an image. By observing all those as a layer as lighter pixels rise and darker pixels fall, the transition also in watershed distinguishes catching basins and catching ridge edges in such an image. This could, therefore, identify or mark primary objects, even contextual points, that best segment using the watershed transition (Herrmann et al. 2018). The marker-controlled segmentation of the tubes is accompanying a simple strategy: Compute the segment's function. A single approach is performed by watershed segmentation managed by markers: the segmentation function estimate. It is indeed an image wherein darkened parts can be segmented: (a) for the foreground, measure labels. Throughout each stage, there appear to be pixel blocks, (b) Assessment of background indications. Pixels in a single object could never be identified, (c) Modify the segmentation feature enough so that it will provide the front and the rear markers with minimal information, (d) Quantify enhanced watershed transformation segmentation feature.

9.4.2 Otsu's Thresholding

Thresholding is the means by which pixels are removed along the front of the background. The Otsu process implemented by Nobuyuki Otsu is one of several ways to reach the maximum limit. The Otsu method is a technique for determining the threshold levels at which there is the lowest difference in weight between some base and the forward pixels. The basic concept ought to be to learn all possible values of the threshold and measure the allocation of the first and the subsequent pixels. Also, it seeks minimal distribution. The formula for assessing the lower-class variance until a certain threshold t is reached is described by: $\sigma^2(t) = \omega_{bg}(t)\sigma_{bg}^2(t) + \omega_{fg}(t)\sigma_{fg}^2(t)$.

This algorithm (Rodrigues et al. 2018) attempts to minimize the internal variance of the two groups of variances by a weighted maximum (background and foreground). The standard gray color is between 0 and 255 (0–1 for floats). Therefore, if another threshold is 100, the backdrop should be all pixels under 100, and the foreground with the same image would be all pixels above it or equal to 100 pixels. Where $\omega_{bg}(t)$ and $\omega_{fg}(t)$ describe the probability of pixel number by threshold t and σ^2 is the variance of the pixel value. P_{all} = the total count of pixels in an image, $P_{bg}(t)$ = the count of background pixels at threshold t, $P_{fg}(t)$ = the count of foreground pixels at threshold t. Hence, the weights are then indicated as: $\omega_{bg}(t) = \frac{P_{bg}(t)}{P_{all}}$, $\omega_{fg}(t) = \frac{P_{fg}(t)}{P_{all}}$. The variance could be determined with the following formula: $\sigma^2(t) = \frac{\sum(x_i - \bar{x})^2}{N-1}$ where x_i is the value of a pixel at i in the class (bg or fg), $\bar{x}$ = the means of pixel values in the class (bg or fg), and N is the number of pixels.

9.5 FEATURE EXTRACTION

Feature selection plays an important role in the processing of images. This method determines the requisite details for processing. That lung could also be detected as conformity or abnormality. These separated structures were used for tumor diagnosis and classification. The separate prominent characteristics in that same article are also the regions major axis length, minor axis length, eccentricity, convex area, fill area, perimeter, solidity, extent, medium strength, actual axis area, current main axis length, and compactness. The goal would be to use the smallest possible steps to properly identify an entity to be unambiguously classified. The precision of the key image as well as the precision of key images and how they are presented (Murillo, BR 2018). That precision of all its main images, and the preprocessing of all its images, depends on the effectiveness of such measurements. Object erosion, like small holes, including noise, may contribute to the bad result obtained and the unavoidably misidentified results. Shapes are what remain after the location, orientation, and size features of an object have been removed (Nasser and Abu-Naser 2019). The following shape characteristics can be classified in boundaries and regions (Rabbani et al. 2018).

9.5.1 DISTANCES: All differences among two separate pixels were the shortest determined (x_1, y_1) and (x_2, y_2), Euclidean $d = \sqrt{(x_1 - x_2)^2 + (y_1 - y_2)^2}$, and Chessboard $d = \max(|x_1 - x_2|, |y_1 - y_2|)$.

9.5.2 AREA: A sequence of pixels in a shape consists of the area. That's the number among all pixels in the whole potion, including its lung tumor. A region is obtained by a scalar evaluation.

9.5.3 CONVEX AREA: The area surrounding a convex hull is the convex surface area. The convex hull shows the tumor section in the palm as the smallest convex polygon. The number of nodes in the convex region is given in this convex polygon. It's a scalar attribute.

9.5.4 PERIMETER: The pixel values on the image boundary appear in the perimeter. This was the total of the tumor area in the lung. For both the perimeters, this is a scale parameter. When the limiting number is x1, . . .,xN, the perimeter is defined as $Perimeter = \sum_{i=1}^{N-1} d_i = \sum_{i=1}^{N-1} |x_i - x_{i+1}|$ Equation: perimeter calculation, distance di is equal to one for four-connected limits as well as one or $\sqrt{2}$ eight-connected boundaries. There has been one pixel in the length of the number of diagonal ties between $N_4 - N_8$ and one $N_8 - (N_4 - N_8)$ connection within the eight-related border. The complete perimeter is thus: $Perimeter = (\sqrt{2} - 1)N_4 + (2 - \sqrt{2})N_8$

9.5.5 CONVEX PERIMETER: Only the convex hull surrounding the element is a convex perimeter about such an element. This means that now the convex limit of a sphere hull is the convex perimeter of its item.

9.5.6 AXIS LENGTH: The element's major axis is (x, y), the maximum line of terminals. Calculating the distance from each boundary pixel pair inside the object limit and determining the pair only with the maximum length of the main pixel end axis (x1, yl) and (x2, y2) (Tripathi et al. 2019).

9.5.7 MAJOR AXIS LENGTH: That largest axis of an image is the pixel difference between both the endpoints of the principal axis $MajorAxisLength = \sqrt{(x_2 - x_1)^2 + (y_2 - y_1)^2}$. The outcome is object length estimation.

9.5.8 MINOR AXIS LENGTH: The small axis is the (x, y) ends of the object's longest line, which are perpendicular to the main axis. The small axis access points (x1, yl) can be identified by calculating the pixel distance from the endpoints of the limit pixels and (x2, y2). $MajorAxisLength = \sqrt{(x_2 - x_1)^2 + (y_2 - y_1)^2}$. Here, the outcome is object width estimation.

9.5.9 COMPACTNESS: Compactness is a relationship between the area of an object and the boundary with the circumference of the same area (Kumar and Bakariya 2021). As this is the smallest object, a circle is selected. A circle has the highest value. The square's compactness is $\frac{\pi}{4}$ $Compactness = \frac{4\pi.Area}{(Perimeter)^2}$ and $Compactness = \frac{(Perimeter)^2}{4\pi.Area}$.

9.5.10 ECCENTRICITY: The calculation decreases structures that are not smooth or that have elliptical shapes. For just a circle, a minimum value again for calculation is just one. The artifacts have more challenging and irregular boundary $Eccentricity = \frac{AxisLength_{short}}{AxisLength_{long}}$. Equation: Eccentricity seems to be the ratio of the small (short) dimension duration to the large (major) axis of the object (Alamar and Cherezov 2018). Consequently, the measure of object eccentricity is provided at 0 to 1. Often, the ellipticity is understood.

9.5.11 STANDARD DEVIATION: It calculates the mean square deviation from the mean of the gray pixel scale value.

9.5.12 SKEWNESS: It is the asymmetry rating of pixel distribution in the designated ROI around the average.

9.5.13 KURTOSIS: It tests a distribution in contrast to a normal distribution's peak as well as flatness.

9.5.14 ENTROPY: It is a measurement of the maximum possible amount of the segmented ROI content.

9.6 METRICS OF PERFORMANCE EVALUATION

These techniques are necessary to know and explain how much they have learned after introducing ML algorithms. Such tests are known as measures of results. Multiple parameters have been implemented in the analysis, each of which finds different components of algorithm efficiency. With each learning task, an appropriate set of measurements is needed for the performance evaluation (Cui et al. 2018). In the article, the author uses several standard metrics to collect useful data, mostly on the efficiency of the algorithm as well as to evaluate related problems for classification. All these are precision, reminder, uncertainty matrix, precision, and a rating of ROC-AUCs. The exams are the wrong way.

9.6.1 PRECISION/POSITIVE PREDICTIVE VALUE (PPV): $PPV = \frac{TP}{TP + FP}$

9.6.2 NEGATIVE PREDICTIVE VALUE (NPV): $NPV = \frac{TN}{TN + FN}$

9.6.3 RECALL/SENSITIVITY/PROBABILITY OF DETECTION/HIT RATE/TRUE POSITIVE RATE (TPR): $TPR = \frac{TP}{TP + FN}$

9.6.4 SPECIFICITY/SELECTIVITY/TRUE NEGATIVE RATE (TNR): $TNR = \frac{TN}{TN + FP}$

9.6.5 FALL OUT/FALSE POSITIVE RATE (FPR): $FPR = \frac{FP}{FP + TN}$

9.6.6 MIS RATE/FALSE NEGATIVE RATE (FNR): $FNR = \frac{FN}{FN + TP}$

9.6.7 FALSE DISCOVERY RATE (FDR): $FDR = \frac{FP}{FP + TP}$

9.6.8 FALSE OMISSION RATE (FOR): $FOR = \frac{FN}{FN + TN}$

9.6.9 F1-SCORE/F-measure/ SØRENSEN—DICE COEFFICIENT/DICE SIMILARITY COEFFICIENT (DSC): $F1 - Score = 2 \times \frac{Precision \times Recall}{Precision + Recall}$

$$F1 - Score = \frac{2TP}{2TP + FN + FP}$$

9.6.10 ACCURACY: $Accuracy = \frac{TP + TN}{TP + TN + FP + FN}$

9.7 DEEP LEARNING ALGORITHMS

As cattle and humans learn naturally, ML algorithms help to speed up the process. These algorithms are used to learn, without using a default machine model, the raw data the computer processes and adapt their performance as the number of trials accessible to learning improves. These models can recognize natural phenomena by giving insight and helping make more informed everyday decisions, like forecasts, medical diagnoses, inventory, trading, energy consumption assessment, and more. Media sites concentrate on the teaching of computers and millions of possible ways to propose movies or songs. This is being achieved by businesses listening to their consumers' shopping patterns (Mobile et al. 2017). It includes two approaches: supervised learning, which builds a framework with knowing input and output data to forecast future results, and unsupervised learning, including input information with hidden patterns or intrinsic structures.

Up to this point, the author has indicated various MLs, and many of these algorithms will have their advantages and limitations, which make them perfect for use in such fields. Enhanced versions of such algorithms are defined as deep learning (DL) algorithms, which have become extremely popular lately. While machine learning approaches derive from biological neural networks and go back several decades, data science contributions are more recent. The author has used simple ML and DL algorithms in this research, and some of its most widely used LIDC-IDRI datasets for the research (Li et al. 2019; Dartmouth-Hitchcock Medical Centre 2019). The output analyses, as well as the findings of this deep study, will assist researchers and specialists in similar fields to gain improved information and experience for the collection.

9.7.1 AlexNet Architecture

The primary concept of AlexNet was developed by Alex Krizhevsky. It is a convolutional neural network (CNN) that was released by Ilya Sutskever and Geoffrey Hinton (Hussein et al. 2019). AlexNet came into popularity during competition in ImageNet. The particular concern of the source manuscript was for its superior performance.

AlexNet is the vital eight-level coevolution network depicted in Figure 9.3. This network effectively identifies the different features from the input images, which can divide 1,000 classes (Wang et al. 2019). The structure would then be revised such that the binary type is classified as benign or malignant. This AlexNet revision categorizes the images correctly. Mostly, in the article, there have been 197 DICOM images; now the input size of 227x227x3 is indeed fc7 layer (fully connected). The interactive imaging training continues to include expertise {\LIDC IDRI 0001\ 0084.png' and more} info, as well as the labels as categorically. (a) AlexNet architecture contains three layers, two normalization layers, two fully connected layers, and one layer of softmax, and consists of five convolution layers. (b) That layer includes coevolution filters and a non-linear ReLU activation feature. (c) For optimum pooling, pooling layers are used. (d) Due to similarly connected layers, the input value is measured. (e) The size of the input is mostly referenced as 224x224x3 because of certain padding, 227x227x3 is the outcome. (f) The total number of AlexNet metrics is 60 million. AlexNet brief info: (a) A system that reached the finals became optimized to precise details-\s1. ReLU is an activation feature; (b) utilized layers of standardization are no longer required; (c) 128 lots size; (4) SGD learning algorithm momentum; (e) high data increase with items such as flipping, jittering, cuts, and standardization of color; (f) System assembly to achieve the best performance.

The particular concern of the source manuscript was it for its superior performance; its depth of a framework was absolutely necessary. That is computer-intensive but had been made necessary through training gratitude to GPUs or graphics-processing elements made necessary through training gratitude to GPUs. AlexNet demonstrates,

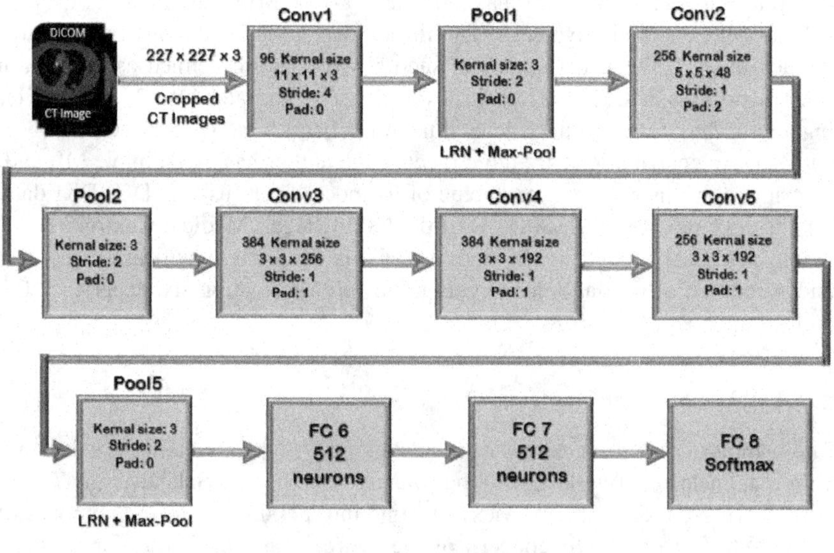

FIGURE 9.3 Architecture of CNN AlexNet.

in the following sentences, the specific layer strategy. The effective purpose, including functional factors of the layer, are therefore disclosed here:

```
layers1 = [imageInputLayer ([227 227 3])
convolution2dLayer (11,128,'Padding',0,'stride',4)
reluLayer
crossChannelNormalizationLayer (5)
maxPooling2dLayer (3,'Stride',2,'Padding',0)
convolution2dLayer (5,512,'Padding',2,'stride',1)
reluLayer
crossChannelNormalizationLayer (5)
maxPooling2dLayer (3,'Stride',2,'Padding',0)
convolution2dLayer (3,384,'Padding',1,'stride',1)
reluLayer
convolution2dLayer (3,256,'Padding',1,'stride',1)
reluLayer
convolution2dLayer (3,256,'Padding',1,'stride',1)
reluLayer
maxPooling2dLayer (3,'Stride',2,'Padding',0)
fullyConnectedLayer (5000)
reluLayer
dropoutLayer
fullyConnectedLayer (1000)
reluLayer
dropoutLayer
fullyConnectedLayer (5)
softmaxLayer
classificationLayer]
options = trainingOptions ('sgdm','MaxEpochs',100, . . .
'InitialLearnRate',0.0001)
convnet = trainNetwork (imdsTrain, layers1, options)
```

Some subsequent classification techniques, such as KNN, Gaussian Naive Bayes (GNB), LDA, and SVM, which have been the machine teaching algorithms, have all been implemented in this paper. Like that of the leading DL segment of CNN, AlexNet is being used to evaluate for the identification of lung cancer.

9.8 CLASSIFICATION

The authors comprise 1,010 malignant as well as benign cases who want to pick a model training set, then test the pattern for accuracy and have frequently sought both supervised and unsupervised learning. However, there is no question of collinearity between the variable as shown in Figures 9.4 (l), 9.4 (m), 9.4 (n), 9.4 (o) and 9.4 (p) to predict cancerous or non-cancerous (Kumar and Bakariya 2021). The correlation between the predictor sector, entropy, and the standard deviation is high (Yiwen et al. 2019). Kurtosis and skewness are also closely related. This collinearity also means that such forecast variables are eliminated.

9.8.1 Gaussian Naive Bayes (GNB)

Naive Bayes is a set of classification methods, most of them by Bayes theory. This is an algorithm party, not one algorithm. A single algorithm is not present here, but many algorithms. It appears that way.

Now, let us start with such a dataset. Consider a fictional set of data that describes lung cancer. Consider the symptoms and conditions that are classified as malignant for the presence of cancer or benign for the healthiness of a person. Bayes' theorem determines the likelihood of such an occurrence happening given the likelihood of a previous event.

Statistically, the theorem of Bayes is as follows: $P(A \mid B) = \dfrac{P(B \mid A)P(A)}{P(B)}$

In which the events are A and B as well as P(B) 0? Practically, since event B is real, we are seeking the possibility of event A. Event B is also called proof. That prior of P(A) is A. The proof is indeed an undefined instance attribute value. P(A|B) is the probability B posterior, i.e., the probability because after proof is shown.

It can now implement the concept of Bayes in the following manner about a certain set of data: $P(y \mid X) = \dfrac{P(X \mid y)P(y)}{P(X)}$

Thus, y is a parameter of class, X is a conditional (size n) function vector in which $X = (x_1, x_2, x_3, \ldots\ldots, x_n)$

$$P(y \mid x_1,..,x_n) = \frac{P(x_1 \mid y)P(x_2 \mid y)...P(x_n \mid y)P(y)}{P(x_1)P(x_2)...P(x_n)}$$

$$P(y \mid x_1,..,x_n) = \frac{P(y)\prod_{i=1}^{n} P(x_i \mid y)}{P(x_1)P(x_2)...P(x_n)}$$

$$P(y \mid x_1,..,x_n) \alpha P(y)\prod_{i=1}^{n} P(x_i \mid y)$$

Thus, it has finally escaped the role of $P(y)$ as well as $P(x_i \mid y)$ estimation. Please be reminded that $P(y)$ is known as probability class, and $P(x_i \mid y)$ is known as the probability situation. Various naive classification algorithms from Bayes vary primarily by their assumptions underlying $P(x_i \mid y)$.

9.8.2 K-Nearest Neighbors (KNN)

In the case of the supervised learning approaches, KNN has been one of the fastest ML techniques. KNN theory hypothesizes that now the scenario is identical or the case accessible and positions the situation in the group somewhat identical to just the categories available. The KNN algorithm collects all existing evidence and categorizes a similarity measure database. This helps new data to be quickly sorted by using the KNN algorithm with a well class. With both classification and regression problems, the KNN algorithm would be used, but especially for classification tasks. KNN is a non-parametric method, meaning that the original data is not taken for granted. Also, it is known as a sluggish learner algorithm, as it will not automatically learn from the training set but retains the set of data and implements the operation mostly on data when it is graded. And in the training process, the KNN algorithm only holds the data and categorizes this into the category like the latest ones if it receives new information.

Mostly based on the following algorithm, the KNN algorithm work could be understood: (a) pick the neighbor's k number; (b) compute k number of neighbors only for Euclidean distance; (c) consider the nearest k neighbors in the Euclidean distance measured; (d) the number of observations for each group is counted by such k neighbors; (e) introduce new information to just the group with the maximum amount of the k neighbor; (f) the framework is up and running.

9.8.3 Linear Discriminant Analysis (LDA)

A strategy for dimension reduction is an LDA. Because the word means reduction techniques, the dimensionality (i.e., dependent variable) in such a sample group is decreased with as much information is preserved as necessary. For example, imagine tracking the correlation between two variables that each color has become a different category. Further, the following steps may be classified as:

1. Measure scattered matrices both within and between groups.
2. The eigenvectors calculate and evaluate the optimal value of dispersion matrices.
3. Pick the top k values and process them.
4. Develop a new matrix with k's map value.
5. Get additional functionality by point multiplication of the data and matrix at step 4.

9.8.4 Support Vector Machine (SVM)

SVM is a highly efficient supervised learning algorithm for the creation as well as classification of regression models. Including both nearly linear and non-linearly separable sets of data, the SVM algorithm could function effectively. With a minimal amount of intelligence, the SVM algorithm can also provide these wonders.

An SVM is a learning machine model that can generalize through classifiers when a collection of marked data is given in the machine training set. This hyperplane, which can differentiate between the different groups, seems to be the main characteristic of SVM. Most hyperplanes would do this; however, the goal is to decide if a hyperplane with the largest margin, which will indicate maximal differences between some two groups, will be able to be counted easily and precisely whether there's a specific set of data that is two.

The SVM theory is as follows: the training set. i.e., disclosed as $D = \{(xj, yi)\}_{i=1}^{L} 1$ with every input data n i $x \in R_n$, and also having $f(x, \{w, b\}) = sign(w.x + b)$ and next to the output associated with $yi \in \{-1, +1\}$. Initially, each input will be mapped to its space of higher dimensional features F, $z = \phi(x)$ which is a non-linear mapping $\phi : R_n \rightarrow F$. When F is linear, unspeakable info, its hyperplane is represented as a $W \in F$ vector and $y_i = (W'.Z_i b) \geq 1 - \xi_i, \forall i$.

It relates to the unidirectional classic recurring network architectures. It implies that the input sequence to such a network is interpreted in the correct direction: it behaves in the right direction. The input background is collected in the previous context of a network. These are problems, therefore, in which only a prior context will not properly learn that problem. It may also be enough to obtain a potential context, for example, a clinical scenario to recognize the local, possibly disturbing portion of an input sequence by providing information about what is yet to come. A future context could only be provided exactly when, at the point of calculation, the input sequences are often fully given. The principle of two-way RNNs allows certain past contexts and future contexts accessible in RNNs (Yang et al. 2017). Consider a classic RNN just with one hidden layer recurring. Now, a second recurrent hidden layer is introduced and linked with the input layer as well as to the output layer, though not specifically by the other hidden layer. Second, frequently in the forward direction, it introduces an input sequence to an input layer. Just the first computing of hidden layer and then all activations are processed with each process (Mayan et al. 2014). Second, the input sequence is displayed backward. Just the second hidden layer works, and it also keeps all its operations. After this, the output layer generates the output sequence by incorporating the historical knowledge that is the first hidden layer, and future information that is the second hidden layer at every stage. The second layer contains a second, hidden layer. Mention that now the backward hidden layer gradient calculation operates in almost the same way; however, at reversal time.

9.9 EXPERIMENTS AND RESULTS

For AlexNet, the following conditions are defined in the MATLAB R2018b GPU frameworks. CoreTM i5–7200U 7th generic variant of CPU @3.1GHZ, Intel(R) HD Graphics 620 RAM 1.00GB adjuster, and DDR4 8GB RAM. For the results of the classifier, as seen in the figures, many criteria are considered. A device of 395 CT fragments was trained to detect pulmonary cancer, and after a DCM file (dot) was transformed into a PNG format, it received a dataset with the LIDC-IDRI. The machine will identify the lung tumor and stagger it with the Matlab script.

Cancer and non-cancer processed images are illustrated in the following figures, as Figure 9.4(a) is an original CT image; Figure 9.4(b) is normalized through Gabor filter CT image; Figure 9.4(c) showed an 800 iteration process for compiling an image; Figure 9.4(d) presented a region-based global segmentation; Figure 9.4(e) elaborated the active contour with interactive mask segmented image; Figure 9.4(f) embodied a histogram of lung CT image volume; Figure 9.4(g) is proven to show a binary's segmented nodule image by the Otsu segmentation approach; Figure 9.4(h) is identified an inverse distance transform of a binary image; Figure 9.4(i) highlighted

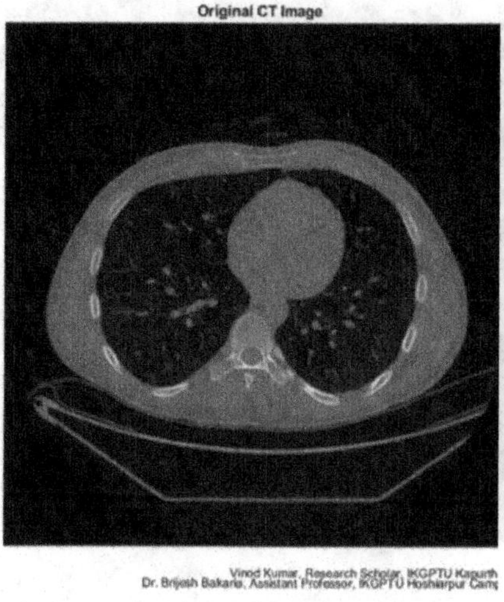

FIGURE 9.4(a) Original CT DICOM Image.

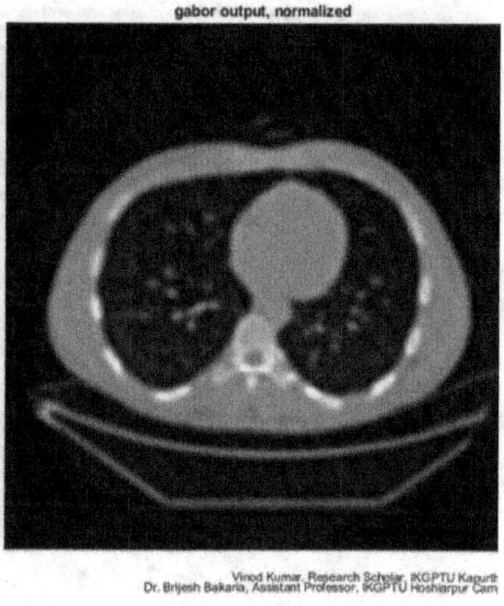

FIGURE 9.4(b) Gabor Filtered Normalized CT Image.

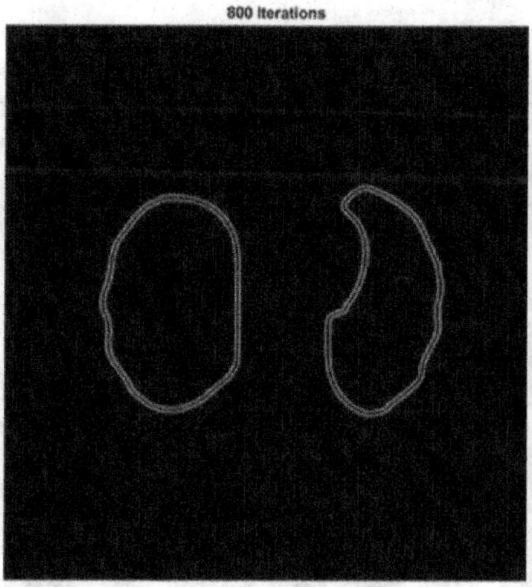

FIGURE 9.4(c) Processed 800 Iteration Image.

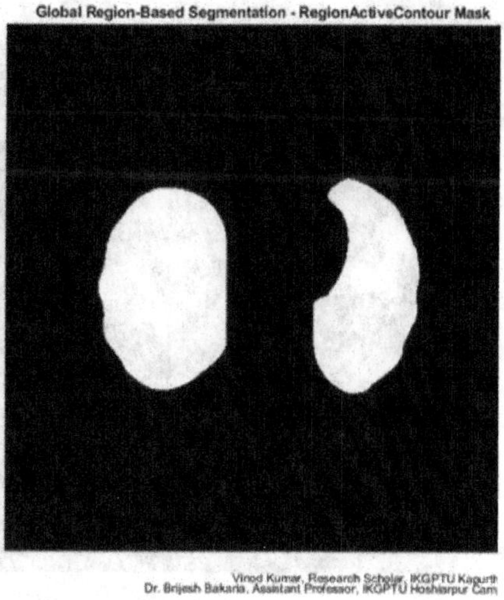

FIGURE 9.4(d) Global Region-based Segmented Image.

a watershed transformed image; Figure 9.4(j) expressed a watershed segmented mask binary image; some possible nodule locations in the image are shown in Figure 9.4(k); Figure 9.4(l) got to show a high false positive T1a stage; Figure 9.4(m) demonstrates the possible appearances of T1a tumors in the CT image; Figure 9.4(n) depicted a number of T1a tumors, i.e., 1,2,3,4, and 5 in CT image; Figure 9.4(o) presents

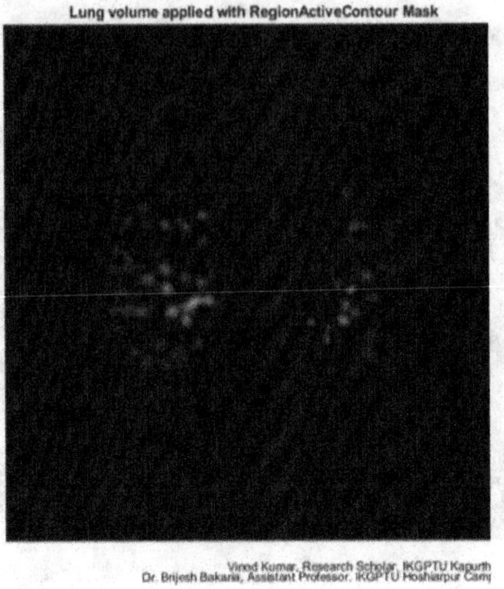

FIGURE 9.4(e) Region Active Contour Mask Image.

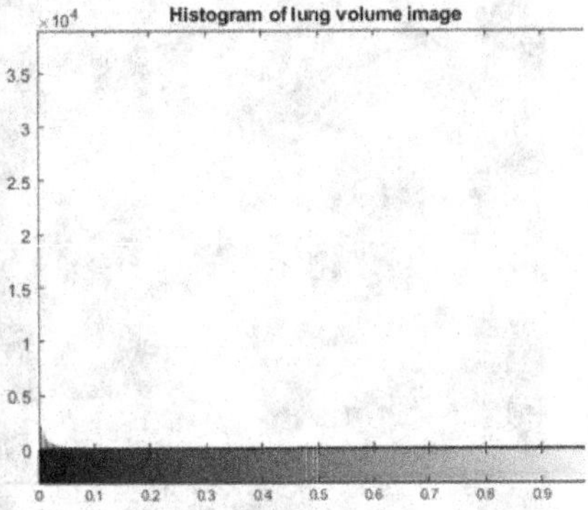

FIGURE 9.4(f) Histogram of Lung Volume (CT Image).

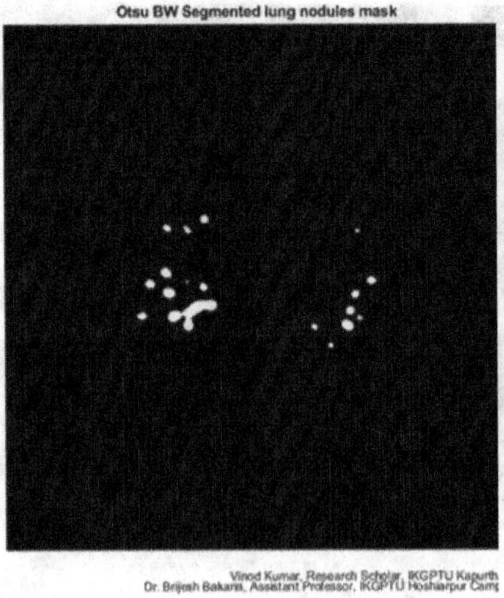

FIGURE 9.4(g) Otsu BW Segmented Nodule Image.

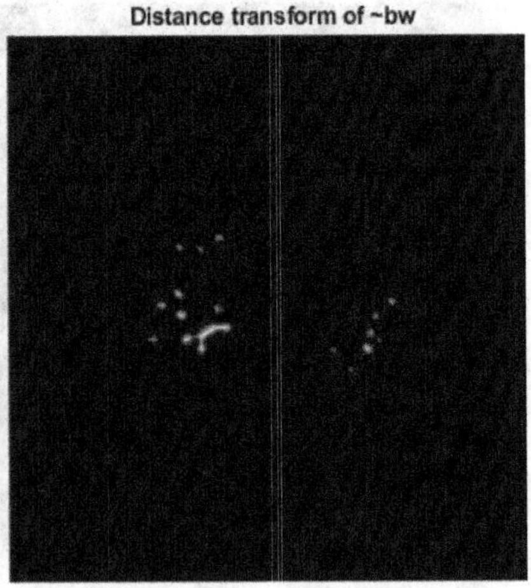

FIGURE 9.4(h) Distance Transform of Inverse BW Image.

probable T1b tumors in the CT image; and A number of T1b tumors, i.e., 1,2,3,4, and 5 in CT image are presented in Figure 9.4(p). The respective masks are used during training in CAD images, and during the trials, the researchers show this map with an example of progress. The lung parenchyma is then applied to this image by the field of the distribution of interest.

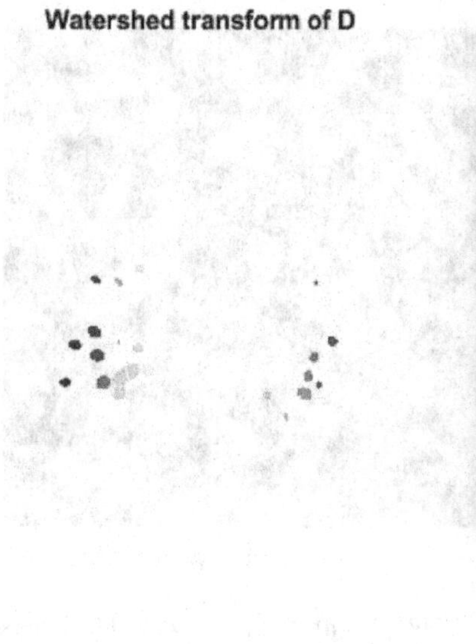

FIGURE 9.4(i) Watershed Transformed Image.

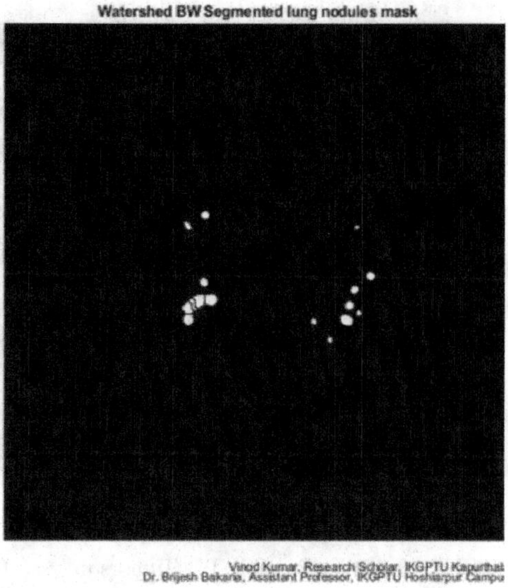

FIGURE 9.4(j) Watershed BW Segmented Mask Image.

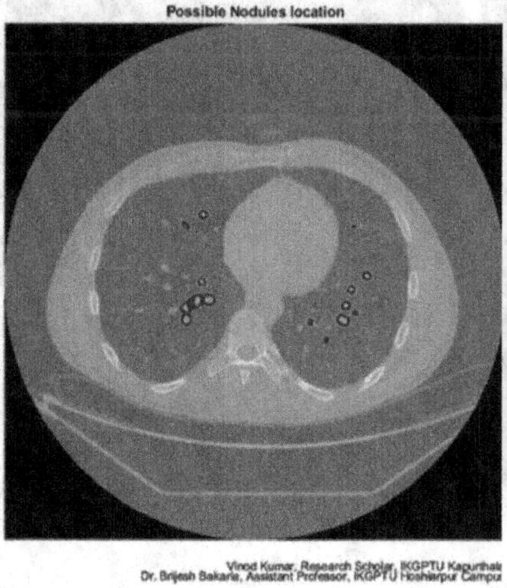

FIGURE 9.4(k) Possible Nodule Locations in Image.

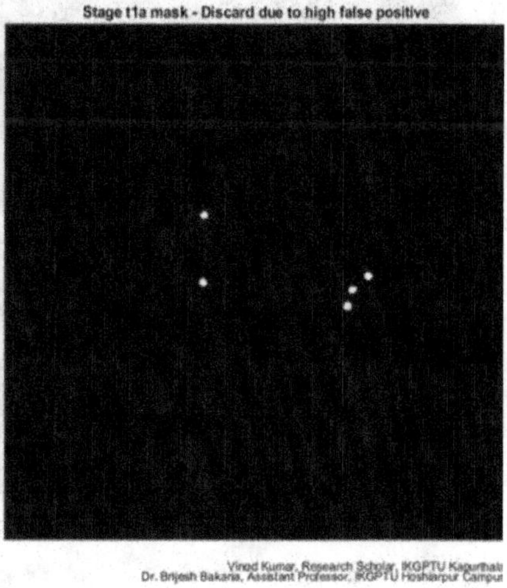

FIGURE 9.4(l) High False Positive T1a Stage.

The dataset used in this research is the LIDC-IDRI (Shah et al. 2020), which deals with the diagnosis of thorax CT scans and predictive lung cancer lesions. This is an outstanding database for the development of the identification and diagnostic strategies for lung cancer (Mayan et al. 2014).

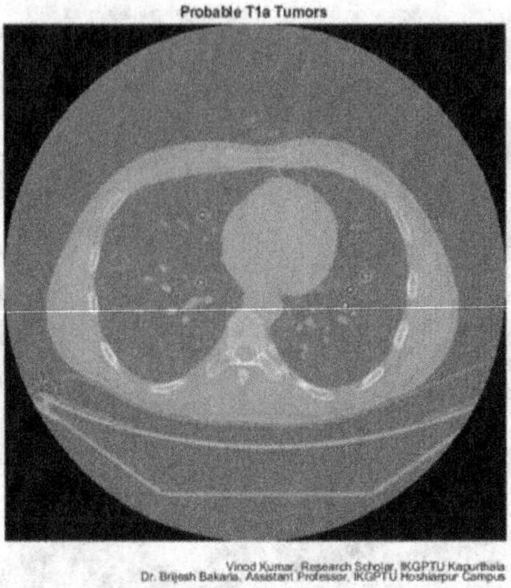

FIGURE 9.4(m) Possible T1a Tumors in CT Image.

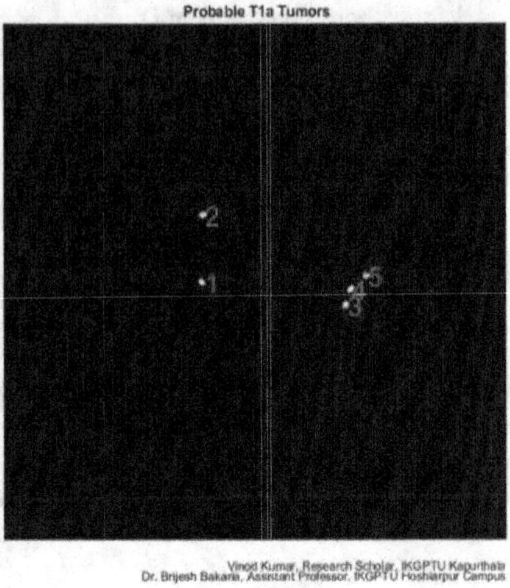

FIGURE 9.4(n) T1a-1,2,3,4, and 5 Tumors in CT Image.

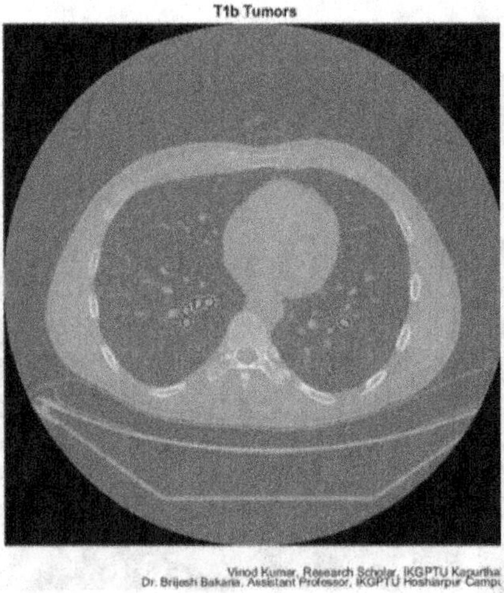

FIGURE 9.4(o) Possible T1b Tumors in CT Image.

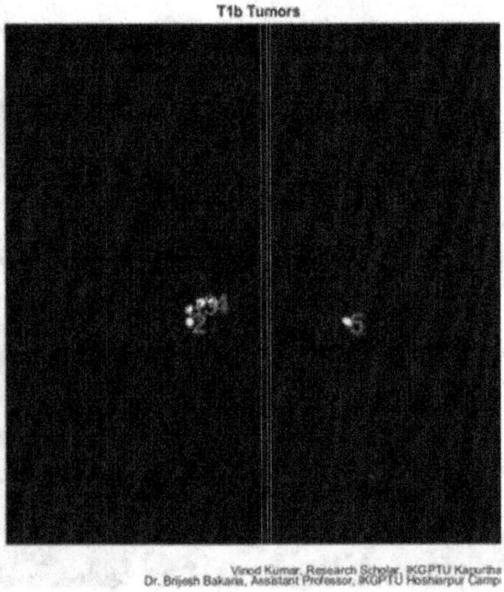

FIGURE 9.4(p) T1b-1,2,3,4, and 5 Tumors in CT Image.

Manual segmentation of the lung parenchyma has also been provided, and investigators have conducted a preprocessing step before experiments that have been carried out on this data packet to extract any information.

Tables 9.1, 9.2, 9.3, and 9.4 display the AlexNet confusion matrix using SVM, KNN, LDA, and GNB classifiers. Four groups can be categorized into the predictive

TABLE 9.1
Confusion Matrix with Advanced Classification Metrics Using AlexNet with SVM.

		Actual Class			
		Positive	**Negative**		
Predicted Class	Positive	TP = 179 (True Positive)	FP = 0 (False Positive)	Precision = TP/(TP+FP) Positive Predictive Value = 100%	Total Test Positive
		FN = 0 (False Negative)	TP = 18 (True Negative)	TN/(FN+TN) Negative Predictive Value = 100%	Total Test Negative
	Negative	TP/(TP+FN) Sensitivity = 100% Total Disease = 179	TN/(TN+FP) Specificity = 100% Total Normal = 18	(TP+TN)/(TP+TN+FN+FP) Accuracy = 100%	

TABLE 9.2
Confusion Matrix with Advanced Classification Metrics Using AlexNet with KNN.

		Actual Class			
		Positive	**Negative**		
Predicted Class	Positive	TP = 177 (True Positive)	FP = 2 (False Positive)	Precision = TP/ (TP+FP) Positive Predictive Value = 98.88%	Total Test Positive
		FN = 0 (False Negative)	TN = 18 (True Negative)	TN/(FN+TN) Negative Predictive Value= 100%	Total Test Negative
	Negative	TP/(TP+FN) Sensitivity = 100% Total Disease = 177	TN/(TN+FP) Specificity = 90% Total Normal = 20	(TP+TN)/(TP+TN+FN+FP) Accuracy = 98.98%	

form: true positive (TP), false positive (FP), true negative (TN), and false negative (FN). In percentiles 50–50, 60–40, 70–30, 80–20, and 90–10, the image volume for the training test data is being implemented.

The positive outcome percentage is accurate; the predicted outcomes have to be positive. False positives are expected to equal the number of true negative results. The positive results are false and are supposed to be negative. The number of negative outcomes is negative and therefore should be negative.

TABLE 9.3
Confusion Matrix with Advanced Classification Metrics Using AlexNet with LDA.

		Actual Class			
		Positive	**Negative**		
Predicted Class	Positive	TP = 171 (True Positive)	FP = 1 (False Positive)	Precision = TP/(TP+FP) Positive Predictive Value= 100%	Total Test Positive
	Negative	FN = 2 (False Negative)	TN = 23 (True Negative)	TN/(FN+TN) Negative Predictive Value= 99.42%	Total Test Negative
		TP/(TP+FN) Sensitivity = 98.84% Total Disease = 173	TN/(TN+FP) Specificity = 95.83% Total Normal = 24	(TP+TN)/(TP+TN+FN+FP) Accuracy = 98.48%	

TABLE 9.4
Confusion Matrix with Advanced Classification Metrics Using AlexNet with GNB.

		Actual Class			
		Positive	**Negative**		
Predicted Class	Positive	TP = 163 (True Positive)	FP = 16 (False Positive)	Precision = TP/(TP+FP) Positive Predictive Value= 91.06%	Total Test Positive
	Negative	FN = 8 (False Negative)	TN = 10 (True Negative)	TN/(FN+TN) Negative Predictive Value= 55.56%	Total Test Negative
		TP/(TP+FN) Sensitivity = 95.32% Total Disease = 171	TN/(TN+FP) Specificity = 38.46% Total Normal = 26	(TP+TN)/(TP+TN+FN+FP) Accuracy = 87.82%	

The author raises training and test sets for 70 percent of AlexNet using the SVM classification, TP at 179, TN at 18, and 0 for both FP and FN, suggesting that 179 of the total exhibits were malignant, and 18 were benign, as shown by Table 9.1.

Again, 177 are malignant and 20 are positive. AlexNet and KNN with the same training ratio are used. Moreover, there are 172 cases of malignancy and 24 cases of benignity classified by LDA. Finally, the authors described malignant and benign as 171 and 26, respectively, with the GNB classifier.

Following the images of the training gap, training is seen for 60 percent, 40 percent, 70 percent, 30 percent, and then for 20 percent, for 80 percent, for 90 percent, and for 10 percent. As for the images for the training gap, 50 percent; the training gap 60 percent; the training gap 40 percent; the training gap 70 percent; the study 30 percent; the training gap 80 percent; and finally 90 percent for training and 10 percent for testing. Tables 9.1, 9.2, 9.3 and 9.4 show the AlexNet confusion matrix. The type of forecaster can be split into four different groups: true positive and true negative (FN). The share of positive results is accurate, and the outlook should be positive. It is calculated to be optimistic about the amount of very bad outcomes.

This implies that the total display is malignant to 6,902 and that the 2,211 conveyors to be benign to 2 are listed in the table. Compute 95.77 percent and 96.65 percent of sensitivity and precision, which contribute to the improvement of performance.

When AlexNet is 70 percent, TP and TN are accordingly 6,578 and 2,103; FP is 108; and FN is 324. That means 6,902 frames are malignant, and 2,211 frames are good, as shown in Table 9.3. In contrast, FP is 108. Sensitivity and specificity are 95.77 percent and 96.65 percent, respectively.

It was not possible to cross-validate scholarly work for a variety of reasons. Due to a large amount of data, it takes too much time to pursue cross-validation approaches in an artificial neural network (ANN) in particular. Secondly, collective performance evaluation has been correctly performed, and an appropriate appraisal model has been established. The technology is not an exact and secure method to test diagnostic systems for computers. It should be as informative for researchers as possible.

The proposed approach is frequently employed rather than cross-checked, and the training and test data were often chosen at random. Despite cross-checking, the proposed approach is used several times, and training and testing information is always subjective. Analysis was carried out ten times in this report. The results show that the final results were ten times applied.

Statistics show the results of such surveys, which included 70% of educational data and 30% of data. Based on their analysis, training and test data are selected. As a result of the analysis in Figures 9.5, 9.6, 9.7, 9.8, 9.9, 9.10, and 9.11, the conclusions were drawn. The accuracy of Alex Net is 93.79 percent as shown in figures 9.7 and 9.8.

Accuracy at 70 percent is 93.79% AlexNet as seen in Figures 9.7 and 9.8, as this reminder is 96.18%.

SVM can easily be compared. The SVM normally takes less time (0.11 seconds with 80 percent training data). AlexNet spends the most time in the usable extraction process (20.48 seconds with 60 percent training data). The SVM is therefore sufficient for assessment.

The SVM should then be evaluated, as shown in Figure 9.11. The training algorithm according to Figures 9.7, 9.8, 9.9, 9.10, and 9.11 is one of the key elements for improving performance. For this analysis, AlexNet was chosen because it is more sensible and reliable by SVM and AlexNet. The study found that 6,902 patients were properly distinguished from 2,211 others. Cancer was correctly diagnosed relative to non-cancer patients with 89.13 percent of SVM patients (healthy individuals and patients with other lung diseases). The automated SVM breath evaluation allows for discrimination and wrinkling of safe products in patients with lung cancer.

An automatic breathing assessment with SVM causes patients with lung cancer to be discriminated against in safe artifacts and patients with multiple conditions to

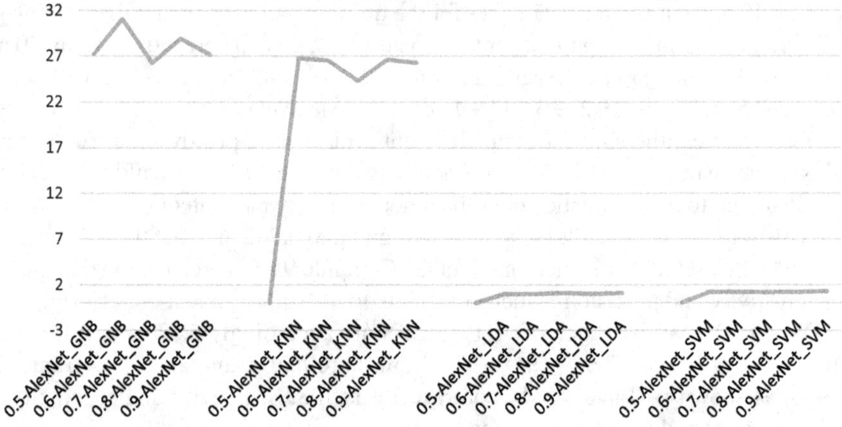

FIGURE 9.5 AlexNet Classification Time Analysis of GNB, KNN, LDA, and SVM.

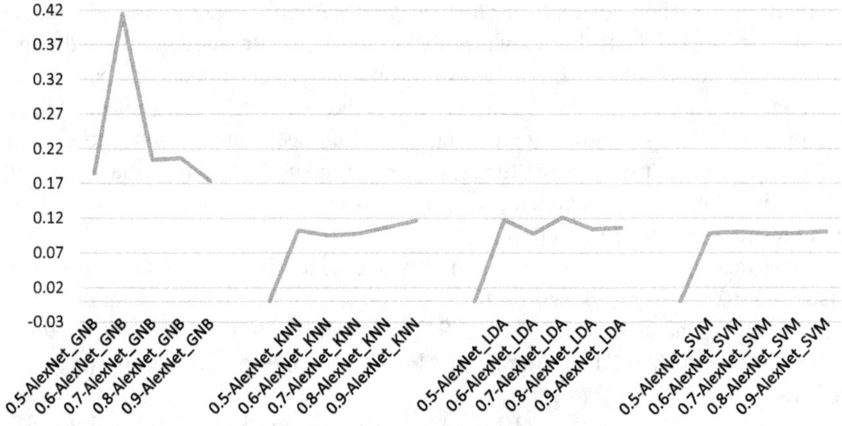

FIGURE 9.6 AlexNet Performance Analysis, Time Analysis of GNB, KNN, LDA, and SVM.

be scrapped. The prejudice can be tolerated between cancer patients and a healthy population, as shown in Figure 9.12.

AlexNet correctly lists 6,902 patients from 2,211 instead of 9,113. Cancer was positive compared to non-cancer patients in 95.47 percent of GNB. There is a degree of segregation between cancer patients and healthy communities.

The 6,379 and 2,734 samples were analyzed, with a training evaluation ratio between 70 percent and 30 percent.

There are 6,902 and 2,211 nodules in the 118 samples. Nodules misclassified as non-nodules and nodules are displayed correctly in table 9.2. Statistically significant sensitivity and specificity values were 95.77 percent and 96.65 percent, respectively. When various test ratios are used, receiver operating characteristic curves reflect the

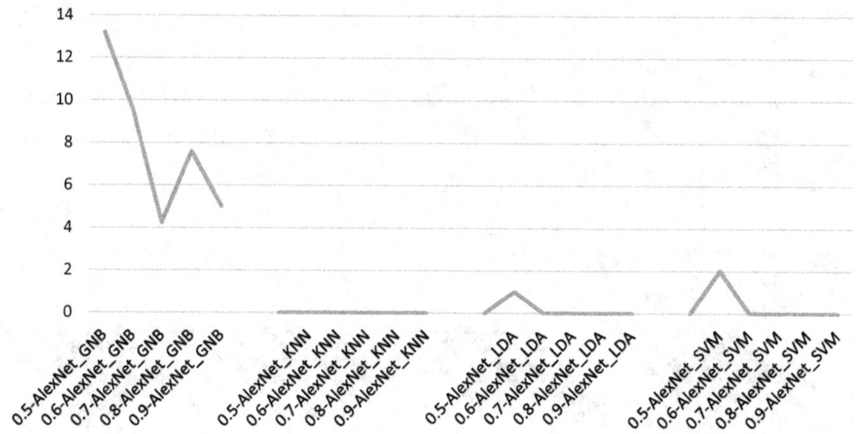

FIGURE 9.7 AlexNet False Alarm Rate Analysis of GNB, KNN, LDA, and SVM.

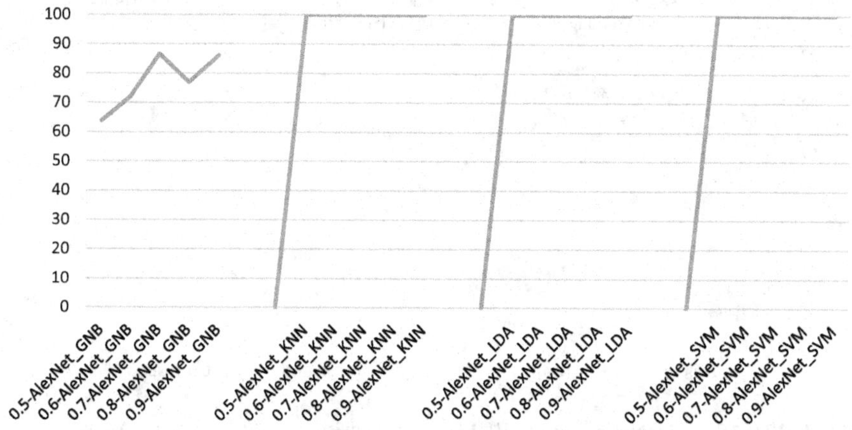

FIGURE 9.8 AlexNet Sensitivity Analysis of GNB, KNN, LDA, and SVM.

total output of the system. It has been determined that ROC is responsible for the correct differentiation between nodules and nodules in AlexNet. The ROC test is one of the best ways to test the effectiveness of classification skills. The ROC curve in TPR FPR is the right curve for testing classification skills.

The research processes and their preparation phase, pulmonary sections, were performed properly. The processes of learning and preprocessing, lung segments, and extraction were performed well.

Following the method with the highest precision, sensitivity, and accuracy, the results were calculated as less reliable. This involves more analyses of the most widely used approaches, such as segmenting datasets. High tumor metabolism and better CT images have been observed. The proposed approach can be used for such diseases as lung cancer, breast cancer, and brain and cardiovascular diseases in this

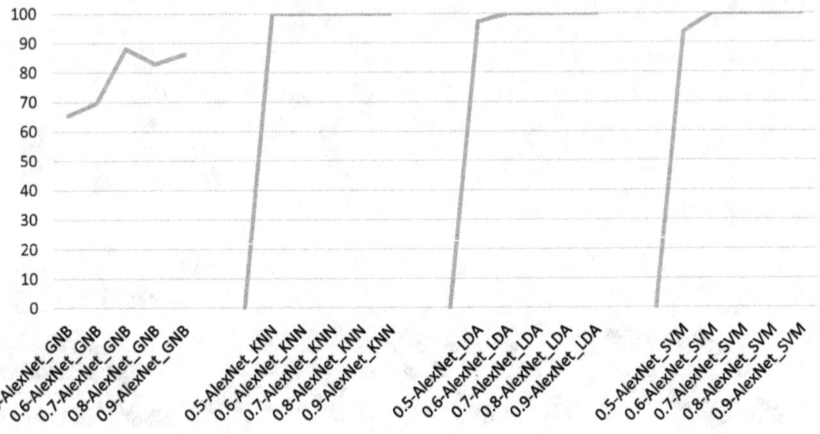

FIGURE 9.9 AlexNet F1-Measure Analysis of GNB, KNN, LDA, and SVM.

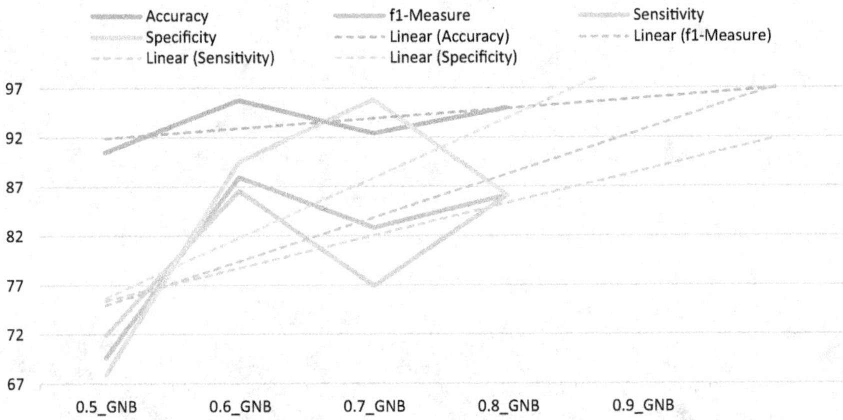

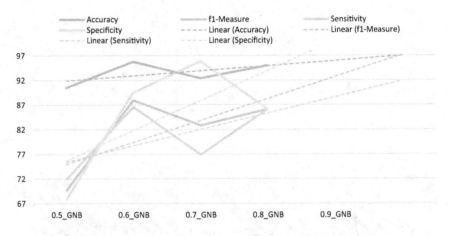

FIGURE 9.10 AlexNet Accuracy, F1-Measure, Sensitivity, and Specificity Analysis of GNB.

imaging scenario. This technique would provide the patients with a much more accurate set of slices.

Only those cuts were then registered. Only a few wounds were reported in the treatment, not all angles. The authors claim that the concept provides advantages, such as limited time spent on images by reducing the use of stored bits, performance improvements, power savings, financial cost savings, time and energy savings, cost scanning, etc. All will be much better served.

This approach enables AlexNet's accuracy to be correctly classified at 96.15 percent by 70 percent, 95.45 percent by AlexNet in 60 percent, and 89.13 percent by SVM in 90 percent of training, as described in Figure 9.9. Even AlexNet is relatively stronger than the others. The likelihood of AlexNet picking its option is therefore higher.

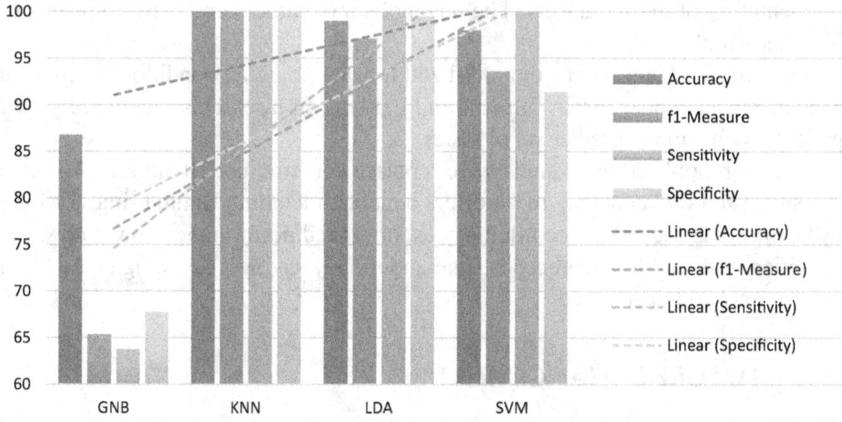

FIGURE 9.11 Accuracy, F1-Measure, Sensitivity, and Specificity Analysis of GNB, KNN, LDA, and SVM at 50–50 TTR.

SVM can easily be compared. It takes less than 0.11 seconds for SVM to train AlexNet (20.48 seconds with 60 percent training data). Figure 9.11 illustrates how SVMs are determined. A key component of the increase in performance is the training algorithm shown in Figures 9.7, 9.8, and 9.9, and 9.10. Overall, 89 percent of patients improved; 6,902 of 2,211 patients successfully distinguished. In SVM patients, 89,13 out of 100 patients are accurately diagnosed with cancer versus healthy people and patients with other lung diseases Several conditions can be identified by an automated SVM breathing evaluation, including lung cancer patients and wrinkled patients, as well as the severity of the wounds in patients with cancer and healthy individuals.

The total number of 9,113 in 2,211 patients identified as 6,902 is properly categorized by AlexNet. Cancer was positive in non-cancer patients in 95.47 percent of AlexNet. The isolation of cancer patients is listed with the healthy population.

There are 9,113 patients in 2,211 who are 6,902 in AlexNet. It covers 95.47 percent of patients who are cancer-free. Cancer sufferers are isolated. In all, 6,379 workout samples were provided, and 2,734 study samples, a ratio of 70 percent to 30 percent. A total of 6,902 and 2,211 nodules are found in 118 samples. Nodules and non-nodules are classified correctly and incorrectly in Table 9.2. Sensitivity and specificity were 95.77 percent and 96.65 percent respectively. ROC curves show different test-to-test ratios for the overall machine output in Figure 9.14.

The AlexNet is calculated so that nodules and non-nodules are properly separated. A ROC chart is an effective method for assessing the distinct ability of classifiers. By deriving the TPR from the FPR, the ROC curve is extracted, demonstrating the AlexNet classifier.

The research and preprocessing stages, lung segments, and extractions were properly performed.

The results were determined to be more accurate, with the highest precision and accuracy according to the system. This involves further study of methods such as the segmentation of main datasets. The metabolism of tumors and better images of CT have been high.

The proposed approach can be utilized to track imaging conditions such as lung cancer, breast cancer, brain tumors, and cardiac disease. This technique provides patients with a much greater set of slices.

Therefore, only some wounds were reported for care, not all angles. The authors ave suggested this concept, and it offers benefits, including limited time for images by minimizing the use of recorded pieces, improved performance, lower power consumption, reduced financial costs, time and money savings, scanning costs, etc. All can be much better handled.

9.10 FUTURE ENHANCEMENT

The scientist has only used a watershed marker segmentation in the proposed method and proposed several more segmentation methods, including comparisons, to improve the possible quality of features. Also, some additional filters and methods for image improvement must be checked. The precision of extra tumor properties is improved by increasing the data entry. Additional rankings, such as AlexNet, GoogLeNet, and others, on AI, should be implemented.

9.11 CONCLUSION

A proposed detection of lung cancer detects the tumor in the lung. A watershed marker is used to preprocess and subsequently segment the CT image. The processes of learning and preprocessing the lung section and extraction were properly conducted. The results were measured as less mistaken with the highest accuracy, sensitivity, accuracy, and others according to the method. This involves more methodology analyses, such as the segmentation of the most important datasets. For extraction of functions, the segmented image is used. The tumor is in the pulmonary system with removed features. The supervised cl for medical d and the unattended d. The tumor is found in the lung with the extracted characteristics. For medical diagnostics, supervised and unattended classifiers are used.

The highest precision rate is 96.14 percent with GNB followed by 95.47 percent with LDA and 89.13 percent with SVM classifiers. The radiologist therefore predicts and enhances the accuracy of the tumor level. The radiologist also studied ML techniques that provide health care professionals and radiologists with diagnostic instruments to select therapies for their patients based on histological characteristics. Lung cancer medical treatment can be helpful by using AI tools. To decide therapies based on histological characteristics, the researchers analyzed ML techniques that provide diagnostic instruments for health care professionals and radiologists. Via AI approaches, medical assistance can be useful for lung cancer and assess certain benefits and existing disadvantages over the entire care period.

CONFLICT OF INTEREST: No conflict of interest.

REFERENCES

Alamar, S. S., and D. Cherezov (2018), Delta radionics improves pulmonary nodule malignancy prediction in lung cancer screening, *IEEE Access* 6: 77796–77806.

Altaf, Fouzia, Syed M. S. Islam, Naveed Akhtar, and Naeem Khalid Janjua (2019), Going deep in medical image analysis: Concepts, methods, challenges, and future directions, *IEEE Access* 7: 99540–99572.

Asan, Onur, Alparslan Emrah Bayrak, and Avishek Choudhury (2020), Artificial intelligence and human trust in healthcare: Focus on clinicians, *Journal of Medical Internet Research* 22, no. 6: e15154. https://doi.org/10.2196/15154. PMID: 32558657; PMCID: PMC7334754.

Bhatia, S., Y. Sinha, and L. Goel (2019), Lung cancer detection: A deep learning approach. In *Soft Computing for Problem Solving,* pp. 699–705. Springer, Singapore.

Chen, Ying, Yerong Wang, Fei Hu, and Ding Wang (2020), A lung dense deep convolution neural network for robust lung parenchyma segmentation, *IEEE Access.*

Cui, Sunan, Yi Luo, Huan-Hsin Tseng, Randall K. Ten Haken, and Issam El Naqa (2018), Artificial neural network with composite architectures for prediction of local control in radiotherapy, *IEEE Transactions on Radiation and Plasma Medical Sciences* 3, no. 2: 242–249.

Dartmouth-Hitchcock Medical Centre (2019), *A New Machine Learning Model Can Classify Lung Cancer Slides at the Pathologist Level.* School of Computer Dublin Institute of Technology.

Deep Prakash, K., W. C. Prasad, A. Alsadoon, A. Elchouemi, and S. Sreedharan (2017), Early detection of lung cancer using the SVM classifier in biomedical image processing. In *2017 IEEE International Conference on Power, Control, Signals and Instrumentation Engineering (ICPCSI),* pp. 3143–3148. https://doi.org/10.1109/ICPCSI.2017.8392305.

Echegaray, Sebastian, Viswam Nair, Michael Kadoch, Ann Leung, Daniel Rubin, Olivier Gevaert, and Sandy Napel (2016), A rapid segmentation-insensitive "digital biopsy" method for radionic feature extraction: Method and pilot study using CT images of non—small cell lung cancer, *Tomography* 2, no. 4: 283.

Gupta, Yubraj, Ramesh Kumar Lama, Sang-Woong Lee, and Goo-Rak Kwon (2020), An MRI brain disease classification system using PDFB-CT and GLCM with kernel-SVM for medical decision support, *Multimedia Tools and Applications* 79, no. 43: 32195–32224. https://doi.org/10.1007/s11042-020-09676-x.

Herrmann, Markus D., David A. Clunie, Andriy Fedorov, Sean W. Doyle, Steven Pieper, Veronica Klepeis, Long P. Le, et al. (2018), Implementing the DICOM standard for digital pathology, *Journal of Pathology Informatics* 9.

Hussein, Sarfaraz, Pujan Kandel, Candice W. Bolan, Michael B. Wallace, and Ulas Bagci (2019), Lung and pancreatic tumor characterization in the deep learning era: Novel supervised and unsupervised learning approaches, *IEEE Transactions on Medical Imaging* 38, no. 8: 1777–1787.

Johora, F. Tuj, M. Jony, and Parvin Khatun (2018), A new strategy to detect lung cancer on CT images, *International Research Journal of Engineering and Technology (IRJET)* 5, no. 12: 27–32.

Kavitha, M. S., J. Shanthini, and R. Sabitha (2019), ECM-CSD: An efficient classification model for cancer stage diagnosis in CT lung images using FCM and SVM techniques, *Journal of Medical Systems* 43, no. 3: 73.

Kim, B. C., S. Sung, and H. Suk (2016), Deep feature learning for pulmonary nodule classification in a lung CT. In *2016 4th International Winter Conference on Brain-Computer Interface (BCI),* 2016, pp. 1–3. https://doi.org/10.1109/IWW-BCI.2016.7457462.

Kim, Catherine S., and Melenda D. Jeter (2021), *Radiation Therapy, Early-Stage Non-Small Cell Lung Cancer.* Stat Pearls Publishing, Treasure Island, FL. Available from https://www.ncbi.nlm.nih.gov/books/NBK459385/

Kuan, K., M. Ravaut, G. Manek, H. Chen, J. Lin, B. Nazir, and V. Chandrasekhar (2017), Deep learning for lung cancer detection: Tackling the Kaggle data science bowl 2017 challenge, *arXiv preprint arXiv:1705.09435*.

Kumar, Vinod, and Brijesh Bakariya (2021), Classification of malignant lung cancer using deep learning, *Journal of Medical Engineering & Technology* 45, no. 2: 85–93. https://doi.org/10.1080/03091902.2020.1853837.

Kumar, Vinod, and Dr. Kanwal Garg (2012), Neural network-based approach for detection of abnormal regions of lung cancer in X-Ray image, *International Journal of Engineering Research & Technology*, ISSN: 2278–0181.

Kumar, Vinod, Ashu Gupta, Rattan Rana, and Kanwal Garg (2015), Lung cancer detection from X-Ray image using statistical features, *International Journal of Computing* 4, no. 6: 178–181.

Layman, Mark, and Hayit Greenspan (2020), Semi-supervised lung nodule retrieval, *ArXiv, abs/2005.01805*.

Li, L., Y. Wu, Y. Yang, L. Li and B. Wu (2019), A new strategy to detect lung cancer on CT images. In *3rd IEEE International Conference on Image, Vision, and Computing, ©2018*. IEEE. 978-1-5386-4991.

Li, Lingling, Yuan Wu, Yi Yang, Lian Li, and Bin Wu (2018), A new strategy to detect lung cancer on CT images. In *2018 IEEE 3rd International Conference on Image, Vision and Computing (ICIVC)*, pp. 716–722. https://doi.org/10.1109/ICIVC.2018.8492820.

Liu, Liang, Jianjiao Ni, and Xinhong He (2018), Upregulation of the long noncoding RNA SNHG3 promotes lung adenocarcinoma proliferation, *Disease Markers*.

Luo, Yi, Daniel McShan, Dipankar Ray, Martha Matuszak, Shruti Jolly, Theodore Lawrence, Feng-Ming Kong, Randall Ten Haken, and Issam El Naqa (2018), Development of a fully cross-validated Bayesian network approach for local control prediction in lung cancer, *IEEE Transactions on Radiation and Plasma Medical Sciences* 3, no. 2: 232–241.

Manikandan, T., and N. Bharathi (2016), Lung cancer detection using fuzzy auto-seed cluster means morphological segmentation and SVM classifier, *Journal of Medical Systems* 40, no. 7: 181.

Mathur, Prashant, Krishnan Sathish Kumar, Meesha Chaturvedi, Priyanka Das, Kondalli Lakshmi Narayana Sudarshan, Stephen Santhappan, Vinodh Nallasamy, et al. (2020), Cancer statistics, 2020: Report from national cancer registry programme, India, *JCO Global Oncology* 6: 1063–1075.

Mayan, K. M., M. Tun, and A. Khaing (2014), Implementation of lung cancer nodule feature extraction using digital image processing. *IJSETR* 3, no. 9: 1610–1618.

Mobile, Aryan, SupratikMoulik, and Hien Van Nguyen (2017), Lung cancer screening using adaptive memory-augmented recurrent networks, *arXiv preprint arXiv:1710.05719*.

Murillo, B. R. (2018), Health of things algorithms for malignancy level classification of lung nodules. *IEEE Access* 6: 18592–18601. https://doi.org/10.1109/ACCESS.2817614.

Nasser, I. M., and S. S. Abu-Naser (2019), Lung cancer detection using artificial neural network, *International Journal of Engineering and Information Systems (IJEAIS)*, 3, no. 3: 17–23. Available from SSRN: https://ssrn.com/abstract=3700556

Patel, Vaibhavi, Samkit Shah, Harshal Trivedi, and Urja Naik (2020), An analysis of lung tumor classification using SVM and ANN with GLCM features. In Singh, P., Pawłowski, W., Tanwar, S., Kumar, N., Rodrigues, J., and Obaidat, M. (eds) *Proceedings of First International Conference on Computing, Communications, and Cyber-Security (IC4S 2019)*, pp. 273–284. Springer, Singapore. https://doi.org/10.1007/978-981-15-3369-3_21

Pelc, Norbert J., and Adam Wang (2020), CT statistical and iterative reconstructions and post processing. In Samei, E., and Pelc, N. (eds) *Computed Tomography*. Springer, Cham. https://doi.org/10.1007/978-3-030-26957-9_4.

Pradhan, Kanchan, and Priyanka Chawla (2020), Medical internet of things using machine learning algorithms for lung cancer detection, *Journal of Management Analytics* 7, no. 4: 591–623. https://doi.org/10.1080/23270012.2020.1811789.

Rabbani, Mohamad, Jonathan Kanevsky, Kamran Kafi, Florent Chandelier, and Francis J. Giles (2018), Role of artificial intelligence in the care of patients with non-small cell lung cancer, *European Journal of Clinical Investigation* 48, no. 4: e12901.

Razzak, Muhammad Imran, Saeeda Naz, and Ahmad Zaib (2018), Deep learning for medical image processing: Overview, challenges, and the future. In *Classification in BioApps*, pp. 323–350. Springer, Cham.

Riti, Yosefina Finsensia, Hanung Adi Nugroho, Sunu Wibirama, Budi Windarta, and Lina Choridah (2016), Feature extraction for lesion margin characteristic classification from CT Scan lungs image. In *2016 1st International Conference on Information Technology, Information Systems and Electrical Engineering (ICITISEE)*, pp. 54–58. https://doi.org/10.1109/ICITISEE.2016.7803047.

Rodrigues, Murillo B., Raul Victor M. Da Nóbrega, Shara Shami A. Alves, Pedro Pedrosa Rebouças Filho, Joao Batista F. Duarte, Arun K. Sangaiah, and Victor Hugo C. De Albuquerque (2018), Health of things algorithms for malignancy level classification of lung nodules, *IEEE Access* 6: 18592–18601.

Shah, Anwar, Javed Iqbal Bangash, Abdul Waheed Khan, Imran Ahmed, Abdullah Khan, Asfandyar Khan, and Arshad Khan (2020), Comparative analysis of median filter and its variants for removal of impulse noise from gray scale images, *Journal of King Saud University – Computer and Information Sciences*. ISSN 1319-1578. https://doi.org/10.1016/j.jksuci.2020.03.007.

Tabish, Tanveer A., MdZahidul I. Pranjol, Hasan Hayat, Alma A. M. Rahat, Trefa M. Abdullah, Jacqueline L. Whatmore, and Shaowei Zhang (2017), In vitro toxic effects of reduced graphene oxide nanosheets on lung cancer cells, *Nanotechnology* 28, no. 50: 504001.

Taher, F., N. Werghi, and H. Al-Ahmad (2016), Rule-based classification of sputum images for early lung cancer detection. In *2015 IEEE International Conference on Electronics, Circuits, and Systems (ICECS)*, pp. 29–32. https://doi.org/10.1109/ICECS.2015.7440241.

Tripathi, Priyanshu, Shweta Tyagi, and Madhwendra Nath (2019), A comparative analysis of segmentation techniques for lung cancer detection, *Pattern Recognition and Image Analysis* 29, no. 1: 167–173.

Wang, Shidan, Alyssa Chen, Lin Yang, Ling Cai, Yang Xie, Junya Fujimoto, Adi Gazdar, and Guanghua Xiao (2018), Comprehensive analysis of lung cancer pathology images to discover tumor shape and boundary features that predict survival outcome, *Scientific Reports* 8, no. 1: 1–9. https://doi.org/10.1038/s41598-018-27707-4.

Wang, Shidan, Donghan M. Yang, Ruichen Rong, Xiaowei Zhan, Junya Fujimoto, Hongyu Liu, John Minna, Ignacio Ivan Wistuba, Yang Xie, and Guanghua Xiao (2019), Artificial intelligence in lung cancer pathology image analysis, *Cancers* 11, no. 11: 1673.

Wu, Jian, Chunfeng Lian, Su Ruan, Thomas R. Mazur, Sasa Mutic, Mark A. Anastasio, Perry W. Grigsby, Pierre Vera, and Hua Li (2018), Treatment outcome prediction for cancer patients based on radionics and belief function theory, *IEEE Transactions on Radiation and Plasma Medical Sciences* 3, no. 2: 216–224.

Wu, Jianrong, and Tianyi Qian (2019), A survey of pulmonary nodule detection, segmentation, and classification in computed tomography with deep learning techniques, *Journal of Medical Artificial Intelligence* 2: 2–8.

Yang, Shuo, Ran Wei, Jingzhi Guo, and Lida Xu (2017), Semantic inference on clinical documents: Combining machine learning algorithms with an inference engine for effective clinical diagnosis and treatment, *IEEE Access* 5: 3529–3546.

Yiwen, X., H. Ahmed, et al. (2019), Deep learning predicts lung cancer treatment response from serial medical imaging. *Clinical Cancer Research*. https://doi.org/10.1158/1078-0432.CCR-18-2495.

10 Cancerous or Non-Cancerous Cell Detection on a Field-Programmable Gate Array Medical Image Segmentation Using Xilinx System

C. Gopala Krishnan, Prasannavenkatesan Theerthagiri, and A.H. Nishan

CONTENTS

10.1 INTRODUCTION

Computer-aided image evaluation has pulled in large concentration commencing each indicator process as well as medicinal researchers because of its ability to

DOI: 10.1201/9781003184140-10

surmount their challenges related to the prejudiced experimentation about infinitesimal images. Characterization of biomedical images acting as a second reader for quantitative tools, it mitigates the consequences of inter and intra booklover inconsistency on analysis and complements the selection (Kharrat et al., 2009). Decisions can be made in a straightforward manner, whereas a computer-aided diagnosis (CAD) system prevents pathologists from killing their instance on picture region (Alfonse and Salem, 2016); the compassionate prostate biopsies percentage in the United States is about 80%, and it was monitored through computerized image analysis, to deal with challenging issues, and it is helpful for the pathologists (Abdel-Basset et al., 2017). Biologists use a huge quantity of diverse cell types by automated tools as biochemical experiments for the identification of cells, and it is very costly (Rajendran and Dhanasekaran 2012). The objective of this research is to employ image processing and machine learning (ML) techniques to implement novel algorithms for computer-aided analysis of biomedical images (Kharrat et al., 2009). The main role is to provide a high interpretation of biological and pathological images and work out automated methods for characteristic extraction, recognition, and categorization (Hebli and Gupta, 2016).

10.1.1 Image Processing In MRI

Inside the human body, tumor formation occurs due to cancer cells. It is identified that nearly 3 million people in India are affected by cancer, out of which 1 million people are diagnosed with new forms of cancer in a survey taken by *Times of India* (Nandi, 2015). It leads to the formation and development of lesions and tumors inside the body, and the consequences are highly pervasive (Islam, Hossain, and Saha, 2017).

Neat experimentation and learning are made by a radiologist to recognize the pathologies embedded within the tissue of the brain, the most convoluted part of the human body (Hebli and Gupta, 2016).

Brain tumors can occur at any age and can often have destructive results. Primary brain tumors can be defined as benign or malignant growths that produce in intracranial tissue. Inside the cerebrum, the brain tumor cells grow quickly (Hebli and Gupta, 2016). A tumor is the wild developing and separation of cancer cells that forms a large mass of tissue inside any part of the human body; it affects the total metabolism of the human body. Generally, the enormous development of cancerous cells within the brain is called a tumor (Aly, 2006). In a benign tumor, active cells are not available, whereas a malignant tumor does have a heterogeneous tumor. Segmentation and classification are employed to find the affected tumor with the help of medical imaging systems (Kharrat et al., 2009).

The most popular methods for growth monitoring, growth analysis and status detection are MRI and probably ML algorithms (Kothari and Indira, 2016). Differences in cell composition and size don't affect the accuracy of the algorithmic program. The effective distribution of gray substance, a white substance that reduces the dimensions and form of cancer, is also combined with artificial intelligence (Apirajitha, Krishnan, Swaminathan et al., 2019).

10.1.2 Brain Analysis

While testing any design in all fields of engineering, data acquisition is an important step. This is salient to comprehend the method and limitations of measurements to create experiments and controls that help to analyze the phenomena of the world (Gomathi and Krishnan, 2020). Data acquisition is a process, allows for collecting signals from measurement sources and digitalizing the signals for storage, analysis and showcasing on another system. It is a process of moving inadvertently accurate data from a sensor or transducer directly into the system. A sensor is a device that acknowledges a physical change, and a transducer is a device that converts energy from one form to another. Data acquisition plays a significant role in image processing (Krishnan, Sivakumar, and Manohar, 2018).

In imaging science, an image is the input information, whereas image processing is a version of the processing of signals. Image processing is a technique that changes a given image into a digital one and performs a few operations for getting a tribute image or to uproot helpful information. For example, in a photography or video outline, its output or response will be an image or arrangement of qualities or the arrangement of qualities or the parameters identified with the image. Often, an image-processing framework considers a picture as a two-dimensional signal and standard signal-processing algorithms are applied to it (Aly, 2006).

It is a briskly developing technology that has various applications in business. Within the engineering and computer science disciplines, the core research area in image processing (Hebli and Gupta, 2016).

10.1.3 Brain Anatomization

To study the interested brain tissue by MRI data, it is necessary to know about the anatomy of the brain (Figure 10.1) and specifically the tissues that are white matter (WM), gray matter (GM), and cerebra spinal cord (CSF) (Islam et al., 2017). CSF is a colorless liquid that surrounds and fills the space in the brain. Pressure in the skull is maintained by CSF. It provides nutrients to the brain tissues and removes the waste from brain metabolism. The brain receives sensory inputs from GM and WM of spinal (Rajendran and Dhanasekaran, 2012). Bundles of axons coated with the sheath of myeline from the nerve tissue. Gray matter consists of a mass of cell bodies and dendrites lined with synapses.

The WM contains more amount of fat, and GM has more water. CSF mainly consists of water (Rajendran and Dhanasekaran, 2014). The different constituents of the tissues give dissimilarities in the MR scan, which differentiates the various tissues found in MRI (Kothari and Indira, 2016).

A divergent growth of tissues in the brain is called a brain tumor. When abnormal cells get multiplies for unidentified reasons, a brain tumor develops (Abdel-Basset et al., 2017). Glioma is the most common primary brain tumor (Bahadure, Ray, and Thethi, 2017). Gliomas occur in glial tissue, which helps and provides cells that can send messages to other parts of the body from the brain (Rajendran and Dhanasekaran, 2012). This tumor can be malignant or benign. It can classify as one of the three most common forms: astrocytoma's, ependymoma's, and mixed gliomas (Anitha and Murugavalli, 2016).

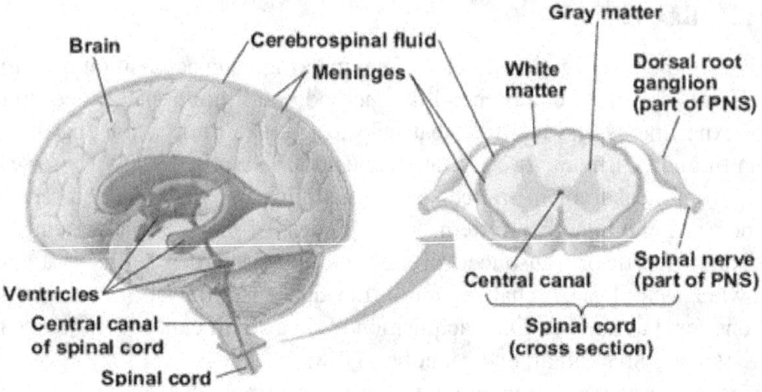

FIGURE 10.1 Anatomy of Brain.

Benign and Malignant Brain Tumors

A benign tumor consists of risk-free cells and has well defined borders. They can be removed completely and do not occur again (Anitha and Murugavalli, 2016). At the initial stages of a brain tumor, it does not spread into the neighbor tissues, but it may lead to pain, brain damage, and death. Malignant brain tumors do not have defined borders (Gupta and Pahuja, 2017). They tend to grow, increase pressure on the brain and can metastasize inside the brain and spinal cord. It is very uncommon for a malignant brain tumor to metastasize further than the central nervous system (CNS) (Nandi, 2015).

Primary and Secondary Brain Tumors

Primary brain tumors occur in the brain. They signify 2.5% of cancer deaths and 1% of cancer cases (Rajendran and Dhanasekaran, 2012). Most of the cancer patients evolve secondary or spread brain tumor when these cancer-causing cells metastasize into the brain from other parts of the body (Islam et al., 2017).

Naming and Grading Brain Tumors

The type of brain tumor describes its origin, how it will grow, and what types of cells it consists of. A tumor occurring in an adult can be graded or staged by how quickly it is growing and how it spreads to other neighboring tissues. Obvious borders are present in low-grade tumors. Some of the poor-quality tumor forms are surrounded by cysts. The low-grade tumor is slow growing. It may spread all over the brain but does not spread to other body parts.

The mid-grade and high-grade tumor is fast growing compared to the low-grade tumor. This tumor causes other health problems. It is very difficult to remove the entire tumor due to its blooming patterns, and the tumor occurs again frequently. A brain tumor consists of many types of cells. These malignant cells are detected under a microscope to determine the grade of the tumor. A tumor that grows to the neighboring tissues is called a penetrating tumor. The World Health Organization (WHO) grading system classifies brain tumors based on the rate of growth into the four categories, grades I, II, III and IV. A grade I tumor is less malignant and slow growing but it is life threatening and it is difficult to do surgery. Grade II tumors are fast growing compared to grade I tumors and have an abnormal appearance in the microscope. These tumors do not harm normal tissues and can occur again. Grade III tumors are normally malignant. They have a high possibility of occurring again. The most malignant tumor is the Grade IV tumor, which can also occur in wide areas of neighboring tissues.

In the medical field, image processing is used for the detection of tumors, CAD, and detection starts with making the images acceptable for further processing, tissue segmentation, features extraction and finding of optimal features, and classifying the tissues for identifying the tumor. Meaningful features learned by the human experts based on their knowledge play a vital role; hence, there is a maximal chance of inappropriate accuracy and difficult task in the learning algorithm.

10.1.4 Preprocessing

In medical image processing, the most essential steps are data acquisition and preprocessing. The quality of images in health care is more important than in other computer vision fields for the following reasons.

In biomedical data mining, for obtaining high-quality data, automatic data processing is required, as a collection of biomedical images are mainly designed in a doctor's manual diagnosis (Krishnan et al., 2018). Most of the data in health care is inconsistent and not standard, as different data sources provide medical images in different formats (Theerthagiri et al., 2021; Krishnan et al., 2021).

Without significant efforts during data preparation, these data cannot be used for data mining, which includes two techniques: image registration and image segmentation (Gupta and Pahuja, 2017). Image segmentation is a method of separating data into a continuous region for representing individual anatomical objects. It is used for many applications in the computer-related medical field in which they can classify different tissues in to separate logical classes. The major problem in image segmentation is about how to separate interested object from background noise. These different forms help to derive information that helps for more data-mining tasks. The amount of rotation and the translation of images in any direction helps to distinguish medical images, and they may also differ in their scale (Prasannavenkatesan, 2021).

Image analysis is essential in preprocessing, especially in the stages of filtering and contrast enhancement (Ruby, Prasannavenkatesan, and Vamsidhar, 2020; Theerthagiri et al., 2021). The results of the segmentation are closely related to these first stages (Krishnan, Robinson, and Chilamkurti, 2020). In fact, these actions allow for the best possible image, with accurate, reliable and excellent representation of the real image. Initially, the pixel range of the MRI image is 0 to 255, and the gray level values should be normalized from 0 to 1. Once normal, the function is extracted for further processing.

Noise Removal

Research shows that noise reduction is essential for digital image processing. The noise described during image acquisition and transmission is high-speed noise. There may be image noise from various sources. Noise can be caused by communication errors or image pressure. There is no image without noise, but some types of images are noisier than others. To screen the noises present in the MRI images and during image smoothing, the technique of noise removal is done (Gupta and Pahuja, 2017).

To get efficient results during image analysis, the image must be of good quality. Due to different kinds of noise corruption of images occur over the channels may occur during image transmission. The most commonly occurring noise is impulse noise. For salt and pepper noise, the corrected pixel may be taking the gray value of 255 or 0. The noise spreads all over the image and causes changes in the information pixel that may lead to thrashing in image information and quality of the image. To eliminate noise from the images, various filters are used. Filters with better edges and image information preservation properties are very important in image filtering. To remove impulse noise, different filters are used. The most efficient filter when the density of the noise is below 20% is standard MF, but when the density of the image is higher than 20% there will be a failure in image information and edges of the image.

The standard median filter is subdivided into WMF and CWMF. AMF is a method to process image noise with short noise density and by increasing the window size that will increase the computation time, but it sometimes results in a blurring effect. DBMS processes the noise signal by finding the presence of impulse noise. Initially, it checks for noise-affected pixels by selecting pixel values between the minimum and maximum range. This is because impulse mainly affects these pixel values.

A pixel value between 1 and 254 is considered an uncorrupted pixel, and there is no need for processing. A pixel that does not lie within that range is a corrupted pixel. At the point when the noise density is high, it takes the median value as well as altered values. In such cases, these encompassing pixel values are utilized for replacing the processed pixel. This should be possible because of the high correlation between neighboring pixels and undermined pixels, and it might prompt great perseverance. DBMF will utilize a specific length window with estimate 3*3, so its preparation time is low compared to different filters. The primary drawback is the streaking impact, which shows up amid high noise intensity, and the undermined pixel is changed with encompassing pixel value and processed. Subsequently, these techniques make it extremely hard to hold data, and edges in images are with high noise intensity. So, a new decision-based median filter is proposed to remove impulse. Based on noise-distributed functions, the window size is chosen. This technique helps us preserve important details of MRI images by eliminating the noise (Zhang and Xu, 2016).

Median Filtering Theory

A nonlinear signal-processing method of measurements is done by the median filter. The median value of the area replaces the noise value. As indicated by their gray matter, the constituent of the duvet is positioned, and to substitute the noise price, the median estimate of the cluster is placed away (Apirajitha et al., 2019). The median filtering output is g(x,y)=med, wherever W is that the two-dimensional price the mask estimate is often odd and also the state of the mask is sq., roundabout, cross, linear and so on.

Performance of the Median Filter in Noise Reduction

The median filter has advanced mathematical analysis for images with nonspecific noise. In Gaussian distribution, the image has zero mean noise and noise variance.

$$\sigma^2 med = \frac{1}{4nf^2(n)} \approx \frac{\sigma_i^2}{n+\frac{\pi}{2}-1} \cdot \frac{\pi}{2} \tag{10.1}$$

Things that determine the median filtering: the mask size and the noise distribution (equation 10.1). Compared to average filtering, the filtering performance is best for random noise reduction.

Fast Computation of the Median Filter

The average calculation determines the complexity of the algorithm. This work provides a statistical map for the growth of average search speed. There are two improvements in the new algorithm compared to the traditional media filter algorithm. First, equalize the number of pixels to N using the historical information of the mask and compare the value of the pixel with the initial average value of the mask. The next step is to minimize the problem of the algorithm.

10.1.5 MRI Techniques

MRI is an imaging technique used to produce images of the inside of our bodies. This depends on nuclear magnetic resonance (NMR). NMR is a spectroscopic system that gets substance and physical data about particles. The first MRI test conducted on the human body took five hours to complete.

The MRI system provides a detailed image of tissues, and it observes and processes the generated signal, while hydrogen atoms are replaced in the strong magnetic field and then excited using a resonant magnetic excitation pulse. Because of the consequence of the atomic turn, hydrogen atoms have a high magnetic moment. While setting these in a strong magnetic moment of the cores in a static attractive field will turn into a string under pressure. These nuclei consist of Larmor frequency, the tension on it determines the resonant frequency of the string. Hydrogen nuclei have a 1.5 T MRI field, and the resonant frequency is 64 HZ.

Image Formation in MRI

In Figure 10.2, the magnetic resonance imaging suite is illustrated. The patient is kept out of the table, and the table is movable. It can be mobilized up, down, front, and back and onwards. To settle the patient in the correct place, movement is automatically done by the MR imaging suite.

The real preferred standpoint of an MRI is that it doesn't utilize radiation, whereas CT scans do. This radiation is perilous if there is frequent contact. As an imaging innovation, MRI imaging has significant growth in the previous ten years; however, it keeps on advancing, and new abilities can without a doubt be produced. MRI gives a strong magnetic field of 0.2 to 3 tesla that may align the protons present on the water molecule of the body. A radiofrequency current is produced and can be measured

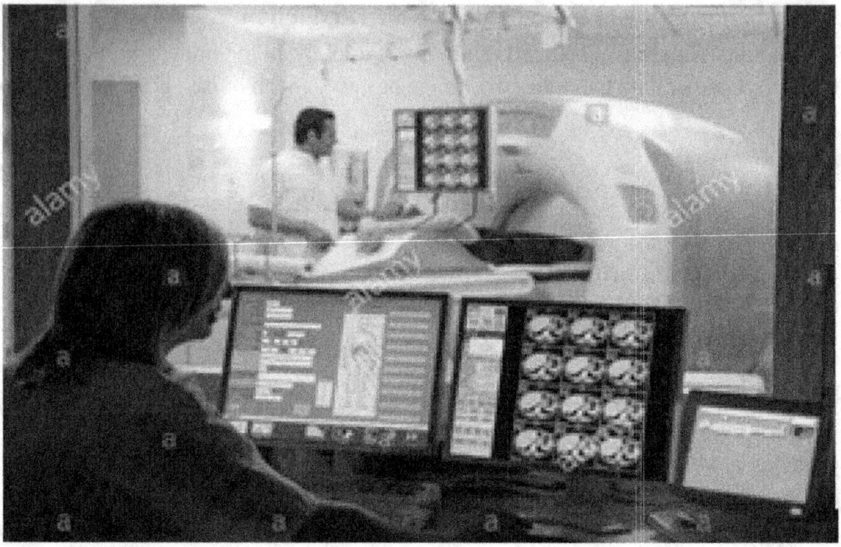

FIGURE 10.2 MRI Imaging Suites.

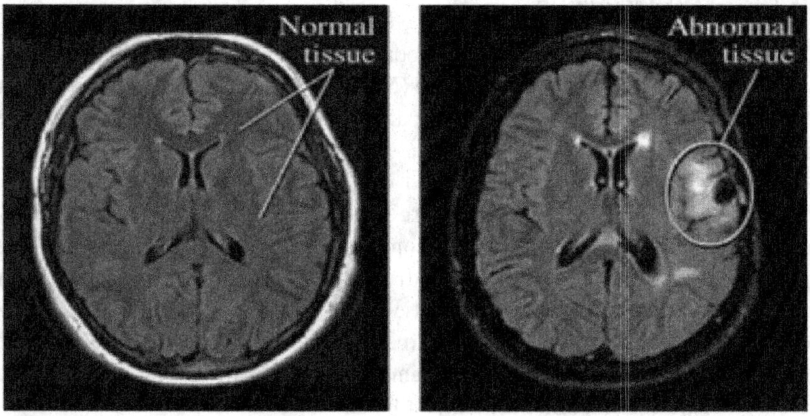

FIGURE 10.3 MRI Brain Image.

by the scanner. In different rates, the proton that is present on tissues returns to the proper state; hence, it is possible to make a difference between different tissues. The picture from the MRI suites is outlined in Figure 10.3.

Magnetic Resonance Image of Brain

MRI is a refined therapeutic imaging system giving delicate human tissues life structures. It delivers three-dimensional learning with high qualifications between delicate tissues. Be that as it may, the quantity of learning is the top measure for manual examination translation, and this has been one of the most critical obstructions inside the successful utilization of MRI. Thus, programmed or self-loader methods of PC-aided picture examination are essential. The division of MRI into totally extraordinary tissue classification, especially dark issue GM, WM, and CSF, is a fundamental assignment. Brain MRI has an assortment of choices, especially accompanying in the first place, MRI is, in principle, piecewise, consistent with a small, low assortment of classes. Also, they require sensibly high qualifications between totally extraordinary tissues. The refinement in MRI relies on the strategy that the picture is gotten by fluctuating radiofrequency and gradient pulses by choosing relaxing timing. It's feasible to get the image with high segmentation images. These two alternatives encourage division.

Types of MRI Scan

MR pictures can be gained by utilizing diverse methods. The subsequent images have diverse properties of the delineated materials. The most well-known weightings are T1 and T2, which feature the properties of T1 relaxation and T2 relaxation separately. Determination of the most suitable weighing is vital for a fruitful segmentation.

T1-Weighted Images

T1 images indicate high differentiation between tissues having distinctive T1 relaxation times. Tissues with long T1 relaxation times emanate minimal signals and along these lines, they will dim in the subsequent images. T1 images air. Borne and CSF have low intensity. T1 images have high differentiation between WM and GM. If the CSF is bright (high signal, it must be a T2-weighted image. If the CSF is dull, it is a T1-weighted image (Figure 10.4). Fluids are mid-dark, dull, water-related issues, and the fat-related tissues are bright. Clear edge can be seen in between the different tissues, with the goal that they are called 'anatomy scans'. The pathology having protrusion or many vessels may be darker compared to conventional tissues, where fatty spots will have high signal intensity.

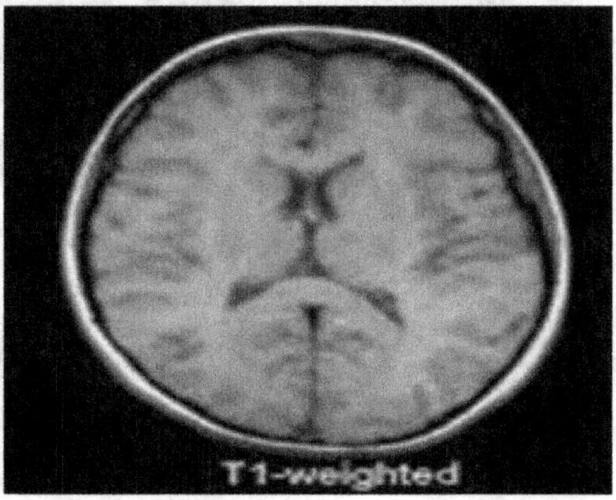

FIGURE 10.4 T1-Weighted Image.

T2-Weighted Images

In T2 images (Figure 10.5), ordinary gray and white tissues are dim and have comparative intensities. CSF is bright, white born, air and fat appear to be dull. As restriction T1 images, T2 images have high qualification amongst CSF and bone.

Flair Images

Fluid attenuated inversion recovery (FLAIR) (Figure 10.6) relates reversal pulse arrangement acclimated the signal from liquids. Fluid arrangement has replaced the PD image. As inside the instance of the inverse reversal recovery arrangements, related imaging succession of the brisk turn reverberate sort is appealing to repay the long obtaining time connected to TR. These grouping region units are routinely utilized in cerebral imaging for edema imaging. By critically choosing the reversal

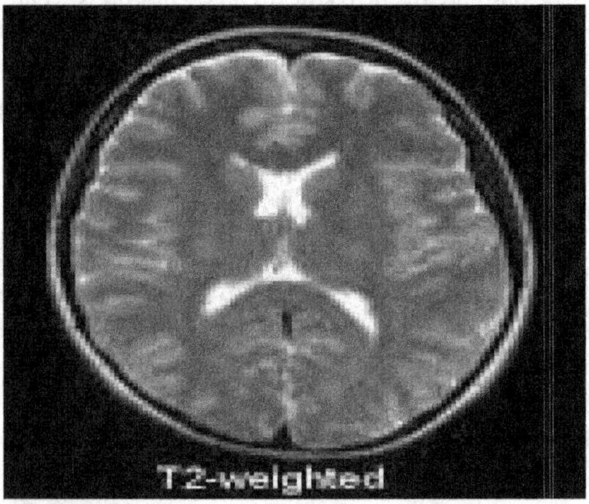

FIGURE 10.5 Weighted Images.

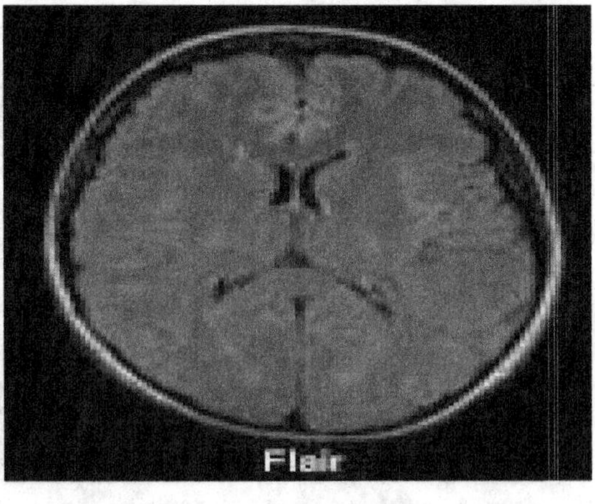

FIGURE 10.6 Flair Image.

time (the time between the reversal and execution beats), the signal from any express tissue will be suppressed. FLAIR imaging of the brain has built up a repetitive tool for measure lesions in patients with examined neurological disorders.

Multi Planar Reconstruction (MPR) Image

Multi planar reconstruction (MPR) images help us utilize the anatomy of the human brain in all possible axes. MPR is nothing but the post-processing reformation of the 3D data sets into 2D slices of random thickness at any angle view. For better resolution, thin slice embodied with the isotropic vowel and overlapping slices with good signals is advantageous.

Contrast enhancement is induced for the aforementioned image sequences to have a clear view of the pathologies present in the brain. Contrast enhancing agent GADD IV (Gadolinium IV) is used extensively in the previously mentioned MRI sequences. Normally, contrast agents favor a clear-cut examination of the pathologies.

MRI Sectional Anatomy of the Brain

Three basic stages of imaging used as part of neuroimaging are a print stage: photos talk about "cuts" in the body. Photos of the arrows were taken opposite the axial plane between the left and right sides (side view). Photographs of the coronal plane were taken against the middle plane, which separated the front from the rear.

In the coronel read, the subsequent regions of the brain are known. The regions are the nerve tract, caudate (head), putamen, septum pellucid, ventricle and lobe. Once viewing a coronel image, we can read the image as if we have a chance to view the substance of the patient. Thus, the spectator's right is that aspect of the brain inside the imaging and according to the left feature of the patient cerebrum. Conversely, the spectator's left is that the right facet of the brain within the imaging and therefore the right features of the patient's brain. The sagittal read of brain MR images shows the intense mass with bone invasion and compression against the membrane cortex. The varied imaging planes are delineated within Figures 10.7 and 10.8.

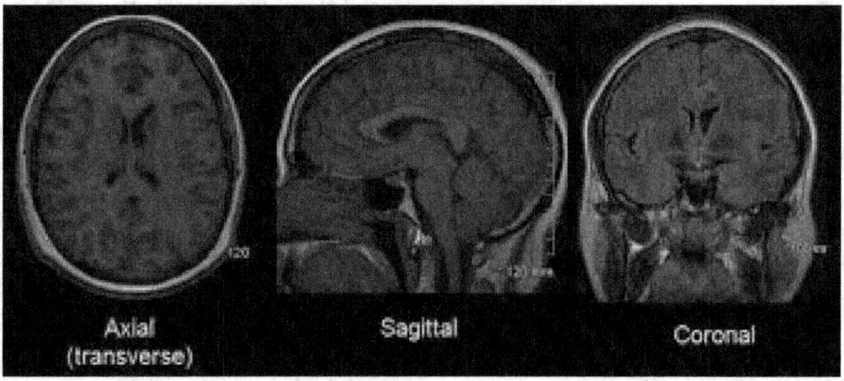

FIGURE 10.7 Imaging Planes of the MRI Brain Image.

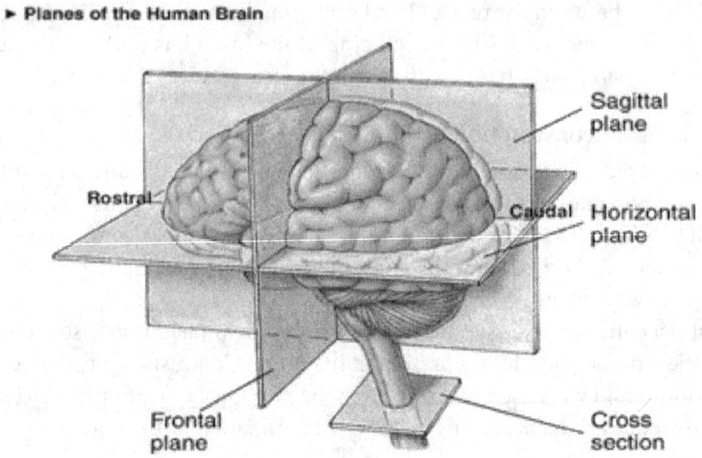

FIGURE 10.8 Spatial Demonstration of the Human Brain.

10.2 EXISTING SYSTEM

In this setting, they propose a new multi-atlas segmentation system (MAS) to visualize MRI of the brain (Figure 10.9). The basic idea of MAS is to separate marked information from different atlases of the brain and combine them into a new image of the brain. Several successful MAS projects have been proposed. However, most of them are designed for general brain imaging, and brain imaging is often a major challenge for them. This is because the tumor causes problems in recording the normal atlas of the brain's shape. To address this challenge, a new low-quality method is used in the first phase of the MAS framework to identify the hidden image of the brain as shown by the MRI of the brain tumor. Unlike the traditional low-scoring method, which produces an image of coverage with a distorted area of the brain, the low-scoring method uses local barriers to obtain a coverage image that maintains a normal region of the brain. Then, in the second stage, the normal atlas of the brain can be recorded on a covered image, which does not affect the tumor. Both steps are performed repetitively to achieve the final division of the tumor's brain image. During repetition, the recovery image in the recovery icon and the recording of the normal brain atlas gradually improve. This proposed method has been compared to the current method in which images of the brain are used with real and artificial MRI tumors. Experimental results show that our proposed method can effectively recover the image and improve the partition status.

Figure 10.10 shows synthetic brain images: 10.10(a) shows iterations 1 and 6 of LRSD + MAS, 10.10(b) depicts iterations 1 and 6 of SCOLOR + MAS, 10.10(c) shows related images of a yellow rectangular composite image. The image shown in 10.10(d) is of a tumor-free brain. For each restored image shown on the upper line (b–c), a similar 3D version is shown on the bottom line. Images recovered with SCLOR + MS show higher quality of the recovered tumor area and better

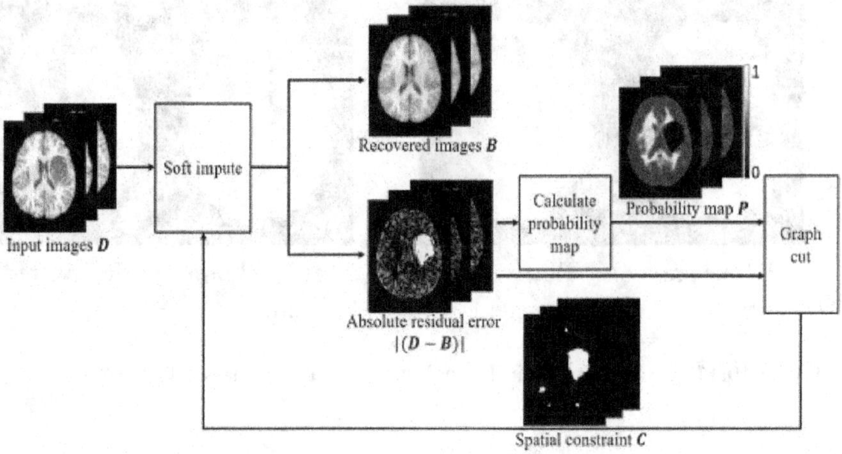

FIGURE 10.9 Flow Diagram of an Existing System.

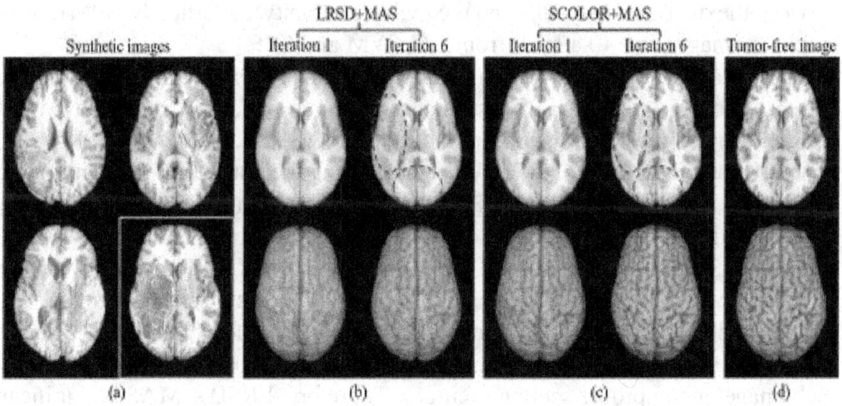

FIGURE 10.10 Images of Artificial Cut-Off Brain Image.

protection of normal areas of the brain than LRSD, mainly in areas where red features elliptical bars.

For the most success, 30 T1 MRI brain images were selected from the BRATS2015 database. Both images were removed due to poor image quality, resulting in a total of 28 experimental images (included in the appendix). Every second picture, there is at least one tumor in the brain. The mean tumor/brain ratio (constant deviation) is 0.059/0.037, which is similar to the size of most tumors in BRATS2015. Atlas 40 used in LRSD + MAS and SCOLOR + MAS is an image of a normal LPBA40 brain. Since there are no images without volume, it is not possible to calculate relationship errors to evaluate images before linking. In our experience, the display quality of recovered images taken with SCROR + MAS is better than LRSD + MAS. Especially in the

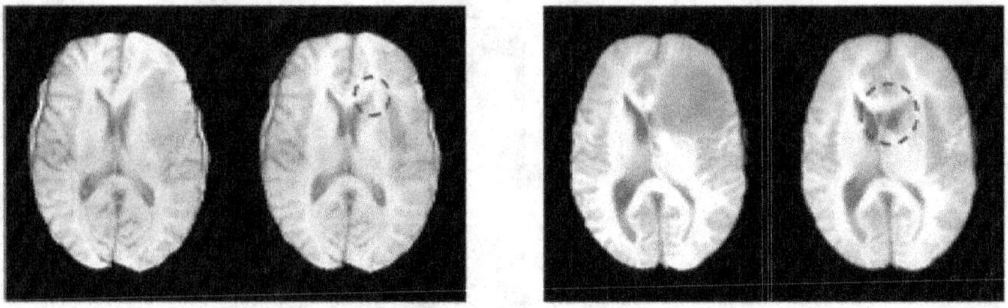

FIGURE 10.11 Example of Brain Tumor Image with Tumor Mass Effect and Similar Recovery Image.

area indicated by the red arrow, it can be seen that Scroll + MS generally restores the crop area more efficiently, without distorting the brain area, compared to LRSD + MAS. The accuracy of the segment is measured by the GM, WM and CSF data index between the distribution results and the region's veracity. Specifically, SPM12 is used in 28 test images and 40 atlases from GM, WM and CSF.

These class results have been manually modified by experts and are considered correct. The class results of 28 test images (without any classification-based image search) are also reviewed using the traditional MAS structure (indicated by ORI + MAS) (Gupta and Pahuja, 2017). When calculating the data code, keep in mind that all changes for GM, WM and CSF in the collection area are ignored when checking the section results. The average code for the 28 test images after each iteration was ORI + MAS, CFM + MAS, LRSD + MAS and SCOLOR + MAS (Krishnan, Rengarajan, and Manikandan, 2015). Combine 28 photos after repeating L, LRSD + MAS and SCOLOR + MAS 5 and 4. First and again ORI + and more. Improving the alignment of atlases and primary images for each crop improves the quality of each image and improves section accuracy. Therefore, LRSD + MAS is significantly better than ORI + MAS in the second iteration.

The misalignment of the right ventricle between the normal brain atlas and the image of the brain tumor creates artificial structures in the retrieval image, as indicated by the red dotted circle in Figure 10.11.

The limitation of our procedure is that the area of the tumor looks relatively different from the normal brain area on the MRI of the brain. Otherwise, no growing spots will be detected, and the color will rot using traditional low-grade methods. Also, scanning for large-scale crop effects affects the process.

10.3 PROPOSED SYSTEM

10.3.1 Concept Involved in Proposed Work

The proposed architecture includes nine blocks for allocating separate blocks of memory starting from preprocessor stage to final algorithm implementations. The preprocessing stage is used to get the samples of value from the **CANCER signal,** which is given by the **CANCER main processor module**. For the denoising feature extraction and classification, we are using the DHT module, **ANFIS module**, and CANCER norm selections modules. The SRAM and mux modules are used to hold the process operations and select the signal conditions, respectively. The proposed modules are used to create the block of the stationary wavelet transform (SWF), which is used for noise reduction of the CANCER signals. Cache memory is high-speed memory that can be used to speed up data processing. A CPU case is a case in which the central processing unit of a computer is used to reduce access to average memory. Cache is a small, fast memory that stores copies of data with frequently used critical memory.

10.3.2 There Are Different Levels of Cache

1. The L1 case is the fastest and is usually integrated into the processor board. The L1 cache typically range from 8 kB to 64 kB and use fast SRAM (fixed memory) instead of the slow and cheap tram (dynamic memory) used in central memory. The Intel Celeron processor uses two temporary 16 kB L1 memories for instructions and data.
2. The L2 caches are between L1 and RAM (L1-L2 RAM processor) and are larger than the initial cache (typically 64 kB to 4 MB).
3. The L3 cache was not recently discovered because its functions were taken over by the L2 cache. The case is located on the motherboard, not on the L3 processor, and is stored between the RAM and L2 cache. If your computer has L1, L2 and L3 memory, data-L1-> L2-> L3-> RAM will be reset. If L1 does not have data, look at L2, L3 and RAM.

The functional block diagram in Figure 10.12 shows the general architecture utilized in FPGA for CANCER signal classifications.

Waves work in time and space and are intermittent. They focus energy on time and space and are suitable for short-term signal analysis. Foyer's change and STFT use waves to analyze the signal, while wave change uses infinite energy waves.

CANCER is the most common process for heart-rate monitoring. CANCER records the electrical activity of the brain through electrodes that are placed on the skin over some time, and from it the contraction (depolarization) and relaxation (repolarization) of the heart can be seen and measured in the form of waves. CANCER is widely used around the world in the medical industry for identifying and the continuous monitoring of several heart diseases and disorders.

The overall objective of CANCER is to obtain information about the function, structure, and condition of the heart. It helps in the detection of several diseases, such as suspected pulmonary embolism, cardiac murmur, cardiac stress testing and so on. The normal CANCER signal reading can be observed in Figure 10.13. The CANCER is explained in terms of five intervals: P, Q, R, S and T. These intervals describe a deflection, that is, brain activity, rhythm and morphology. A normal CANCER wave with a normal heart rhythm consists of a P wave, a QRS complex and a T wave. The P wave represents the continuous decline of the left and right atria. The QRS reflects

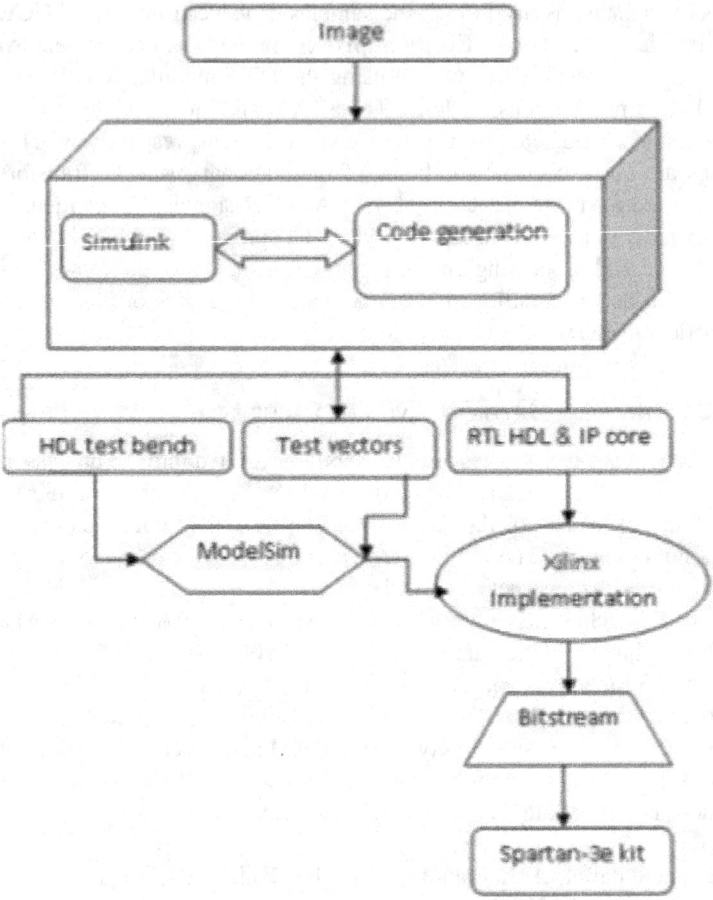

FIGURE 10.12 Functional Block Diagram of FPGA in CANCER Signal Classification.

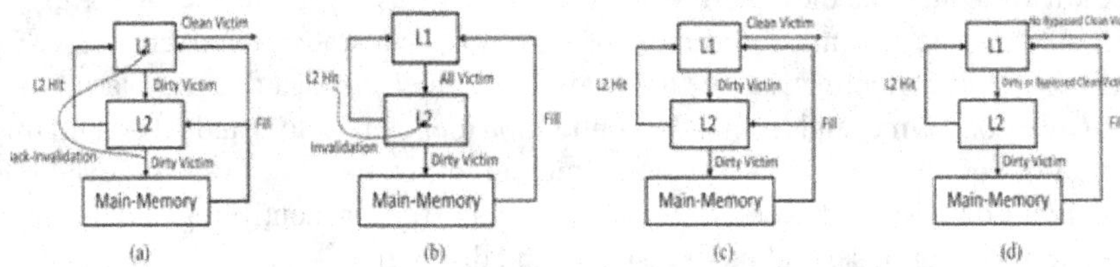

FIGURE 10.13 Memory Utilization in FPGA for CANCER Victim Identifications.

the depolarization of the left and right ventricles. It lasts for 70–110 milliseconds in general; the heartbeat has the largest amplitude of the CANCER waveform, which can be seen in Figure 10.14. The T wave represents the ventricular repolarization and about 300 milliseconds extension after the QRS wave complex. The positioning of the T wave is mainly dependent on brain activity.

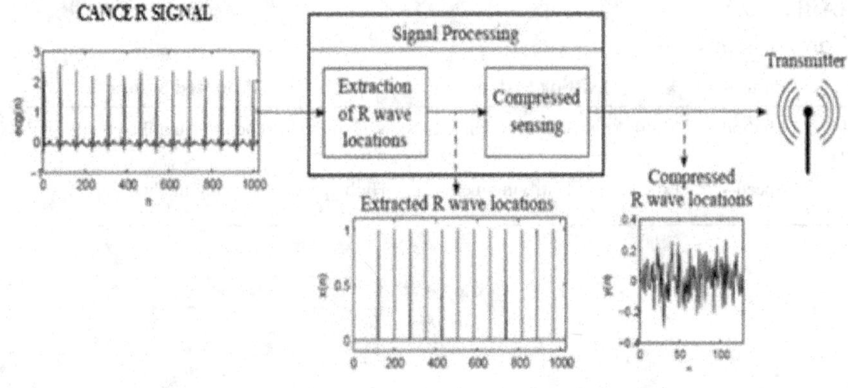

FIGURE 10.14 CANCER Signal Analyses.

10.3.3 RBF Neural Network

The radiation functional network (RBF) solves the problem of curve fitting, which is a broader approach. In this case, training is equivalent to finding a dispersion surface in a multidimensional space that best fits the motion data, which is measured according to pre-selected data criteria. There are three layers called input, hidden/primary layer and output point. Each hidden unit represents the function of a single radial base with the position and width of the integrated center. Such hidden units are sometimes called centrodes or nuclei. Each output unit summarizes the hidden units by weight. Exercise is usually done in two stages, after which the width and center are determined, then the weight.

The project uses a new RBF neural network proposed by (Krishnan, Julie, and Robinson, 2020). The following steps illustrate the algorithm for creating RBF neural networks. First, segregate the entire data into two parts: training data (V) and test data (Vt). Then craft a replica of V called Vc. It is used to select the center of the hidden cover. Then randomly select the VKX model as the center of the cover. Find all models in the neighborhood of xk, that is, all models with a distance of xk are fewer than δ. Those patterns form the nucleus. Next, examine the clarity of the core, that's the relationship involving the number of standards in the identical basic classes and the entire number of standards in that core. But if it exceeds a certain limit, its purification is possible.

Otherwise, as long as the purity is not more than 1, the basic one will be qualified. When a certified core is developed, all samples of this core cover are transferred from V. Continue building the core until all samples are transferred from V. The choice of all centers will not be between Vc. Thus, the algorithm allows a large overlap between nougat of the similar classes. A minute overlap between seeds improves normality in cases that do not overlap. However, too much overlap (between different classes of pips) leads to less precision. The overlap between PPS in the same class does not compromise health and at the same time helps to improve the granularity and sturdiness about the RBF network. Table 10.1 compares the accuracy and speed of the MAS and RBFNN methods. Figure 10.15 depicts the accuracy comparison of both methods.

TABLE 10.1
Comparison Table.

S. No	Parameter	Existing System	Proposed System
1	Method	Multi-Atlas Segmentation (MAS)	Radial Basis Function Neural Network (RBFNN)
2	Accuracy	89%	93%
3	Speed	Low-speed identifying method	High-speed identifying method

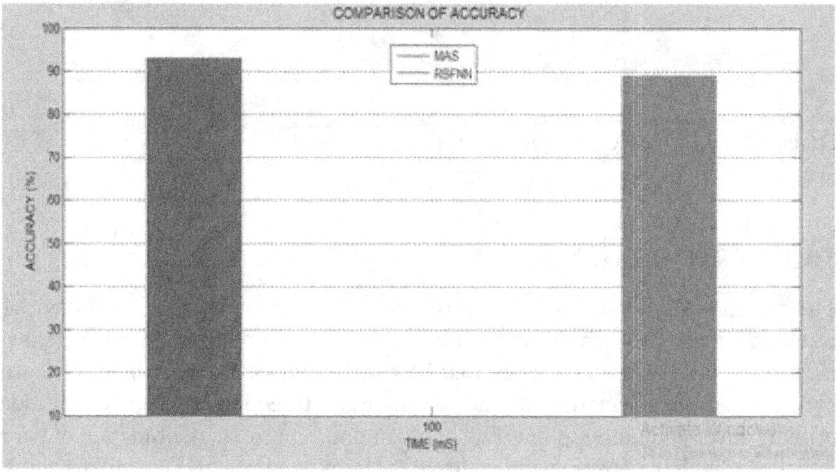

FIGURE 10.15 Comparison Chart.

To get the nucleus to the correct value, put the standard deviation of the entire pattern of movement at the starting worth of (δ0). Once you have the entire course, define the standard deviation for each core, as there is a center on each core. Finally, the weight between the hidden core and the output is obtained using linear square lines.

Evaluate the proposed RBF neural network; the approach allows huge overlap between pips about the identical classes. This helps this overlay set-up to exclude "noise" as well as improve the accurateness of classification surrounded by patterns belonging to different layers, as shown in Model A (where A is "noise"). If you make your core part according to the traditional methods, Model A will extract the smallest particles. However, if large overlaps are allowed between pips of the same class, a larger core is formed. This eliminates the effects of large basic Model A (noise). In addition, they could diminish the quantity of concealed cores and develop network-building performance.

10.4 EXPERIMENTS AND RESULTS

The proposed structure was validated in three applications: fetal magnetic resonance imaging, 2D distribution of different organs and distribution of 3D portions of brain tumors with T1c imaging (T1c) and FLAIR. For these applications, we first tested the accuracy of hidden objects that were not in the training set.

10.4.1 Architecture Design and Analysis

The SRAM has only one input signal and two control signals, clk (clock) and en (enable). The address register is used to tell the RAM where data can be going to write. ANFIS cancer analysis is a disruptive technology that can revolutionize product performance in many areas, from consumer electronics and personal computers to automotive, medical, military and aerospace equipment.

Figure 10.16 shows the CANCER classification and extraction module, which is designed by using the Xilinx 12.3 synthesizer tool. For portable and mobile applications, you can eliminate the MCP (multi-chip package), provide a single memory subsystem, reduce system power consumption and extend battery life. On a private computer, you can replace high-speed caches for running high-speed programs, flash for unstable caches, and SR SRAM for PSRAM and DRAM.

With a recording time of 2 nanoseconds (ns), STT RAM is as fast as SRAM. SRAM currently has a recording time of 1–100 ns, depending on the technology used. In terms of cell size, STM RAM is much larger than SRAM. When the STT-RAM reaches the 32 nm technology node, the cell is under DRAM or NOR memory.

Figure 10.17 shows the cache memory design for high-speed CANCER analysis. Cache memory is high-speed memory that can be used to speed up data processing. A CPU cache is used by a computer's central processing unit to reduce access to average memory. A cache is small, fast memory that stores copies of data in places with frequently used critical memory.

The L1 cache can be deactivated to save power over the ANFIS cycle. The L1 instruction box is typically designed with high-performance (HP) cells to achieve low-access delays at high standard power-consumption costs. The static power of the L1 instruction temporary storage may be greater than or equal to the standard power of the L2 cache. Therefore, despite the relatively small size, it is very important to reduce the standard power consumption of the instruction cache.

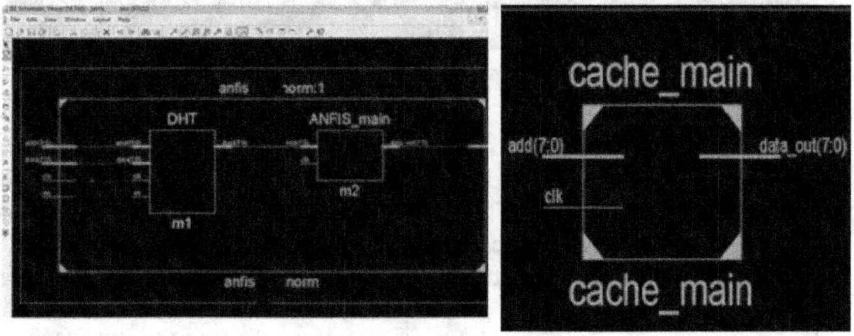

FIGURE 10.16 ANFIS CANCER Classification Block with DHT Module and Cache Memory.

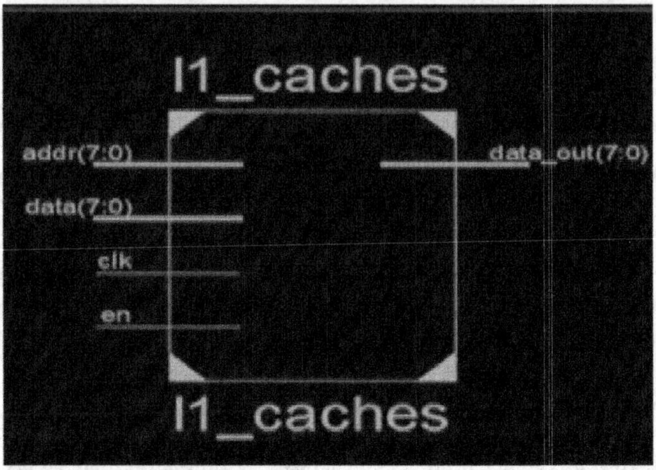

FIGURE 10.17 L1 Cache for ANFIS Main.

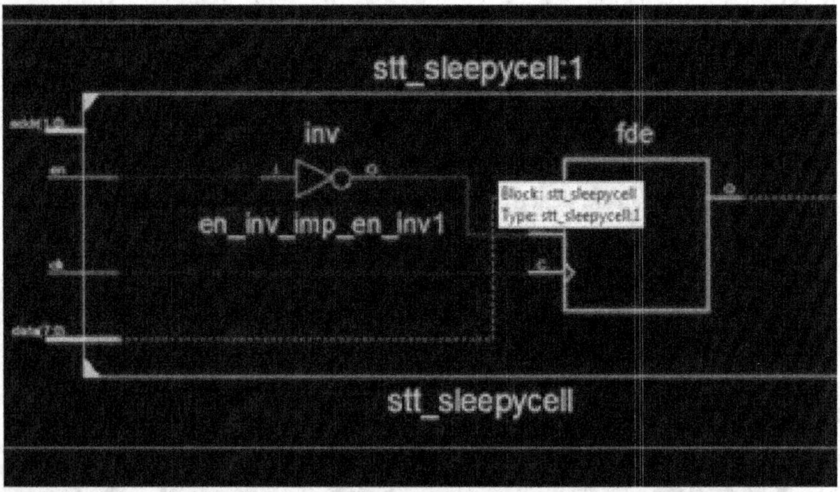

FIGURE 10.18 Stt_sleepy Cell.

Stt_sleepy cell L1 is used to reduce the static allocation of cache power according to DHT specifications (Figure 10.18). Using this concept, the STT-RAM cache is an ideal target for using power-to-heat technology. Traditionally, the SRAM card has been plagued with a trade-off between maintaining the status quo and minimizing leakage.

Due to variations in functions, the supply voltage of the SRAM cell level must be carefully adjusted. With a generator, the STD RAM cache can leak simultaneously and significantly reduce workloads.

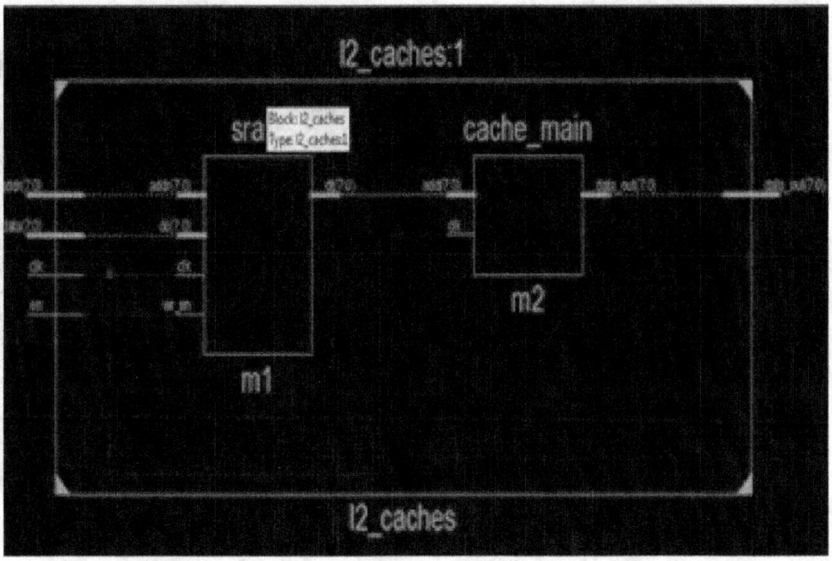

FIGURE 10.19 L2 Loop Basic Cache Memory.

Cache L2 uses low-power (LP) cells with integrated static loop chips (Figure 10.19). This L2 design saves the cancer processor time and can significantly reduce power consumption while minimizing the impact on performance and cache power.

When the cancer processor performs a loop, the L2 instruction box prompts you to repeat. If you can add the entire body of the loop by adding a small buffer to the loop cache, you can follow the instructions in the loop body in the buffer instead of the L1 statement.

This allows the L1 instruction cache to be disabled when it repeats less than the capacity of the loop cache, which consumes a significant portion of program execution.

Figure 10.20 intimates the group of architecture in CANCER analysis of Spartan 3E FPGA, which includes nine blocks for allocating separate blocks of memory starting from the preprocessor stage to the final algorithm implementations. The preprocessing stage is used to get the samples of value from the CANCER signal, which is given by the **CANCER main processor module**. For de-noising feature extraction and classification, we are using the **DHT** module, the ANFIS module (Hebli and Gupta, 2016) and the CANCERnorm_selections modules. The SRAM and mux modules are used to hold the process operations and select the signal conditions, respectively. The proposed modules are used to create the block of the stationary wavelet transform (**SWF**), which is used for noise reduction of the CANCER signals.

Figure 10.21 represents the inner synthesized architecture of loop aware static selections of CANCER signal classification, i.e., CANCER normal and above normal distribution result. Figure 10.22 intimates the different levels of CANCER signal analysis, which starts from the CANCER sampling signal to CANCER normal/up to normal analysis. Figure 10.23 illustrates the timing and area analysis of proposed the architecture.

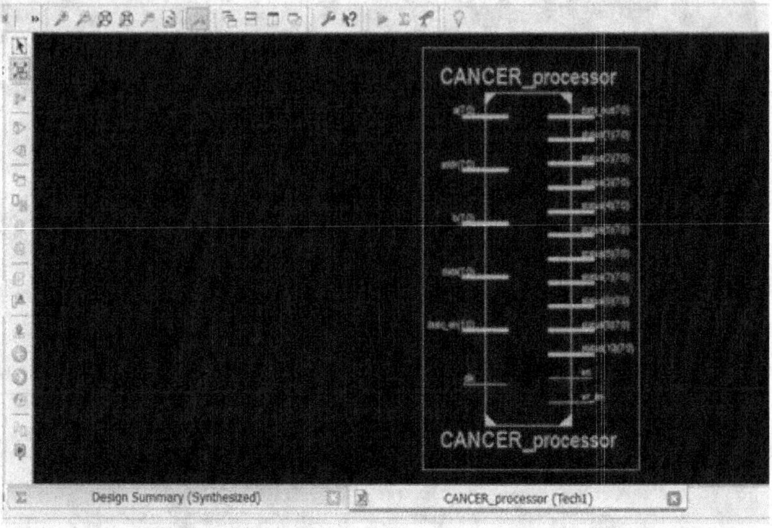

FIGURE 10.20 Final Architecture of CANCER_in Spartan 3E FPGA.

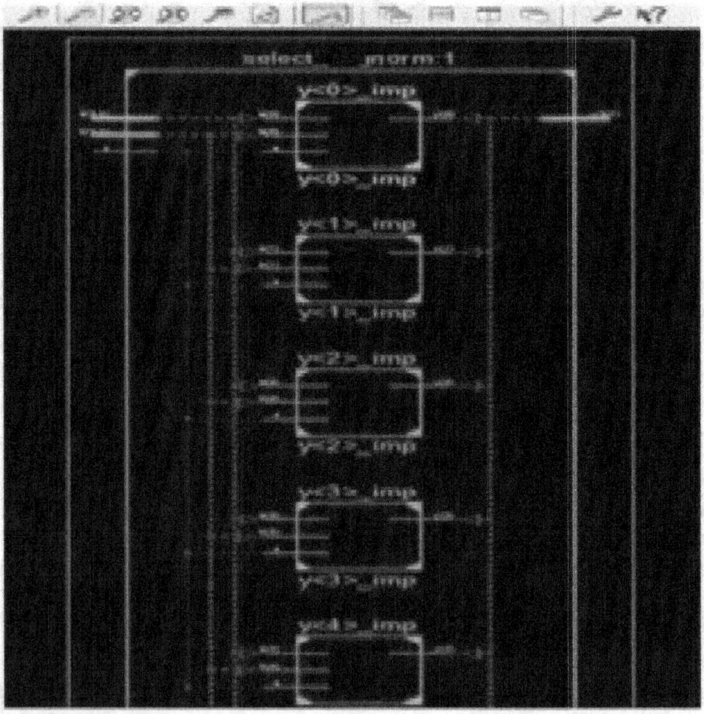

FIGURE 10.21 Inner Architecture for Classification of CANCER Signal.

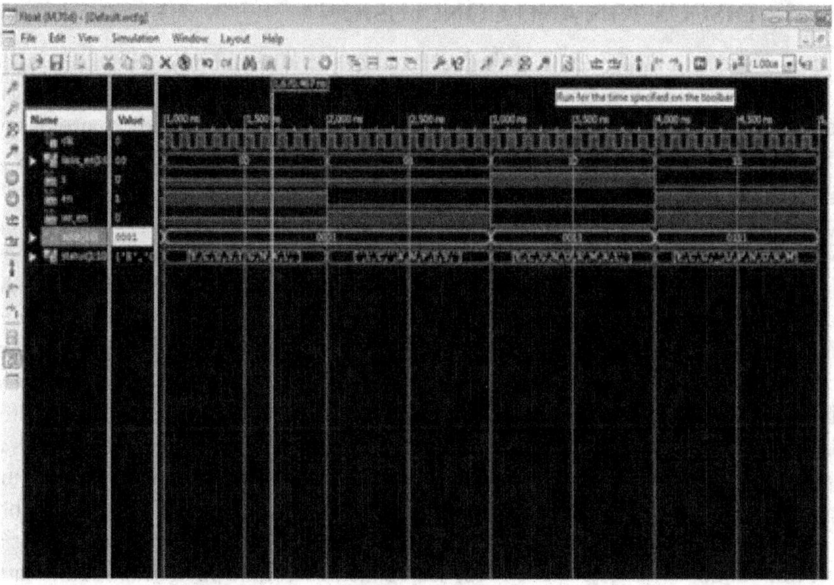

FIGURE 10.22 Simulation Results of the Proposed Architecture.

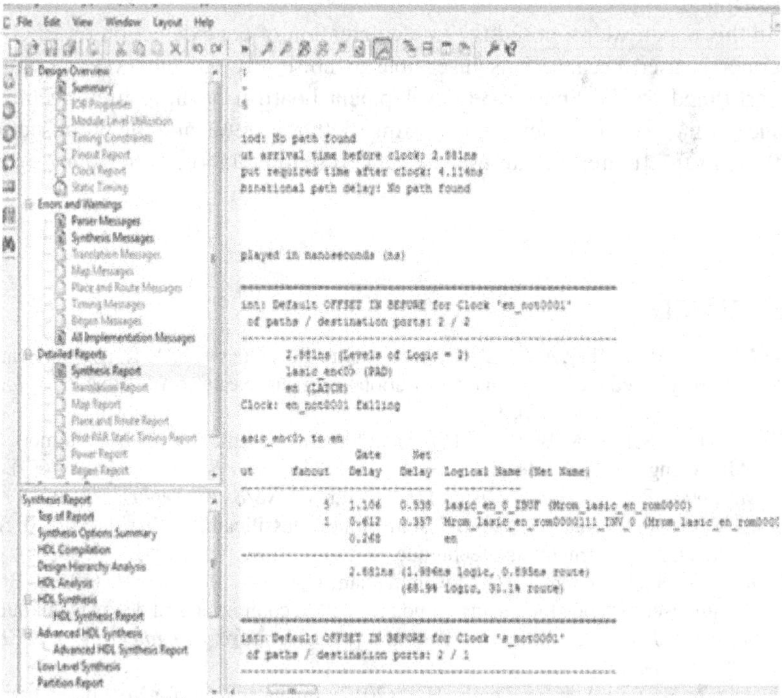

FIGURE 10.23 Timing and Area Analysis of Proposed Architecture.

10.5 DISCUSSION AND CONCLUSION

Tumors grow in the brain. The lingual band of the human body, formed by uncontrolled growth and timely cell extraction, is known as a tumor or nodule. The normal metabolic function of the human body is affected by crops. A medical professional could lead patients to live with remarkable health. Additionally, cancerous cell growth along with the tumor cells in the brain is considered a tumor. Stable cancer cells did not develop into benign brain tumors, and contrasting malignant tumors were never analyzed in active cancer cells. To trace out the presence of tumor cells in the part of the tissue under analysis is possible with the assistance of medical imaging techniques automated with ML processes. MRI is a well-known technique for tumor analysis, the process of detecting tumors after the identification of tumor cells. Predicting the affected area using MRI is a slow process, and this analysis is suitable for improving existing automated interaction models. ML algorithms help doctors identify areas affected by tumors. The dissimilarities in the mass and shape of the tumor cells do not affect the prediction rate of the systems. Accurate extraction of the region of interest in terms of the white and gray matter of brain tissues and infected tumor location is made with the retrieval of feature vector and minimization of search space with less exactness value in tumor identification. Information-based visualization techniques drive significant character in medical examination for health issues. In clinical fields, to identify different tissue levels and detect abnormalities, MRI is helpful, and a dimensionality-rooted resolution report is generated.

In this paper, we proposed the implementation of edge detection on FPGA. This operator is used because it is insensible to noise and reduces system complexity. It performed on the Spartan-3e development board to achieve high accuracy and reduced time. The accuracy of an existing method is 89%, and the proposed method is 93%. In the future, we can use the cuckoo search algorithm to improve accurate results.

REFERENCES

Abdel-Basset, M., Fakhry, A. E., El-Henawy, I., Qiu, T., & Sangaiah, A. K. (2017). Feature and intensity based medical image registration using particle swarm optimization. *Journal of Medical Systems*, *41*(12), 1–15.

Alfonse, M., & Salem, A. B. M. (2016). An automatic classification of brain tumors through MRI using support vector machine. *Egyptian Computer Science Journal*, *40*(3).

Aly, M. (2006). Face recognition using SIFT features. *CNS/Bi/EE Report*, *186*.

Anitha, V., & Murugavalli, S. (2016). Brain tumor classification using two-tiered classifier with adaptive segmentation technique. *IET Computing. Vis*, *1*, 9–17.

Apirajitha, P. S., Krishnan, C. G., Swaminathan, G. A., & Manohar, E. (2019). Enhanced secure user data on cloud using cloud data centre computing and decoy technique. *International Journal of Innovative Technology and Exploring Engineering (IJITEE)*, *8*(9), 1436–1439.

Bahadure, N. B., Ray, A. K., & Thethi, H. P. (2017). Image analysis for MRI based brain tumor detection and feature extraction using biologically inspired BWT and SVM. *International Journal of Biomedical Imaging, 2017*, 1–12.

Gomathi, S., & Krishnan, C. G. (2020). Malicious node detection in wireless sensor networks using an efficient secure data aggregation protocol. *Wireless Personal Communications, 113*(4), 1775–1790.

Gupta, A., & Pahuja, G. (2017, August). Hybrid clustering and boundary value refinement for tumor segmentation using brain MRI. In *IOP Conference Series: Materials Science and Engineering* (Vol. 225, No. 1, p. 012187). IOP Publishing, Secunderabad, India.

Hebli, A. P., & Gupta, S. (2016, November 20). Brain tumor detection using image processing: A survey. In *Proceedings of 65th IRF International Conference*. International Research Forum for Engineers and Researchers (IRF), Pune, India.

Islam, A., Hossain, M. F., & Saha, C. (2017, September). A new hybrid approach for brain tumor classification using BWT-KSVM. In *2017 4th International Conference on Advances in Electrical Engineering (ICAEE)* (pp. 241–246). IEEE, Dhaka, Bangladesh.

Kharrat, A., Benamrane, N., Messaoud, M. B., & Abid, M. (2009, November). Detection of brain tumor in medical images. In *2009 3rd International Conference on Signals, Circuits and Systems (SCS)* (pp. 1–6). IEEE, Medenine, Tunisia.

Kothari, A., & Indira, B. (2016). An overview on automated brain tumor segmentation techniques. *International Journal of Computer Trends and Technology, 40*, 2231–2830.

Krishnan, G. C., Nishan, A. H., Prasannavenkatesan, T., Jeena Jacob, I., & Komarasamy, G. (2021). Two dimensional and gesture based medical visualization interface and image processing methodologies to aid and diagnose of lung cancer. In J. Nayak, M. N. Favorskaya, S. Jain, B. Naik, & M. Mishra (eds.), *Advanced Machine Learning Approaches in Cancer Prognosis. Intelligent Systems Reference Library* (Vol. 204). Springer, Cham.

Krishnan, C. G., Rengarajan, A., & Manikandan, R. (2015). Delay reduction by providing location based services using hybrid cache in peer to peer networks. *KSII Transactions on Internet & Information Systems, 9*(6).

Krishnan, C. G., Robinson, Y. H., & Chilamkurti, N. (2020). Machine learning techniques for speech recognition using the magnitude. *Journal of Multimedia Information System, 7*(1), 33–40.

Krishnan, C. G., Sivakumar, K., & Manohar, E. (2018, December). An enhanced method to secure and energy effective data transfer in WSN using hierarchical and dynamic elliptic curve cryptosystem. In *2018 International Conference on Smart Systems and Inventive Technology (ICSSIT)* (pp. 1–7). IEEE, Tirunelveli, India.

Nandi, A. (2015, November). Detection of human brain tumour using MRI image segmentation and morphological operators. In *2015 IEEE International Conference on Computer Graphics, Vision and Information Security (CGVIS)* (pp. 55–60). IEEE, Bhubaneswar, India.

Prasannavenkatesan, T. (2021). Forecasting hyponatremia in hospitalized patients using multilayer perceptron and multivariate linear regression techniques. *Concurrency and Computation: Practice and Experience*. Springer, e6248.

Rajendran, A., & Dhanasekaran, R. J. P. E. (2012). Fuzzy clustering and deformable model for tumor segmentation on MRI brain image: A combined approach. *Procedia Engineering, 30*, 327–333.

Rajendran, A., & Dhanasekaran, R. J. P. E. (2014). Brain tumor segmentation on MRI brain images with fuzzy clustering and GVF snake model. *International Journal of Computers Communications & Control, 7*(3), 530–539.

Ruby, A. U., Prasannavenkatesan Theerthagiri, D. I., & Vamsidhar, Y. (2020). Binary cross entropy with deep learning technique for Image classification. *International Journal, 9*(4).

Theerthagiri, P., Jeena Jacob, I., Usha Ruby, A., & Yendapalli, V. (2021). Prediction of COVID-19 possibilities using k-nearest neighbour classification algorithm. *International Journal of Current Research and Review, 13*(6), 156.

Zhang, S., & Xu, G. (2016). A novel approach for brain tumor detection using MRI Images. *Journal of Biomedical Science and Engineering, 9*(10), 44–52.

11 A Deep Learning and Multilayer Neural Network Approach for Coronary Heart Disease Detection

Seema Rani, Neeraj Mohan, Surbhi Gupta, Priyanka Kaushal, and Amit Wason

CONTENTS

11.1 INTRODUCTION

The heart is a hollow muscle organ used to pump blood through the circulatory system by repeated contraction and dilation. We can also describe the heart as a combination of nerves and muscles used to pump blood in the human body. The heart renders approximately 50% of the cardiac myocytes (Steinhauser 2011). Unexpected death may be the outcome of several defects or failures of the heart, such as myocardial

DOI: 10.1201/9781003184140-11

infarction, when the cardiac muscle is permanently damaged. Myocardial infarction is made of three words: "myo," meaning muscle; "cardial," meaning heart and "infarction," meaning the death of nerves because of blood-supply deficiencies.

11.1.1 Structure and Working of Heart

The heart has four chambers, on both atria and on the right ventricles. The heart is a beefy organ that constantly drives blood throughout the body. There are two different pumping systems and the blood cannot move easily between them. The function of the right atrium is to bring blood into the heart. In Figure 11.1, the top left shows the right atrium.

When the blood supplies oxygen to our tissues and body parts, the blood itself gets deoxygenated. Subsequent to this, blood is flowing down into the ventricle of the right atria, from the right atria via the right atrioventricular valve, through the upper and lower vena cava. Valves are a part of the heart that act as a kind of locking system that stops blood from flowing in reverse. The blood is pulled up to the respiratory system via the pulmonary arteries and goes from the right ventricle through the pulmonary valve.

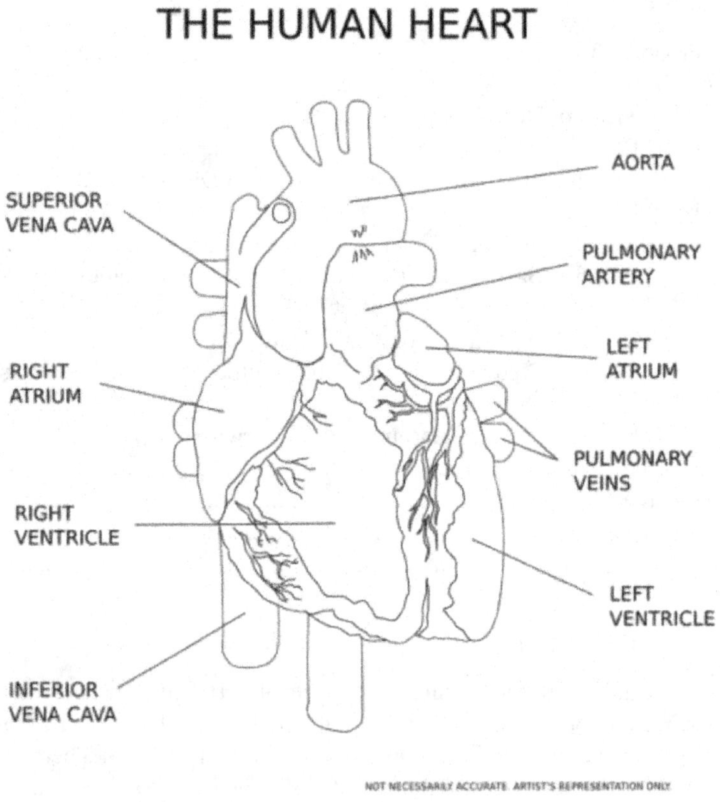

FIGURE 11.1 Structure of the Human Heart.

In the lungs, our blood is oxygenated. The rectal ventricle contraction is insufficient to send the oxygenated blood around after the lungs, so for the second pump the heart must return the oxygenated blood. When the oxygenated blood in the pulmonary veins reaches the heart, it passes to the left ventricle and continues through the bicuspid valve.

Figure 11.1 shows how the left ventricle muscle (depicted in pink) is denser than the right ventricle muscle. The reason behind this is the extra force that is needed to pump blood around the whole body, not just the lungs.

Blood passes by the aortic valve via the left ventricle and leaves the heart from the aorta. The aorta is an important blood vessel that adjusts the blood from the heart to other parts of body. The aorta has elastic properties that permit it to spread into the blood under high pressure, which results in an inoperative contract. Once the blood has returned through the vena cava, it is deoxygenated and the journey starts again (Iaizzo 2009).

We use the artificial neural network (ANN) combined with deep learning (DL), to detect cardiovascular disease. The following paragraphs provide a brief introduction to this approach.

11.1.2 Types of Heart Diseases

Cardiovascular diseases (CVD) involve both the heart and blood vessels. The growth of ischemic heart diseases (IHD) is due to cardiovascular disease, which includes such myocardial infarctions as heart attacks and angina. Coronary heart disease (CHD) develops plaque (a waxy substance) within the coronary arteries. Blood rich in oxygen is supplied to the heart by these arteries. The disease is called atherosclerosis when the plaque starts to grow up in these arteries. The plaque develops over a number of years. This plaque may tighten or rupture over time. The function of the coronary arteries is reduced due to hardened plaque, reducing the flow of oxygenated blood to the heart. Blood clots also develop because of plaque breakage. Blood circulation through a coronary artery is blocked due to large coagulation. The broken plaque hardens and shrinks the coronary arteries. The segment of the heart muscle starts to die if blood flow is not restored soon. A cardiac infarction can lead to severe issues and even death without prompt treatment. Heart attacks are a worldwide prevalent cause of death. The following are some of the prevalent symptoms of cardiac attack:

1. **Chest pain:** This is the most prevalent heart attack symptom. When a person has a blocked artery or a heart crash, he or she can experience pain, strain or chest pressure.
2. **Heartburn, nausea, stomach pain and indigestion:** These cardiac symptoms are frequently ignored. They are more common in women than in men.
3. **Arm pain:** Starting from the chest, pain goes to the arms, mainly to the left.
4. **Sensation of dizziness and slightly faint:** These symptoms result in loss of balance.
5. **Fatigue:** Doing simple tasks results in a sense of tiredness.
6. **Sweating:** The act of releasing liquid from the skin by the sweat glands.

Other prevalent cardiovascular diseases include peripheral artery disease, hypertensive heart disease, stroke, venous thrombosis cardiac failure, heart rheumatism, aortic aneurysm, congenital heart disease and valvular cardiac disease

11.1.3 Risk Factors Causing Heart Disease

Different types of heart disease are there, but still there are some common risk factors that help us to decide whether a person is at risk of acquiring heart disease. Here, we will describe the common risk factors of heart disease to clearly understand the problem. After that, we will discuss the specific conditions of heart disease.

There are various risk factors that can lead to heart disease; some can be kept under control by living a healthy lifestyle and taking precautions, but still some risk factors are out of a person's control. Some of the uncontrollable risk factors that may lead to heart disease include:

1. **Gender:** Males are more prone to heart disease than females. Because of this, females often believe they cannot acquire heart disease. But heart problems are a major cause of death for females, just as they are for males. Hence, males and females should both take proper precautions to avoid heart disease.
2. **Age:** It is very natural that with increasing age the risk of getting heart disease also increases. Among the total number of human deaths due to coronary heart disease, 80 percent of victims fall in the age group of 65 years and older. In fact, as the age increases, females have a higher chance of having a severe heart attack than males.
3. **Family History:** If an individual has a family history of high blood pressure, diabetes or any other kind of heart disease, the chances of that person acquiring heart disease increases. Persons with biological relatives who suffered from a heart attack before the age of 55 have an even higher chance of suffering from heart disease. The ethnic background of the person's family can also affect the chances of heart disease. For example, African Americans are at higher risk for high blood pressure, so the threat of heart disease also rises. If a family history of heart disease exists, there is an increased chance that an individual will develop a heart disease, but this does not mean that individual certainly has a heart disease.
4. **Obesity:** High blood pressure is often experienced by overweight individuals because their hearts have to work harder to pump blood. They are also more likely to have high cholesterol levels, which may result in blockages in the path of blood flowing to the heart. Obesity also increases the chance of having high blood sugar (diabetes), which is another major factor causing heart disease. Among the finest ways to control obesity and related medical problems are exercising regularly and eating a healthy diet. Regular check-ups by a physician are also required to take care of the problems caused by obesity.
5. **High Cholesterol:** Cholesterolis a fatty molecule and therefore a major component of a healthy body and an essential component of a healthy cell membrane. High cholesterol in the blood, however, increases the risk of heart

disease. High concentrations of cholesterol and other fatty substances may trigger atherosclerosis, in which fatty plaques build up on the blood vessel walls and limits blood flow to the heart, eventually triggering a heart attack. There are two different forms of cholesterol: LDL (low-density lipoprotein) is bad cholesterol, and HDL (high-density lipoprotein) is good cholesterol. High concentrations of LDL boosts your risk of heart attack. High levels of HDL lead to more protection against heart attacks. The concentrations of cholesterol depend on the age, gender, heritage, nutritional choices and level of exercise of an individual. Exercise and nutritional changes, such as avoiding saturated and trans fats, helps in reducing LDL cholesterol. Exercise is best way of improving your HDL cholesterol.

6. **Smoking:** Smoking is a significant risk factor for heart attacks. Among other health effects, it makes the blood more clottable and, as a result, blood pressure increases and threatens the heart.
7. **High Blood Pressure:** When blood pressure is uncontrolled, the risk of heart failure rises because it becomes more and more difficult for the heart to pump blood into the body as blood pressure increases. The exertion of the overloaded heart is more extensive, like with all other muscle stress; the heart's walls are thickened, and the general size of it increases. When the heart walls thicken, the volume of the heart chamber is significantly reduced, so every time the heart beats, less blood is being pumped.
8. **Diabetes:** As stated, diabetes, a blood sugar regulatory disease, is an important cardiac disease risk factor. The risk of having heart disease in a diabetic individual is equal to the risk in someone who has had a prior heart attack. Diabetes patients are more at risk for cardiac disease unless their blood sugar is well controlled. Diabetics must also regulate their blood pressure and concentrations of cholesterol. In fact, a diabetic's cholesterol goal is just as low as a person who has suffered a prior cardiac attack.
9. **Other factors:** Stress, alcohol consumption and depression are all associated with increased heart risk. Stress leads to over-eating, and smoking and excessive drinking cause higher blood pressure and obesity. Research shows that the threat of heart disease can be reduced by daily moderate alcohol consumption (one day, one drink). It may be an addictive drug and is a source of empty calories (i.e., limited in value for nutrition).

11.1.4 Techniques Used for Detection of Heart Disease

Research has resulted in the development of several algorithms for data mining. These algorithms are used in a dataset directly to create some models or from that data set to draw important findings and inferences. There are some common data-mining algorithms, such as ID3, decision tree, random forest, regression, k-means and many more.

The following sections discuss them as techniques for the detection of heart disease.

1. Decision tree
2. K-means algorithm

3. ID3 algorithm
4. Support vector machine (SVM)
5. C4.5
6. Artificial neural network (ANN)
7. Naive Bayes (NB)
8. CART
9. Random forest
10. Regression
11. J48
12. Association rules
13. A-priori algorithms
14. Fuzzy logic

11.1.5 Motivation

According to the World Health Organization, every year 17.9 million individuals, or 31 percent of the world's population, die from cardiovascular syndromes. These diseases are due to unhealthy diets, physical inactivity, tobacco smoking and excessive intake of alcohol. These result in high blood pressure, obesity and higher blood glucose (Subhadra 2019).

The increasing price of healthcare nowadays leads to the biggest issues in the world. As the global population increases, the medical sectors face numerous difficulties in identifying and treating patients' needs. Better efficiency in use of medical resources will decrease healthcare costs. An intelligent system must be intended to assist diagnosis of the disease to end the risk of diagnosis. By modeling human operations on a PC, this scheme can be modeled.

Usually, diagnosis is based on indications, conditions and physical treatment of the patient. By studying and experiencing a variety of patients, almost all physicians can predict heart disease. For adequate diagnosis, however, human intelligence alone is not sufficient. By using an intelligent system, the number of problems, such as less precise outcomes, less experience and time-dependent performance, during diagnosis will decrease, as shown in Figure 11.2.

11.1.6 Architecture of Artificial Neural Network

Figure 11.3 illustrates the fundamental architectural view of the artificial neural network (ANN). Artificial neurons consist of ANNs. Circles represent the processing node, and arrows represent the links to other neurons. A popular architecture of ANN, i.e., the multilayer perceptron, arranges the neurons into layer (Sivanandam et al. 2009).The input layer contains a sequenced set of predictor variables (a vector). Values of each input layer neuron are distributed to all hidden layer neurons. There is a connecting weight along every relationship between the input and the hidden neuron to ensure that the hidden neuron is obtained as a product of the input value and connecting weight. Each hidden neuron collects the total sum of its weighted inputs, and a nonlinear function applies to the total. The output of this specific hidden neuron is the result of the function. The output neuron is connected to every

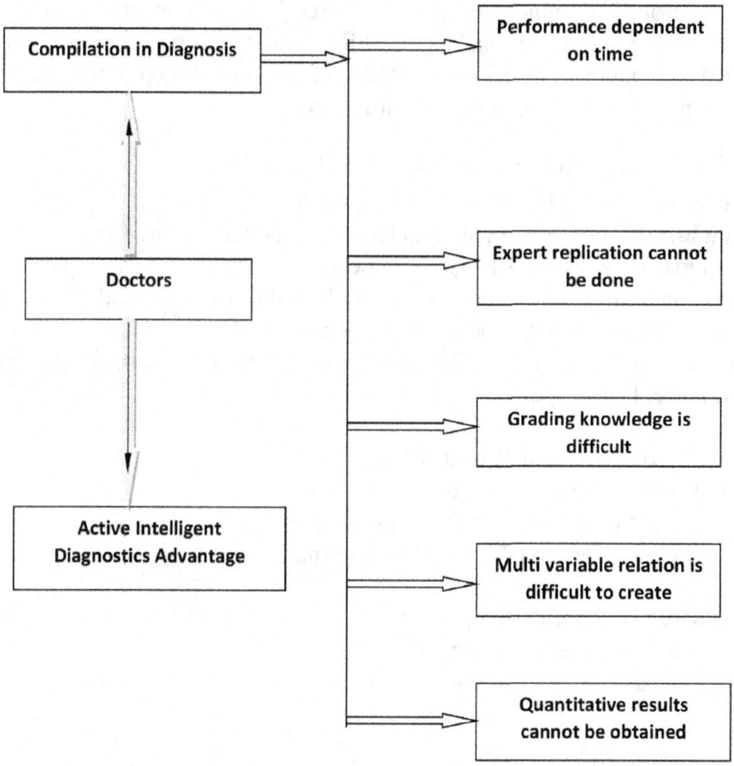

FIGURE 11.2 Diagnosis Complexities of Doctors.

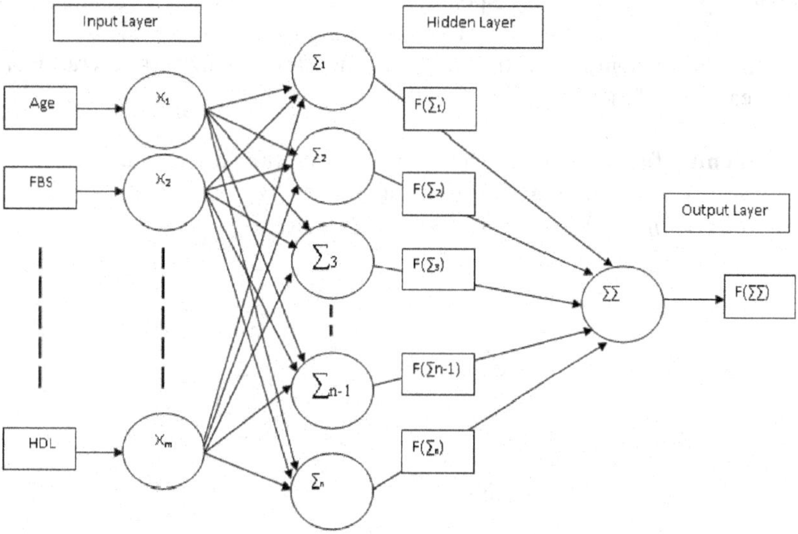

FIGURE 11.3 Architecture of Artificial Neural Network.

hidden neuron. The connecting weight is linked by each connection between the mid neuron and the output neuron. The resulting neuron uses the weighted input sum in the final step, and on the input weighted sum nonlinear function is used. The result of this function is the output for the entire ANN.

11.1.6.1 Basic Artificial Neural Network Terminology

Important terms used for ANN are discussed here.

Weights: To solve a problem, this term is used by the neural net.

A simple network of neurons is indicated in Figure 11.4. W1 and W2 are the weights containing information. They can be either fixed or random. Set weight to zero, or some methods can be used to calculate it. Weight initialization in a neural net is an important criterion. The changes in weight indicate the overall neural net performance. From Figure 11.4,

X_1= Neuron activation 1(input signal)
X_2= Neuron activation 2(input signal)
W_1= Weight of neuron 1 connected to output
W_2= Weight of neuron 2 connected to output

The parameters for calculating net input "Net" are used. Net neuron input signal is computed using the given formula:

$$Net = X_1 W_1 + X_2 W_2$$

From the calculated net input, applying the activation functions, the output may be calculated.

Activation Function: The activation feature is applied to calculate the neuron output response. To get the response, the total of the weighted input signal (Net) is activated. Linear and nonlinear active functions can be provided. A multilayer network utilizes nonlinear activation functions. Here, we discuss several linear and nonlinear active functions.

- **Identity function:** This function is given by f(a)= a, for all a.
- **Binary step function:** This function is given by
  $f(a) = 1 if\, f(a) > 0 else$ f (a)= 1 if f(a)>0 else 0

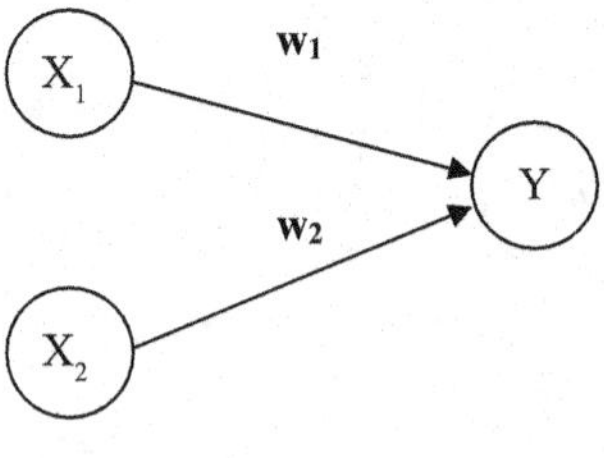

FIGURE 11.4 A Simple Neural Net.

- **Sigmoid functions:** The hyperbolic and logical curves are often used in multilayer networks, such as back propagation.
  - Binary sigmoid (logistic) function is given by f(a)=1/ (1+exp(-a))
  - Bipolar sigmoid (hyperbolic tangent) function is given by f (a) = (1-exp (-a))/ (1+exp (-a))

  Here, x is the net input to the neuron.

11.1.6.2 Types of Neural Network Training

In general, neural network training techniques are divided two categories.

- **Supervised training:** This is a simple model in which the networks calculate an answer to each input and then compare it to the objective value. When the calculated response differs from the target value, network weights are modified in accordance with a learning rule.
- **Unsupervised training:** These networks learn through the identification of specific features of the problems, for example, a self-organizing map.

11.1.6.3 Advantages of Neural Networks

The benefits for classification of the neural networks are:

1. The weights make neural networks more stable.
2. Neural networks using the training data are capable of learning and performing tasks.
3. The neural networks can generate their own information organization during study.
4. Neural networks can carry out simultaneous calculations. Many hardware devices are designed and manufactured to benefit from ANNs.
5. In noisier environments, neural networks are more robust.

11.1.6.4 Pseudo Coding to Train Multilayer Neural Networks

Varieties of training algorithms are available in neural networks to find solutions to the problem at hand in an efficient manner. Let us introduce the back propagation algorithm, which is used to train a multilayer neural network.

Back Propagation Training Pseudo Coding

Initialize all synaptic weights with small random values; typical values lie between –1 and 1.

```
Repeat
For every input vector in the training data sample
    Place the input vector in the network
//  Propagate the input vector forward through the network:
        for each network layer
            for each neuron in the layer
                1. Calculate the weighted sum values of the input to the node
```

```
            2. Add the threshold value to the sum
            3. Calculate the activation level by applying a nonlinear activation
               function for the neuron
         end
      end

//  Propagate the errors backwards to the hidden network layer
      for each output layer neuron
         calculate the error value
      end

      for all hidden layers
         for each neuron in the layer
            1. Calculate the neuron's signal error
            2. Update each neuron's weight in the network
         end
      end

//  Calculate overall Error
      Calculate the Error Function
   end

while (maximum number iterations < specified) and
      (Error function is >specified)
```

11.1.6.5 Deep Learning

Deep structured learning is an automatic way to teach computers naturally. Deep learning (DL) behind driverless cars is an important technique that enables them to identify red lights or to separate the footpath from roads. It is the best approach to voice recognition on customer appliances, such as free speech machines, phones, tablets etc. Deep structured learning has been a big focus. Results have been reached that were previously not possible.

DL features can reach the latest accuracy, and sometimes more than human efficiency. Many numbers of labeled layered and neural network systems are used for training models. The mathematical models learn to execute identification work directly with pictures, text and sound.

DL, initially postulated in the 1980s, has only recently been useful for two main reasons:

1. DL demands a considerable quantity of data to be labeled. For example, it takes millions of images and footage recorded across thousands of hours to develop self-driving systems.
2. DL needs considerable computer ability. This high-performance GPU is designed for deep structured learning, with a parallel architecture. When linked to cluster or cloud computing, it reduces training times in a deep-seated learning network.

11.1.7 Benefits of Deep Learning Neural Networks

There are several benefits of using DL neural networks for detection of CHD. We can use different techniques for the detection of CHD, such as:

1. Electrocardiography
2. Echocardiography
3. Telesonography

Electrocardiography is the method of generating an electrocardiogram (ECG), which records the electrical activity of the heart by using electrodes positioned on the skin. An electrocardiogram is also generated by a voltage versus time graph.

Echocardiography is a popular method for cardiac imagery, which captures ultrasound videos of discrete cardiac views, internal structures and motions. For the evaluation of the function and heart morphology and to diagnose heart problems related to movement deformities, this technique is preferred as a major tool. It permits rapid assessment of cardiac size, structure, function and hemodynamics (Spencer et al. 2013). This technique is the best approach for cardiac imaging. The development by technological advances of portable ultrasound systems has made echocardiography suitable for different situations, for example, health missions to developing countries. Remote assessment of echocardiograms using a cloud-computing environment may be helpful in expediting care in remote areas (Singh et al. 2013).

Telesonography facilitates the guidance of an innovative ultrasound sonographer by the remote specialist in acquiring the ultrasound imagery as generating and interpreting ultrasound images are highly operator dependent (Ferreira 2015). It is a subtype of telemedicine. Many scientists have experimented with streaming apps in actual time, which transmits the cinematic feed to offsite equipment by an ultrasound machine linked to a computer. Investigators have also utilized a variety of commercial video-streaming devices to transmit real-time streaming ultrasound images over either the internet or satellite transmission, with potential remote guidance potentially conducted by telephone or without a video camera depiction of the user (Pian et al. 2013). Most experts think that the video quality largely depends on the ambient Internet speed rather than the type of video calling system used (Kim et al. 2017). The results show that remote management and interpretation can be performed on mobile devices connected to low-bandwidth networks. However, owing to slow internet connectivity, transmission speeds and image quality would be lower. As internet connectivity will only improve worldwide, future developments need to focus on the human factors to optimize tele-sonographic interactions (Crawford et al. 2012). Images were sent for verification of point-of-care diagnosis to two expert echocardiographers in the United States reading on a workstation. However, evaluations were not significant, and the clinical importance of transmitted pictures remains unchanged (Choi et al. 2011).

The traditional methods described require expertise in the same field and thus are difficult to use. Also, there is a tendency toward error, as these are all purely dependent upon the skill of the person involved in these techniques. If the person performing these tests is not skilled, it may lead to misdiagnosis of the disease.

Whereas, machine learning (ML) algorithms are independent of the skills of the person involved in performing the test. Input and output layers and, at most, one hidden layer are used by some traditional ML algorithms. Four different methods of transmission and communication for both the feasibility of transmission and image quality were used, and more than three layers are used for DL, including input and output. So far, DL-based applications have provided positive feedback; however, due to the sensitivity of healthcare data and other challenges, we should look at more sophisticated DL methods that can deal with complex healthcare data efficiently. ML is the best strategy for diagnosis, prediction, management and other associated elements of health management when applying DL techniques in cardiovascular studies (Liteplo et al. 2011).

11.2 RESEARCH GAPS

After the study of the literature survey, we found some gaps in the study, which can be listed as follows:

- A lot of work has done in the field of neural networks, and comparative studies have also been made in this field. However, there is a scope of comparison of effective neural network computational techniques for the detection of CHD.
- Conventional systems cannot provide efficient real-time outcomes. Deep convolutional neural networks (CNNs) can enhance the performance of this smart computing scheme.
- Work has done to analyze and identify the parameters that can be helpful for heart disease diagnosis. However, there is no such fixed data set for the computational and detection of coronary cardiac disease.
- By analyzing the neural network techniques that were proposed in the literature, it might be observed that ANN techniques are helpful in the detection of certain diseases. However, there is a scope of implementation of these neural network techniques for the prediction of CHD depending upon the available data set.
- The comparative analysis of neural network techniques has been performed. However, the comparative analysis of the recent methods of neural networks for the identification of CHD needs to be covered.

11.3 RESEARCH CHALLENGES

There are a number of possibilities and challenges in finding the disease in the early stages. When conducting research in any area, one may face various problems and challenges in that field. Similarly, during this study, we may face the following challenges:

1. As the neural networks work on the dataset, that, in turn, depends on the values, which are required to be collected from one or the other source. We need to create a large dataset, as the larger the dataset the greater the accuracy of the neural network will be. So, the first challenge for this study is

to fix the parameters, which are useful for the diagnosis of CHD. Then, the collection of patient data is the second most important challenge to be taken care of during this study.

2. Another important challenge in this study is to program and code the neural network for this detection. Part of the challenge of coding a neural network is structuring. One needs to determine how many layers to use, how many nodes per layer to use etc.
3. Once the aforementioned challenges are met, we need to produce an effective solution for detection of CHD. After the development of this technique, we are left with another challenge: running the proposed technique on the dataset and figuring out whether the produced technique is better than the existing techniques. Collection of the dataset in large amounts is also a big challenge in the medical field, and the accuracy of all these classification algorithms can be further enhanced by increasing the size of the dataset as well as by using ensemble methods such as bagging and boosting (Bhargav 2018).

11.4 CONCLUSION

After going through the previous description and research gaps, it is concluded that there is a lot of scope of research in the field of cardiac disease detection using ML. As human beings, doctors can misdiagnose diseases, so in the near future, a smart system may be established that can diagnose CHDs in the early stage. That smart system will need data sets of patients with some influential parameters, such as age, high blood pressure, high cholesterol, obesity, gender, smoking, diabetes etc. The person needs to have all required tests from laboratories. The collected data must undergo preprocessing and other methods as per the requirement of the research work. The process will be including an experimental evaluation of the data extraction methods to find out the most accurate way of extracting the region of interest (CHD). Based on the data set provided as input, our system will decide whether the person has CHD or not.

REFERENCES

Bhargav, K. S., Thota, D. S. S. B., Kumari, T. D., & Vikas, B. 2018. Application of machine learning classification algorithms on hepatitis dataset. *International Journal of Applied Engineering Research* 13, no. 16, 12732–12737.

Choi, B. G., Mukherjee, M., Dala, P., Young, H. A., Tracy, C. M., Katz, R. J., & Lewis, J. F. 2011. Interpretation of remotely downloaded pocket-size cardiac ultrasound images on a web-enabled smartphone: Validation against workstation evaluation. *Journal of the American Society of Echocardiography* 24, no. 12, 1325–1330.

Crawford, I., McBeth, P. B., Mitchelson, M., Ferguson, J., Tiruta, C., & Kirkpatrick, A. W. 2012. How to set up a low cost tele-ultrasound capable videoconferencing system with wide applicability. *Critical Ultrasound Journal* 4, no. 1, 13.

Ferreira, A. C., O'Mahony, E., Oliani, A. H., Araujo Júnior, E., & da Silva Costa, F. 2015. Teleultrasound: Historical perspective and clinical application. *International Journal of Telemedicine and Applications* 2015, 1–11.

Ghwanmeh, S., Mohammad, A., &Ibrahim, A. 2013. Innovative artificial neural networks-based Decision Support System for Heart Diseases Diagnosis. *Journal of Intelligent Learning Systems and Applications* 5, 176–183.

Iaizzo, Paul A. ed. 2009. *Handbook of cardiac anatomy, physiology, and devices.* Springer International Publishing, New York, DOI: 10.1007/978-3-319-19464-6

Kim, C., Jin, H., Seung, K. B. Hyuk, C. J., Jeong-Hun, S., Kim, T. H., & Ho, C. J. 2017. Can an offsite expert remotely evaluate the visual estimation of ejection fraction via a social network video call? *Journal of Digital Imaging* 30, no. 6, 718–725.

Liteplo, A. S., Noble, V. E., & Attwood, B. H. C. 2011. Real time video streaming of sonographic clips using domestic internet networks and free videoconferencing software. *Journal of Ultrasound in Medicine* 30, no. 11, 1459–1466.

Pian, L., Gillman, L. M., McBeth, P. B., Xiao, Z., Ball, C. G., Blaivas, M., Hamilton, D. R., & Kirkpatrick, A. W. 2013. Potential use of remote telesonography as a transformational technology in under resourced and/or remote settings. *Emergency Medicine International* 2013, 1–9.

Singh, S., Manish, B., Puneet, M., David, A., Shantanu, S. P., Rhonda, P., Anne, D. L., et al. 2013. American society of echocardiography: Remote echocardiography with web-based assessments for referrals at a distance (ASE-REWARD) study. *Journal of the American Society of Echocardiography* 26, no. 3, 221–233.

Sivanandam, S. N., Sumathi, S., & Deepa, S. N. 2009. *Introduction to neural networks using MATLAB 6.0.* Tata McGraw Hill Publishing Company Ltd, New Delhi, Eighth reprint.

Spencer, Kirk T., Kimura, Bruce J., Korcarz, Claudia E., Pellikka, Patricia A., Rahko, Peter S., & Siegel, Robert J. 2013. Focused cardiac ultrasound: Recommendations from the American society of echocardiography. *Journal of the American Society of Echocardiography* 26, no. 6, 567–581.

Steinhauser, Matthew L., & Lee, Richard T. 2011. Regeneration of the heart. *EMBO Molecular Medicine* 3, no. 11, 701–712.

Subhadra, K., & Vikas, B. 2019. Neural network based intelligent system for predicting heart disease. *International Journal of Innovative Technology and Exploring Engineering (IJITEE)* 8, no. 5, 484–487. ISSN: 2278–3075.

12 Energy-Efficient Green Cities

A Mechanism for Nature-Based Solutions for Future Cities

Prof. (Dr.) Pavnesh Kumar, Chandan Veer, and Raushan Kumar

CONTENTS

12.1 INTRODUCTION

The United Nations (2021) has projected that 60% of the world's population will be living in cities by 2030. Cities of the world are believed to be economic powerhouses

DOI: 10.1201/9781003184140-12

and contribute 60% of the global GDP, but they are also responsible for 70% of the total carbon emissions globally. Rapid urbanization of our cities provides the path for urban development, but it also minimizes the goal of sustainable urban development for green cities. The shift from rural to urban cities mainly in developing states is due to poverty, employment opportunity and other socioeconomic factors. Developing states, like India, are experiencing massive infrastructural development due to increasing population sizes and rapid urbanization. The reduction of green spaces in urban cities has grabbed the attention of environmentalists. Due to the degradation of urban spaces, the quality of urban ecosystems has drastically shrunk. The absence of these spaces from our environment generates poor air quality, the disruption of urban biodiversity, and poor human well-being. Throughout the world, the so-called modern city infrastructure affects the health of the people who live around urban cities.

12.1.1 Green City

Despite many awareness programs, the availability of green city and spaces are poor due to the high costs associated with their development (Franco et al. 2021). Green cities are those that promote energy efficiency and the use of renewable resources and have a significant collaboration between environment and economic objective. Green cities have clean air, water, parks, and streets. Due to the availability of green spaces, the mortality rate in green cities is also improved, and the spread of infectious disease outbreaks is much lower. Green cities are also adopting resiliency in the face of natural disasters and other natural instability phenomena. Despite its many definitions, a city is also considered green when it encourages green behavior, such as the use of energy-efficient public transport and adopting a green lifestyle, and the incidence of disease outbreak is low.

12.1.2 Green buildings

Green buildings may be described as efficiently using natural resources, such as energy, water, light and air. The construction and design of green buildings reduces and eliminates negative impacts on our climate and environment (Franco et al. 2021). Green buildings also include the use of renewable energy sources, such as solar and wind power. The provision of green buildings is different from region to region and is based on climatic conditions; culture; economic conditions and environmental, social and cultural impacts. So, a green building provides protection to the people, reduces waste and pollution, improves health and productivity and provides holistic development of the environment.

12.1.3 Green Spaces

Green spaces may be defined as areas that are not occupied by buildings Gozalo et al. (2019). These areas generate economic, social and ecological benefits. One economic benefit of green spaces is that they act as tourist destinations for people who will generate revenue. A social benefit is that they provide good air quality that will attract people to spend time in the spaces, and an ecological benefit of green spaces is that they provide a sustainable urban ecological balance. In sum, green spaces represent areas that can be public or private and that provide many social, environmental, physical and ecological benefits.

12.1.4 Energy Resilience and Infrastructure

Population growth, rapid urbanization and climate change put pressure on urban infrastructure. Due to massive urbanization and new building development, the existing building stocks were left untouched. A lot of carbon emission occurs in the process of demolishing an existing building and constructing a new one. According to the International Energy Association (IEA), the building sector alone accounted for 39% of carbon emissions in 2019. The IEA predicts GHG (greenhouse gas) emissions from buildings will more than double in the next 20 years.

Due to overwhelming support from various government officials and policymakers, the demand for reliable, affordable and environmentally sustainable energy is growing, but these demands will shear on energy infrastructure resources and financing. The growing demand of policies and investment in sustainable energy infrastructure and the decisions and investment in resilient infrastructure made today will lock us into our energy future for many years. The energy sector must not only be robust but also resilient.

12.2 HOW CAN INDIA PROMOTE ENERGY-EFFICIENT GREEN CITIES?

12.2.1 Adopt and Promote Green Infrastructure Mechanisms

Adopting and promoting green infrastructure is necessary. The increasing amount of carbon emissions from existing concrete creates pressure among environmentalists to think about better alternatives for present-day problems. Green concrete could one such option. A PhD student, Shin Yin from James Cook University of Australia, developed green concrete and won the Australian Innovation Challenge award in the manufacturing, construction and innovation category in December 2015 (Kane 2016). Shin Yin developed a plastic-waste-converted concrete that can reduce carbon emissions by 50%. Another Australian Company, Wagners, developed Earth Friendly Concrete (EFC) made from blast furnace slag (waste from iron production) and fly ash (waste from coal-fired power generation) (Wagner 2021). The advantages of EFC over normal Portland cement concrete include durability, lower shrinkage, earlier strength gain, higher flexural tensile strength and increased fire resistance. The pavement at Toowoomba Wellcamp Airport (formerly known as Brisbane West Wellcamp Airport) is one example of the use of EFC, and it is known as being one of the greenest airports in the world. This project alone saves 8,800 tons of carbon dioxide emissions.

12.2.2 Disaster Risk Reduction

Disaster risk reduction is a plan that aims to reduce the damage caused by natural hazards, such as floods, droughts, earthquakes and cyclones (United Nations 2021). It is a systematic effort to analyze and minimize the risks associated with such disasters. There is a need for a proper channel of decisions about where to build schools, factories, dams and lakes and how much to invest in disaster surveillance.

Through the Jawaharlal Nehru National Urban Renewal Mission (JNNURM), cities are being provided with an opportunity to respond to climate change and upgrade basic infrastructure services (JNNURM 2021). With an objective and mission focused on the planned development of cities, accelerating the flow of financing in urban sustainable infrastructure and the renewal and redevelopment of cities will be a better option to reduce and minimize disaster risks.

12.2.3 Intelligent and Green Transportation Network

The social and economic activity of any country depends on its transportation sector. Making an unobstructed and sustainable transportation service is one of the keys for energy-efficient infrastructure. India is mainly dependent on nonrenewable sources of energy, but its share of renewable sources of energy has also risen to 9% IBEF (India & News 2019). According to data from IBEF (India & News 2019), coal comprises of 59.9% of India's total nonrenewable sources. The transportation network and use of transport system is much poorer and not updated according to current and future requirements. Poor public transport, a large population and the unavailability of green transport contribute to India's road pollution. But the Indian government seems to be more focused on enhancing green transport, and it recently imposed green taxes on vehicles older than eight years (Nair 2021). Roy and Raju (2021) discussed the recent union budget that provides for greening the public transport and combating pollution. Solar energy shows the highest increase in renewable energy in the country, and it will be best option for the transportation sector to minimize and combat risks and create a resilient infrastructure.

12.3 LITERATURE REVIEW

In developed and developing countries, infrastructure development significantly contributes to energy consumption and the amount of carbon emissions. The national vision of almost every country is to have environmental sustainability, and it is an indispensable aspect for the future development of the country.

Artmann et al. (2019) found that with the end of the 21st century, 90% of the total population will shift in the urban areas. So, it becomes necessary to develop urban infrastructure in a sustainable manner for future generations. Green cities can play an important role in urban sustainable development.

Ahvenniemi et al. (2017) defined a sustainable city as one that establishes the relationship between social, economic and environmental aspects by combining these components. While other researchers focus on only one aspect, Wu et al. (2019) emphasize pollution, waste generation and usage of water and energy. Similarly, Rode et al. (2011) combine social equity and a balanced environment.

Urbanization in Qatar has brought several challenges, such as an increase in the consumption of water and electricity, an increase in solid waste and the deterioration of air quality (Youssef 2017). So, to mitigate future challenges, the time for sustainable urban development is now.

According to Dhingra and Chattopadhyay (2016), every smart and sustainable city has certain goals to be achieved in a flexible, authentic, innovative, attainable and durable manner, such as:

- Improvement in quality of life of the people.
- Proper economic growth and equal employment opportunities.
- Adaption of environmentally responsible and sustainable methods of development.
- Efficient infrastructure, transportation, water and energy supply, proper drainage and other facilities.
- Resilience to climate change and environmental issues.
- Practical regulatory and local governance.

Zaina et al. (2016) expressed that, in order to protect the historic importance of the city, it is necessary to combine modern architecture and traditional community living.

Jong et al. (2015) conceptualized a smart city as the use of technology in favor of systems and services of people.

Marsal-Llacuna and Segal (2016) evaluated a smart city as it considers experiences of environmentally friendly cities that comprise quality and sustainability of life along with technological involvement.

As per Carrillo et al. (2014) and Kondepudi (2014), there are four features of smart and sustainable cities: (a) sustainability, (b) quality of life, (c) urban aspects and (d) intelligence. These are investigated under four main headings: (a) society, (b) economy, (c) environment and (d) governance.

Meinert (2014) expressed that there should be enough space for the environmental factors to change while planning for future development; at the same time, the plan should not be too far into the future that it seems irrelevant in the present scenario.

Wiedmann et al. (2014) stated that the vision for country growth should synchronize with economic growth, social growth and the conservation of its natural habitat.

Yigitcanlar (2015) and Lee et al. (2014) explained smart-eco cities as those which are environmentally healthy, use advanced technologies, have industries that are economically productive and environmentally safe, have harmonious culture and have a physically attractive living environment.

Hiremath et al. (2013) explained sustainable urban development as aiming to create a balance between urbanization and the protection of the environment with an aim to have equality in income, employment residence, social infrastructure and transportation in urban areas.

As per Deakin and Al Waer (2012) and Townsend (2013), smart cities evolved because of the use of technology in the fields of health and safety, transportation, energy usage, education and urban governance.

Lazaroiu and Roscia (2012) stated that smart cities reflect a technological community, are internally connected, have sustainable comfort and are alluring and safe.

Caragliu et al. (2011) explained a smart city as one that has such characteristics as enhanced administration, an efficient economy, attention on business intended for urban development, inclusion of urban residents in social services, attention to high-tech and creative industries and synchronization of all these activities toward social and environmental sustainability.

Fuerst and McAllister (2011) compared green and non-green properties from the supply-and-demand perspective. They concluded that green features will lead to residential transaction price premium. The promotion and regulation of green development can be done through mandatory government regulation and voluntary industry involvement. This has resulted in mandatory building codes, like minimum energy and water consumption and residential electricity consumption.

Nam and Pardo (2011) categorized smart cities under three categories: (a) technology, (b) population and (c) institution.

Schaffers et al. (2011) pointed out that a smart city should have strong broadband network and the creation of applications through large-scale participation.

Eichholtz et al. (2010) researched the impact of energy star–labeled and LEED-certified commercial buildings on financial performance. They propounded that economic benefits, such as 3% higher rental rates, are associated with green certified commercial buildings in comparison to non-green buildings.

Gudes et al. (2010), Cocchia (2014) and Lara et al. (2016) defined smart cities as a broader concept that includes several sub-concepts, such as smart urbanization, smart environment, smart economy, smart energy usage, smart transportation and so on.

Harrison et al. (2010) explained the functioning of a smart city as one that uses city statistics for smooth traffic management, efficient energy usage, security and optimal use of municipal services.

Costa and Kahn (2009) found that energy-efficient buildings in California sell at a premium price. Compulsory green building programs can drastically change the market scenario and stimulate the demand for green properties.

Jacobsen and Kotchen (2009) concluded that change in energy code requirements in terms of electricity and natural gas consumption can lead to a 4% decrease in electricity consumption and a 6% decrease in natural gas consumption.

Miller et al. (2008) studied green features in commercial buildings and identified the benefits associated with energy-efficient investment and environmental certifications.

Lee et al. (2008) and Jong et al. (2015) proposed that metropolitan areas are mostly involved in improving urban infrastructure and services with a view to have better social, economic and environmental ambience along with enhancing the attractiveness and competitiveness of cities.

Simons and Saginor (2006) identified that most of the previous research was done on environmental attributes focusing on contamination and residential property values.

Komninos (2002) and Yigitcanlar (2015) propounded that efforts to improve the existing scenario of the cities is a step toward intelligent cities, which is the antecedent of smart cities.

Boyle and Kiel (2001) appraised 30 hedonic price studies on air, water quality and multiple pollution roots. This study found that environmental variables impact residential property values.

Bynum and Rubino (1998) stated that green building programs are a system of green development. They suggested that all builders should be obliged to follow green building practices and that this will bring dramatic shifts in sustainable construction practices and will accelerate the performance of green development.

Reichert et al. (1992), Smolen et al. (1992), Flower and Ragas (1994) and Benjamin and Sirmans (1996) discussed consumer negotiations for price discounts on residential properties if they found poor air quality, poor water quality and undesirable land use.

In general, smart city means the extensive use of information and communication technology to build a conceptual model for urban development with the integration of human and technological capital.

12.4 OBJECTIVES OF THE STUDY

1. To suggest measures for less and safer energy consumption in urban areas.
2. To suggest measures for promotion of green areas in urban spaces.
3. To explore the possibilities of a viable mass transport system.
4. To suggest measures for proper water consumption and conservation.
5. To suggest measures to improve solid waste collection and disposal system.
6. To suggest measures for the reduction of air pollution.
7. To suggest an environment-based framework for green infrastructure development.

12.5 RESEARCH METHODOLOGY

An exhaustive literature review regarding the title and its related concepts has been done. Secondary data, inclusive of quantitative and qualitative data, has been analyzed. The sources of information are from various research publications, published newspapers, online and printed journals, books, magazines and websites (Fuerst & McAllister 2011; Youssef 2017; Zaina et al. 2016; Artmann et al. 2019). The information was collected from libraries and websites. The literature was cross-checked and validated to give the latest information.

12.6 MEASURES TO BE IMPLEMENTED FOR EFFICIENT GREEN CITIES

12.6.1 Intelligent Waste Management

The increase in population has created a challenge in raising awareness about waste management. Improper waste management leads to many health issues. The garbage dumped in the trash bin contains materials like industrial waste, community- or household-generated waste etc. This garbage creates unhygienic, and sometimes toxic, conditions. To achieve the optimal solution, sensor-based dustbins can be an efficient option. Sensors placed in the dustbins work with the help of a GSM module connected to the internet, which sends notifications to the respective authority regarding the status of the bin. The sensor helps in several ways, such as the sorting of garbage and indicating whether the bin is full, and all this information is periodically sent to the respective agency. A mobile app would inform the residents about empty dustbins nearest to current location. Similarly, the nearest cleaner will get an update or notification about filled dustbins. This will help in city garbage management, along with efficient garbage collection.

12.6.2 Water Consumption and Conservation

It is well known that three-fourths of the earth is filled with water, but only 3% is suitable for drinking or for other various purposes. In a district in California, the municipal corporation has incorporated an innovative policy for water management. For this, they adopted three tools: authority tools, incentive tools and capacity-building tools. In authority tools, the governing body implements a directive to restrict residential daily water usage by use of efficient technologies. In incentive tools, the consumers are given a choice to save more water, and rebates are given to them on their water bills. In capacity-building tools, those who are not aware of the efficient use of water are provided training and adequate knowledge.

There are two methods of rainwater harvesting (Chao-Hsien et al. 2015):

1. Surface runoff harvesting: This is an indirect method of rainwater harvesting in which rainwater is stored in a small, constricted area, like a well, pit, shaft etc., and infiltrated under the soil through them. This is much needed in urban areas where rainwater is wasted.
2. Rooftop Rainwater Harvesting: This method is less expensive and very effective if implemented properly. In this method, rainwater is captured from a roof's catchment area, and the collected water is either stored in tanks or diverted to an artificial recharge system.

For example, in Chennai city, there are 201 ponds and 63 temple tanks available. The Municipal Corporation of Chennai has mandated to ensure that every household establishment has working rainwater harvesting. As per Chennai Corporation Commissioner G. Prakash, the rainwater harvesting system reaches 80% of the city and is working effectively.

Rainwater harvesting and utilization of rainwater by accumulating it on rooftops can be a solution for gardening, domestic usage and for groundwater recharge. Recycling and reuse of gray water reduces the concentration of organic matter. With the help of proper treatment and mechanisms, gray water can be transformed for safe use. Leadership in Energy and Environment Design (LEED) certification should be made compulsory for green building development, new construction and maintenance of existing structures.

12.6.3 Green Buildings: A Way Toward Earth-Friendly Buildings

Suhendro (2014) stated that the cement industry is responsible for 8–10% of the world's total carbon dioxide emissions. During crushing and heating, gases are released into the environment, and this leads to serious concerns about the environment. According to the World Green Building Council (WGBC 2021), a green building means that it can create a positive impact on our environment and eliminate negative impacts. A green building consists of materials that are nontoxic and sustainable along with efficient use of energy and resources. The process of green building doesn't lead to environment destruction.

Challenges in green building:

- Myths about the high costs involved.
- Limited awareness.
- Inadequate government policies.
- Lack of skilled manpower.

While the whole world is still facing the COVID-19 pandemic, India has committed to reduce its GHG emissions by 35%, as buildings are the major player responsible for 40% of all carbon emissions in the world. Union budget 2021 has attracted investment in green infrastructure and sustainable building that would spur the economy and build green spaces (Padmanabhan 2021).

Steps to building a green and sustainable building:

- Understanding the climatic conditions of the city: The study of climatic condition is very much needed to build a sustainable building. We need to study the environmental parameters of the region, the availability of sunlight, surrounding water bodies, altitude and range. For example, in hot regions, solar energy is a sufficient renewable source. In a country like India, where there is a different climate in every region, there exists the potential to utilize many different sources of renewable energy.
- Depends on locally designed houses: Every region has their own locally designed houses with their available resources. In India, the design of a house in Jammu Kashmir is different from a house in Delhi due to the climatic conditions. Today, it does not matter whether you are in Jammu or in Delhi; glass-facade buildings are popping up across the country. Environmentalist Narain (2013) says that in glass-facade buildings, the glass throws back heat into the atmosphere, which causes warming. According to the National Environmental Engineering Institute, the temperature can go up by 17 degrees Celsius in glass-facade buildings (*Times of India* 2013). This concept has evolved from the West, where in most cases, the region will experience cold climatic conditions for a period of time. So, the use of glass in buildings should be avoided in India, as it traps heat and raises the temperature, especially in hot weather regions.
- Efficient lightning: Efficient use of lightning is required to reduce the dependency on nonrenewable resources. We need to identify the artificial lightning requirement for any building and provide the minimum required lightning level. The world is moving from incandescent bulbs to LED lights, and all this has happened very fast and is still evolving. Lights with very high efficacy should be adopted. Besides efficient lightning, green buildings

should be installed with higher rating appliances and equipment and with a procurement policy. For example, in an institutions, a procurement policy should be developed that only five-star-rating air conditioners should be fitted, and policies must be executed properly. These passive systems of energy are the part of active systems.

- Green Concrete: Green concrete is defined as concrete that uses waste materials, and its production doesn't have any negative impacts on the environment. Shin Yin's award-winning green concrete helped Queensland, Australia–based company Fibercon develop Fibercon RMP47, a recycled polypropylene fiber from industrial plastic wastes. It can reduce carbon dioxide production by 50%, and plastic wastes are converted into concrete. Wagners' EFC is another example of green concrete (Wagner 2021).

12.6.4 Smart Transportation: A Key Success for a Smart City

The world is moving fast. Kochar (2018) stated that cities are growing and an urban population is rising. It is to be predicted that the urban world population will grow by 66% by 2050. Our urban cities are not ready to accommodate the burden of all the existing and upcoming infrastructure facilities. The United Nations has initiated the United Smart Cities program to improve the efficiency, mobility and sustainability of urban cities with the help of public–private partnerships (PPPs) (United Nations 2021).

Imagine a cleaner, safer, more efficient, silent and emissions-free city without compromising our everyday life and health. This imagination would turn into reality if we adopted smarter transportation systems. Grantthornton (2016) stated that the future is electric because of zero emissions and low noise. We need to develop our transport system in a smarter way to tackle air pollution, road accidents, congestion

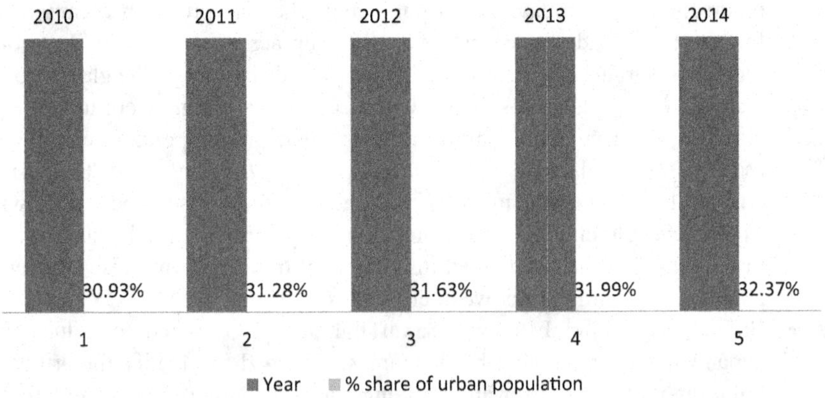

FIGURE 12.1 Indian Share of Urban Population to Total Population.

Source: World Bank 2021

and climate change. To improve the efficiency and capacity of urban cities, silent emissions-free buses should be increasing to meet the demand for public services. Dedicated corridors for metros and monorails are also needed in every smart city to reduce the burden of the urban population on the road. Urban transportation plays a major role in shaping the way we live, and it also contributes major part of the GDP of any country. The Indian automotive sector is a major contributor to the country's GDP, and it is expected to grow about 12% by 2026 and is currently close to 6.7%. The Indian share of the urban population to the total population was 32.37% in 2014 and 30.93% in 2010 (World Bank 2021) (Figure 12.1).

Challenges for smart transportation in India:

- Inadequate public transport.
- Congestion on road.
- Inadequate electronic toll collection.
- Multilevel parking systems.
- Emissions of GHGs.

Suggestions for smart transportation in India:

- Electric vehicles: Electric vehicles are the only solution for fuel efficiency and reducing GHGs from the environment. In India, vehicles are mostly running on petrol, diesel and CNG in some parts of the country.
- Alternative fuel: Due to rapid urbanization, carbon footprints forcing the need for adoption of clean, efficient and sustainable energy. Besides petrol and diesel, vehicles powered with CNG, biofuels and solar power fuels should be a better option to support energy-efficient green cities.
- Adoption of energy standards: To regulate air pollutants from motor vehicles, the Government of India follows the Bharat stage (BS) emissions standard. BS is based on the European standard and the Indian government regularly upgrades it. Most of the European countries now follow Euro VI (equivalent to BS VI), but in India, urban cities are still updating the BS IV. So, there is a need to frequently adopt the energy standards that will help in reducing carbon emissions.

12.7 FINDINGS

The human pursuit to control nature through its practices has brought severe damage to it in the present situation. The threat we have brought to our ecosystem has a silent impact on every sector. There are a few threats that require prioritized attention and action to resolve them. The growing population is in continuous fear due to incidences of natural disasters. Both the industrially developed and growing economies need to come up with a common proposal to meet the ecological challenges. Some of the common challenges the world is facing are:

- The earth's natural system is deteriorating, such as forest depletion, ozone layer depletion and air pollution due to urbanization.

- Increase in temperature due to emissions of GHGs, which trap the sun rays and results in global warming is due to dependency on traditional energy sources.
- Lack of innovation and dependency on traditional practices in construction and unplanned township setup.
- Dependency on fossil fuels for transportation activities causes a severe rise in air pollution.
- The improper utilization of groundwater has worsened the situation all around the globe. The contamination of groundwater due to land pollution has created a challenge for access to fresh water and air.
- Energy requirements and dependency on hydro and thermal power has caused serious environmental problems.

12.8 SUGGESTIONS

- Every individual on this planet will have to take the responsibility to minimize their own carbon footprint by becoming sensible to the environment through their practices.
- We need to provide reliable and sustainable energy requirements and minimize the dependency on traditional means of energy.
- The government should make a legal provision to have green spaces in every city.
- Safeguard the prevailed biodiversity without disturbing it and then develop the urban areas along with urban forests.
- Convert the abandoned areas into urban parks or urban green spaces.
- Develop effective natural drainage systems, like rivers, wells and ponds, to mitigate natural disasters and become resilient to the environment.
- Vegetation and plantation of trees, shrubs and mangrove areas within a city to improve soil health and recharge the groundwater and enhance the biodiversity of the city.
- Any infra project should first be evaluated on the parameters of ecosystem rather than only on socioeconomic benefits. Governments or policymakers should frame a provision for Biophilic cities."
- Governments should develop mass transportation systems based on low- or zero-emissions technologies, green fuels and green power–based locomotives.
- Develop agricultural activities near the consumer market to reduce the energy consumption in transportation, shorten the food supply chain and initiate the promotion of rooftop production of foods.
- Use ICT and green technologies in public and private practices as much as possible.
- Assess the present city biodiversity level and then frame the biodiversity strategy and action plan to preserve the biodiversity.
- Frame legal directives to promote green buildings and provide financial assistance in the projects through green funding gateways.

12.9 CONCLUSION AND FUTURE SCOPE

Environmental challenges arise mainly from population increase, fast urbanization, high vehicle dependency and irregular and irresponsible lifestyles. The achievement of urgent measures to mitigate environmental fade effectively, efficiently and responsibly have shown a path to reinvent the traditional practices into eco-friendly practices. In due course, sustainability and sustainable development have become popular terms for urban policy developers and practitioners.

Green cities are an integrated approach comprising smart technologies and nature-focused practices. They not only consider economic, social, technological and political aspects but also keep environmental priority on top. Green cities are a way toward a climatically adapted and resilient approach to urbanization, as they include sustainable construction activities, structure and environmental engineering. Green cities are the future for infrastructure development and a long-lasting environment.

BIBLIOGRAPHY

Ahvenniemi, H., Huovila, A., Pinto-Seppä, I., & Airaksinen, M. (2017). What are the differences between sustainable and smart cities? *Cities*, 60, 234–245.

Ali Abdallah. (2020). The future of green cities in 2040: A case study on the capital city of Qatar, Doha. *Humanities and Social Sciences*, 8(6), 170–176. doi:10.11648/j.hss.20200806.11

Ali Abdul Samea Hameed. (2020). Green cities and sustainable urban development: (Subject review). *I. J. Of Advances in Scientific Research and Engineering-IJASRE* (ISSN: 2454–8006), 6(11), 31–36. https://doi.org/10.31695/IJASRE.2020.33929

Artmann, M., Kohler, M., Meinel, G., Gan, J., & Ioja, I. (2019). How smart growth and green infrastructure can mutually support each other—A conceptual framework for compact and green cities. *Ecological Indicators*, 96, 10–22.

Benjamin, J. D., & Sirmans, G. S. (1996). Mass transportation, apartment rent and property values. *Journal of Real Estate Research*, 12(1), 1–8.

Boyle, M., & Kiel, K. (2001). A survey of house price hedonic studies of the impact of environmental externalities. *Journal of Real Estate Literature*, 9(2), 116–144.

Bynum, R. T., & Rubino, D. L. (1998). *Handbook of alternative materials in residential construction*. New York: McGraw-Hill.

Caragliu, A., Del Bo, C., & Nijkamp, P. (2011). Smart cities in Europe. *Journal of Urban Technology*, 18(2), 65–82.

Carrillo, J., Yigitcanlar, T., Garcia, B., & Lonnqvist, A. (2014). *Knowledge and the city: Concepts, applications and trends of knowledge-based urban development*. New York: Routledge.

Chao-Hsien, L., En-Hao, H., & Yie-Ru, C. (2015). Designing a rainwater harvesting system for urban green roof irrigation. *Water Science & Technology: Water Supply*, 15, 271. doi:10.2166/ws.2014.107.

Cocchia, A. (2014). Smart and digital city: A systematic literature review. In *Smart city* (pp. 13–43). Berlin: Springer.

Costa, D., & Kahn, M. E. (2009). *Towards a greener California: An analysis of household variation in residential electricity purchases*, Working Paper. Los Angeles, CA: UCLA.

Darshana, S., Patil, A., Devthane, S., & Rathod, N. (2018). Clean city green city-smart garbage control system: Survey. *International Journal for Scientific Research and Development*, 6(10), 376–379.

Deakin, M., & Al Waer, H. (Eds.). (2012). *From intelligent to smart cities*. New York: Routledge.

Dhingra, M., & Chattopadhyay, S. (2016). Advancing smartness of traditional settlements-case analysis of Indian and Arab old cities. *International Journal of Sustainable Built Environment*, 5(2), 549–563.

Flower, P. C., & Ragas, W. R. (1994). The effects of refineries on neighborhood property values. *Journal of Real Estate Research*, 9(3), 319–338.

Franco, M., Pawar, P., & Wu, X. (2021). Green building policies in cities: A comparative assessment and analysis. *Energy and Buildings*, 231, 110561.https://doi.org/10.1016/j.enbuild.2020.110561

Freytag, T., Gössling, S., & Mössner, S. (2014). Living the green city: Freiburg's Solarsiedlung between narratives and practices of sustainable urban development. *Local Environment*, 19(6), 644–659.

Fuerst, F., & McAllister, P. (2011). Green noise or green value? Measuring the effects of environmental certification on office values. *Real Estate Economics*, 39(1), 1–25.

Fuerst, F., McAllister, P., Nanda, A., & Wyatt, P. (2016). Energy performance ratings and house prices in Wales: An empirical study. *Energy Policy*, 92, 20–33.

Götz, G., & Schäffler, A. (2015). Conundrums in implementing a green economy in the Gauteng City-region. *Current Opinion in Environmental Sustainability*, 13, 79–87.

Gozalo, G. R., Morillas, J. M. B., & González, D. M. (2019). Perceptions and use of urban green spaces on the basis of size. *Urban Forestry & Urban Greening*, 46, 126470.

Gudes, O., Kendall, E., Yigitcanlar, T., Pathak, V., & Baum, S. (2010). Rethinking health planning: A framework for organising information to underpin collaborative health planning. *Health Information Management Journal*, 39(2), 18–29.

Harrison, C., Eckman, B., Hamilton, R., Hartswick, P., Kalagnanam, J., Paraszczak, J., & Williams, P. (2010). Foundations for smarter cities. *IBM Journal of Research and Development*, 54(4), 1–16.

Hiremath, R. B., Balachandra, P., Kumar, B., Bansode, S. S., & Murali, J. (2013). Indicator-based urban sustainability: A review. *Energy for Sustainable Development*, 17(6), 555–563.

Jacobsen, G. D., & Kotchen, M. J. (2009). *Are building codes effective at saving energy? Evidence from residential billing data in Florida*, Working Paper. New Haven, CT: Yale University.

Jong, M., Joss, S., Schraven, D., Zhan, C., & Weijnen, M. (2015). Sustainable—smart—resilient—low carbon—eco—knowledge cities; Making sense of a multitude of concepts promoting sustainable urbanization. *Journal of Cleaner Production*, 109, 25–38.

Komninos, N. (2002). *Intelligent cities: Innovation, knowledge systems, and digital spaces*. New York: Taylor & Francis.

Komninos, N. (2016). Smart environments and smart growth: Connecting innovation strategies and digital growth strategies. *International Journal of Knowledge-Based Development*, 7(3), 240–263.

Kondepudi, S. N. (2014). *Smart sustainable cities analysis of definitions. The ITU-T focus group for smart sustainable cities*. Washington: United Nations.

Lara, A., Costa, E., Furlani, T., & Yigitcanlar, T. (2016). Smartness that matters: Comprehensive and human-centred characterisation of smart cities. *Journal of Open Innovation*, 2(8), 1–13.

Lazaroiu, G. C., & Roscia, M. (2012). Definition methodology for the smart cities model. *Energy*, 47(1), 326–332.

Lee, J. H., Hancock, M. G., & Hu, M. C. (2014). Towards an effective framework for building smart cities: Lessons from Seoul and San Francisco. *Technological Forecasting and Social Change*, 89, 80–99.

Lee, S. H., Han, J. H., Leem, Y. T., & Yigitcanlar, T. (2008). Towards ubiquitous city: Concept, planning, and experiences. *IGI Global*, 2, 148–169.

Liaw, C. H., & Tsai, Y. L. (2004). Optimum storage volume of rooftop rain water harvesting systems for domestic use 1. *JAWRA Journal of the American Water Resources Association*, 40(4), 901–912.

Marsal-Llacuna, M. L., & Segal, M. E. (2016). The Intelligenter method (I) for making "smarter" city projects and plans. *Cities*, 55, 127–138.

Meinert, S. (2014). *Field manual: Scenario building*. European Trade Union Institute.

Miller, N., Spivey, J., & Florance, A. (2008). Does green pay off? *Journal of Real Estate Portfolio Management*, 14(4), 385–399.

Nam, T., & Pardo, T. A. (2011). Conceptualizing smart city with dimensions of technology, people, and institutions. In *Proceedings of the 12th annual international digital government research conference: Digital government innovation in challenging times* (pp. 282–291). ACM.

Reichert, A. K., Small, M., & Mohanty, S. (1992). The impacts of landfills on residential property values. *Journal of Real Estate Research*, 7(3), 297–314.

Rode, P., Burdett, R., & Soares Gonçalves, J. C. (2011). Buildings: Investing in energy and resource efficiency. In *Towards a green economy: Pathways to sustainable development and poverty eradication* (pp. 331–373). United Nations Environment Programme.

Schaffers, H., Komninos, N., Pallot, M., Trousse, B., Nilsson, M., & Oliveira, A. (2011). Smart cities and the future internet: Towards cooperation frameworks for open innovation. In *The future internet assembly* (pp. 431–446). Berlin: Springer.

Simons, R. A., & Saginor, J. D. (2006). A meta-analysis of the effect of environmental contamination and positive amenities on residential real estate values. *Journal of Real Estate Research*, 28(1), 71–104.

Smolen, G. E., Moore, G., & Conway, L. V. (1992). Economic effects of hazardous chemical and proposed radioactive waste landfills on surrounding real estate values. *Journal of Real Estate Research* 7(3), 283–295.

Suhendro, B. (2014). Toward green concrete for better sustainable environment. *Procedia Engineering*, 95, 305–320.

Townsend, A. M. (2013). *Smart cities: Big data, civic hackers, and the quest for a new utopia*. New York: WW Norton & Company.

Wiedmann, F., Salama, A. M., & Mirincheva, V. (2014). Sustainable urban qualities in the emerging city of Doha. *Journal of Urbanism: International Research on Place making and Urban Sustainability*, 7 (1), 62–84.

Wu, Z., Chen, R., Meadows, M. E., Sengupta, D., & Xu, D. (2019). Changing urban green space in Shanghai: Trends, drivers and policy implications. *Land Use Policy*, LXXXVII, 104080.

Yigitcanlar, T. (2015). Smart cities: An effective urban development and management model? *Australian Planner*, 52(1), 27–34.

Yigitcanlar, T. (2016). *Technology and the city: Systems, applications and implications*. New York: Routledge.

Yigitcanlar, T., & Dizdaroglu, D. (2015). Ecological approaches in planning for sustainable cities: A review of the literature. *Global Journal of Environmental Science and Management*, 1(2), 71–94.

Yigitcanlar, T., Kamruzzaman, M., Buys, L., Ioppolo, G., Sabatini-Marques, J., da Costa, E. M., & Yun, J. J. (2018). Understanding 'smart cities': Intertwining development drivers with desired outcomes in a multidimensional framework. *Cities*, 81, 145–160.

Youssef, H. (2017). *The future of sustainable urban development in Qatar*. QGBC.

Zaina, S., Zaina, S., & Furlan, R. (2016). Urban planning in Qatar: Strategies and vision for the development of transit villages in Doha. *Australian Planner*, 53 (4), 286–301.

Websites:-

EFC Home | Wagners. Wagner. (2021). Retrieved 10 February 2021, from www.wagner.com.au/main/what-we-do/earth-friendly-concrete/efc-home/.

Glass facades harm environment—Times of India. The Times of India. (2013). Retrieved 26 January 2021, from https://timesofindia.indiatimes.com/home/environment/developmental-issues/Glass-facades-harm-environment/articleshow/22619283.cms.

India & News, I. (2021). Share of renewable energy rises to 9 per cent | IBEF. *Ibef.org*. Retrieved 26 January 2021, from www.ibef.org/news/share-of-renewable-energy-rises-to-9-per-cent.

Jaipurmc.org. (2021). Retrieved 26 January 2021, from www.jaipurmc.org/PDF/Auction_MM_RTI_Act_Etc_PDF/JNNURM_Projects_SP.pdf.

Kane, A. (2021). Making concrete green: Reinventing the world's most used synthetic material. *The Guardian*. Retrieved 10 January 2021, from www.theguardian.com/sustainable-business/2016/mar/04/making-concrete-green-reinventing-the-worlds-most-used-synthetic-material.

Kochar, M. (2018). Smart transportation: A key building block for a smart city | Forbes India Blog. *Forbes India*. Retrieved 16 December 2020, from www.forbesindia.com/blog/who/smart-transportation-a-key-building-block-for-a-smart-city/.

Nair, R. (2021). Government mulls imposing green tax on vehicles older than eight years—Mercom India. *Mercom India*. Retrieved 29 January 2021, from https://mercomindia.com/government-mulls-imposing-green-tax/.

Narain, S. (2013). A green building is not glass. *Business-standard.com*. Retrieved 26 December 2020, from www.business-standard.com/article/opinion/a-green-building-is-not-glass-113040700348_1.html.

Nation, U. (2021). *Cities*. United Nations Sustainable Development. Retrieved 16 February 2021, from www.un.org/sustainabledevelopment/cities/.

News, C. (2020). Chennai: 102cm of rain since October 1, but jury out on water harvested | Chennai News—Times of India. *The Times of India*. Retrieved 26 February 2021, from https://timesofindia.indiatimes.com/city/chennai/102cm-of-rain-since-oct-1-but-jury-out-on-water-harvested/articleshow/79685758.cms.

Padmanabhan, G. (2021). Budget 2021 expectations: Sustainable building could make a major impact. *The Financial Express*. Retrieved 13 February 2021, from www.financialexpress.com/budget/budget-2021-expectations-sustainable-building-could-make-a-major-impact/2180766/.

Roy, H., & Raju, T. (2021). Railways budget: Budgetary provisions for transport to encourage green transport? *The Financial Express*. Retrieved 5 February 2021, from www.financialexpress.com/budget/railways-budget-budgetary-provisions-for-transport-to-encourage-green-transport/2187919/.

Smart Transportation—transforming Indian cities. Grantthornton.in. (2016). Retrieved 8 February 2021, from www.grantthornton.in/globalassets/1.-member-firms/india/assets/pdfs/smart-transportation-report.pdf.

Understanding Risk. Undrr.org. (2021). Retrieved 26 January 2021, from www.undrr.org/building-risk-knowledge/understanding-risk.

United Smart Cities (USC)—United Nations Partnerships for SDGs platform. Sustainabledevelopment.un.org. (2021). Retrieved 17 February 2021, from https://sustainabledevelopment.un.org/partnership/?p=10009.

Urban population (% of total population)—India | Data. Data.worldbank.org. (2021). Retrieved 15 February 2021, from https://data.worldbank.org/indicator/SP.URB.TOTL.IN.ZS?end=2019&locations=IN&start=1960&view=chart.

What is green building? | World Green Building Council. World Green Building Council. (2021). Retrieved 26 January 2021, from www.worldgbc.org/what-green-building.

13 Improving Suspicious URL Detection through Ensemble Machine Learning Techniques

Sanjukta Mohanty, Arup Abhinna Achary, and Laki Sahu

CONTENTS

13.1 INTRODUCTION

Nowadays, the usage of the Web has increased tremendously because it provides enormous services in terms of e-learning, e-shopping, e-banking etc. So, technological advancements, like new functionalities and advanced features of web browsers, lead to different kinds of threats, attacks and frauds by invaders through malicious web

DOI: 10.1201/9781003184140-13

URLs. A single click of a malicious URL by unsuspecting users permits the attacker to enter the victim's machine and execute malware automatically, which results in a loss of sensitive information, the theft of money etc. The ultimate intention of hackers is to trap victims through mail, web searches and web links from other web pages to their desired URL with just a single click. So, it is crucial to know which URL components and meta-data are responsible for propagating the malware in websites (Ma et al. 2009). The host name generally contains an average of two or three dots, and the path component usually contains zero dots. So, a path that contains more than one dot would be considered suspicious. If the potential victim could know that a particular URL is dangerous prior to performing any action, the problem could be solved to some extent. For this purpose, researchers developed blacklisting approaches that they embedded into toolbars, search engines, appliances etc. (Ma et al. 2009). But the disadvantages of this approach are incorrect listing and cloaking. The major issues with blacklisting is that it fails to detect suspicious URLs in the early hours of an attack because of its slow updating process (Yoo et al. 2014). As the different security threats grow exponentially, it is very tedious to identify security breaches. So, it is very important to adopt a robust technology for identifying a malicious web link in the ever-growing web environment.

In this chapter, we design a framework for detecting suspicious URLs by considering URL features without requiring the content of the web page. We distinguish malignant web pages based on different discriminative and effective URL features, including lexical features, HTTP header information–based features, host-based features, geographical features and network features whose predictive power is high and improves performance significantly. Moreover, our approach uses both batch machine learning (ML) algorithms and ensemble machine learning classifiers (EMLCs) to identify suspicious URLs. EMLCs use multiple weak learners that are trained on different training examples to enhance the model performance effectively (TRAGHA 2019). We have compared the batch ML algorithms with ensemble models experimentally and ascertained that the ensemble approach outperforms the batch ML classifiers. We have extended our previous approach (Mohanty et al. 2020), where we used only some batch learning classifiers and a few URL features to identify URLs as either malignant or safe and obtained the classifier; the random forest (RF) model achieved the highest accuracy at 95%. Our proposed approach is evaluated against a training dataset that contains some safe and some malicious URLs and shows that the ensemble techniques obtain a TPR (true positive rate) of 0.98, FPR (false positive rate) of 0.01 and accuracy 98.66%, precision of 0.95, recall of 100%, F1 score of 96% and AUC of 0.982.

13.2 BACKGROUND

In this section, we have presented a brief introduction of URLs, describing their different structural components. Next, the ML concepts, along with different ML classifiers, are explained.

13.2.1 Uniform Resource Locator (URL)

URL refers to the global address of a resource or document. A URL can be divided into four components: a scheme, a host, a path and a query string. A scheme represents the protocol (http, https, ftp, mailto, file etc.) to be used for accessing the

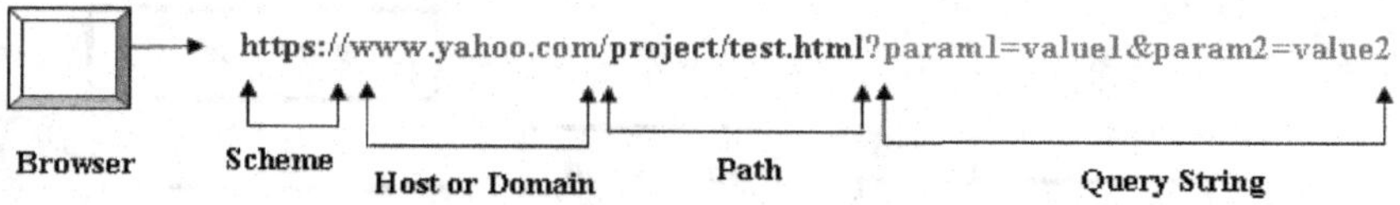

FIGURE 13.1 Different Components of the URL.

information from the internet and to establish secure and smooth communication. A host or domain name (www.yahoo.com) represents the destination location for the URL. A host name consists of subdomain, second-level domain and top-level domain. In Figure 13.1, the term 'www' represents the subdomain, the text 'yahoo' indicates the domain name and the term 'com' represents the top-level domain. The components of domain names are separated by dots and preceded with a double slash. The path name includes the files and directory that represents the location of files in the server and are separated by slashes (/project/test.html). A query string is placed after the path and gives a string of information to the file for some purpose and contains the parameter list, which includes name and value pair (param1=value1). Each query string is preceded by a question mark and separated with an ampersand (&). The components of the URL are represented in Figure 13.1.

13.2.2 Machine Learning

Machine learning is a learning technique in which a machine can learn on its own without being explicitly programmed. It is an application of artificial intelligence (AI) that provides the system the ability to learn automatically, to perform their job skillfully and improve from the experience. The objective of learning is to generate a model that receives the input and produces the required output. It can be broadly classified into three categories according to the kind of data they require.

Supervised learning: In supervised learning, a labeled or training dataset is used to make prediction. The training dataset includes both input and output data, and together they constitute a training example.

The output vector consists of labels or tags for each training example present in the training data (Mohammed et al. 2016). The model or the classifier is trained with the labeled data, and if the training accuracy is acceptable, it is tested with a new dataset or unknown data to generate the class labels. This is represented in Figure 13.2. Two important techniques that come under the supervised ML algorithm are classification and regression.

Classification: Classification falls under discrete categories or predicts the discrete class labels. Here, the inputs are divided into two or more classes. The division of inputs into two classes constitute binary classification, and more than two classes forms categorical classification. The application of the classification model includes fraud URL detection (input is the URL and the classes are malicious URL or benign URL), spam mail filtering (input is the email messages and class is spam or not spam), sentiment analysis and scorecard prediction of any examinations (pass or fail) etc. In our problem domain, we have used the binary classification technique to predict suspicious URLs.

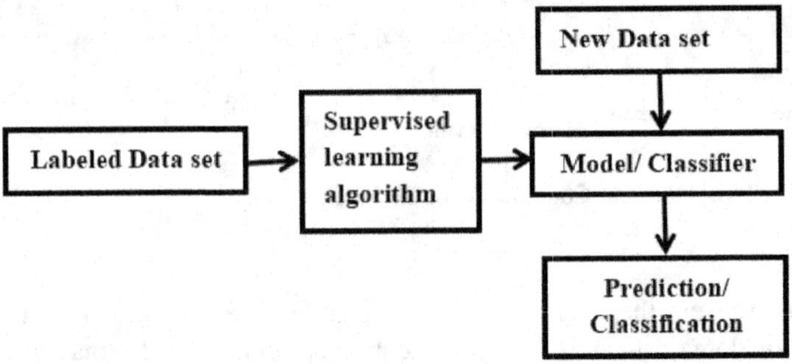

FIGURE 13.2 Working Principle of Supervised Learning Algorithm (IBM Developer 2020).

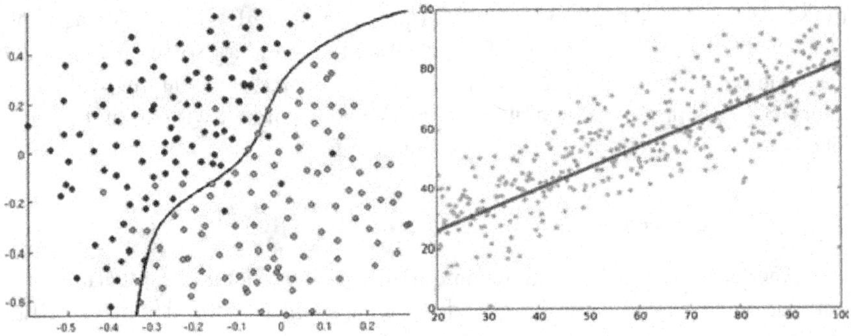

FIGURE 13.3 Classification and Regression (The Genius Blog 2020).

Regression: Regression is the task of predicting continuous quantity or it is a real valued number, such as salary, weight, pressure etc., that exists in a continuous space. The objective is to model the relationship between the dependent and independent variables. The regression technique can be used for making both predictions and inferences.

The classification and regression techniques are depicted in Figure 13.3. The solid line in classification represents a linear boundary that separates the two classes as two different types of dots. In regression, the straight line is drawn through the data points to generate a best-fit line that models the relationship between two variables (dependent and independent variables).

Unsupervised learning: Unsupervised learning is a type of ML algorithm used to draw inferences from datasets consisting of input data without labeled responses. The idea is to find a hidden structure from the unlabeled data.

Reinforcement learning: Reinforcement learning trains the machine to be able to make specific decisions. Here, the machine is exposed to an environment where it uses a trial-and-error method to train itself repeatedly. This machine learns from past experience and tries to capture the best possible knowledge to make accurate decisions.

In this chapter, we have adopted binary classification models of supervised ML techniques because our objective is to classify the URL as suspicious or benign. We have considered the supervised batch ML classifiers, such as SVM (support vector machine), Naive Bayes (NB) and some ensemble classifiers, such as bagging and boosting techniques, for malicious URL detection.

13.2.2.1 Batch Machine Learning Classifiers

SVM: SVM sorts the data into two distinct categories as malicious and benign by drawing a hyperplane or decision boundary in N-dimensional space (N is the number of features) between the two opponent classes (malicious and benign) in such a way that the margin length will be maximized to reduce the high chance of misclassification; this is represented in Figure 13.4. SVM is effective in cases where the number of dimensions is higher than the number of samples, but its training time increases with large datasets, which negatively affects performance. SVM can be formed as (Kazemian et al. 2015):

$$h(x) = b + \sum_{n=1}^{N} yi \propto iK(x, xi) \tag{13.1}$$

$h(x)$ represents the distance of the decision boundary,
b indicates the bias weight,
$\propto$ is a coefficient that maximizes the margin of correctly classified data points on the labeled data,
N represents the number of features,
K is the kernel function and
x represents the feature vector.

The value of h(x) is restricted to predict a binary label for the feature vector x. First, the model is trained by initial kernel function $k(x,x_i)$, then α is computed, which

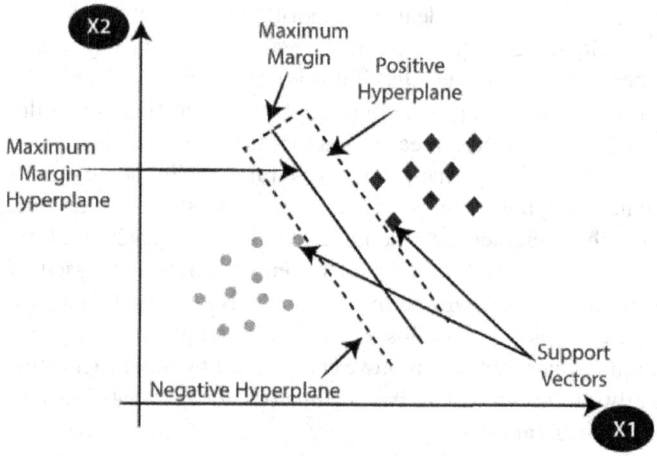

FIGURE 13.4 Support Vector Machine (Javapoint Tutorial 2020).

maximizes the margin length of correctly classified data points on the training data. SVM is the most common method for classifying malicious URLs (Sahoo et al. 2017).

NB: NB is a classification model that works on the principles of conditional probability as given by Bayes' theorem and assumes all the features are independent of each other. This is a very simple model and easy to implement and requires less training data than SVM. It can handle both the discrete and continuous data and is also highly scalable with number predictions and data points. This algorithm is not sensitive to irrelevant data. The NB model can be depicted as Let P(A | B) indicate the conditional probability of the feature vector given a class, the independence assumption implies that $P(A \mid B) = \prod_{k=1}^{n} p(A_k \mid B)$, where n is the number of features. So, the probability that a feature vector A is a malicious URL (Sahoo et al. 2017) can be calculated as:

$$(B = 1 \mid A) = \frac{p(A \mid B = 1)}{p(A|B = 1) + p(A \mid B = -1)} \tag{13.2}$$

13.2.2.2 Ensemble Machine Learning Classifiers

In addition to the batch learning algorithms discussed in Section 13.2.2.1, we have also adopted EMLCs for solving the binary classification issues of malicious and benign URL class. The EML technique combines the individual model together to improve the reliability, stability and predictive performance of the ensemble model. It is a very successful learning model that boosts the performance of the ensemble model intensively. EMLCs are categorized into two types, bagging and boosting.

Bagging: In bagging, the training set is divided into different subsets, which are used with replacement to the same classifiers of the ensemble (Bai et al., 2016). Here, the estimators are run independently or parallel to produce the output. Then, the prediction of individual estimators are aggregated through voting to form a final prediction. Bagging methods work best with strong and complex models (Bai et al., 2016) but fails with the stable learning algorithm.

Boosting: Boosting connects the weak learners sequentially to achieve a strong learner. The different boosting algorithms are as follows:

AdaBoost algorithm: The AdaBoost (adaptive boosting) algorithm trains base weak estimators iteratively. In each iteration, it applies weight w_i = 1/N to each of the training data. In Step 1, the estimator is trained with the original data. For each successive iteration, the weight of the samples is modified and again reapplied to the estimators with reweighted data (Scikit Learn Organisation 2020). If the samples are predicted correctly by the boosted model, weights will be decreased, and if misclassification occurs, weights will be increased. Therefore, weak learners are bound to focus on the examples that are lost or missed in the previous ones in the sequence. As a result, this algorithm can improve errors made by the earlier estimators sequentially. Finally, all the successive base estimators are aggregated using weighted voting to predict the final prediction.

Gradient Boosting (GB): This is an EML technique that uses boosting techniques to join individual models sequentially. It finds the residuals or errors of previous estimators and further sums them to find the final output. This process is repeated till the error becomes zero or the stopping criterion is reached (Towards data science 2020). GB repeatedly leverages the pattern in residuals and strengthens a model with weak prediction. Modeling is stopped when residuals do not have any pattern that can be modeled. As the GB adds many models in series and each model has an aim to minimize the error percentage of the previous one, it makes the boosting method highly efficient and increases the model's accuracy.

Random forest (RF): RF classifies the problem of malicious URL prediction more accurately and provides a stable prediction. It builds multiple decision trees (DTs) and merges them to correct the over-fitting habit of DTs. It uses the principle of bagging, where all the DTs are used as parallel estimators. The result of the RF is calculated on the majority vote of the results obtained from each DT. Unlike DTs, RF randomly selects the samples from the training data with replacements and selects random features for each DT in a random forest. It runs efficiently on large datasets and its training time is less.

13.3 RELATED WORK

In our research study, we have reviewed several static ML techniques pertaining to malicious URL prediction using structural URL features and web page content features of URL sequentially.

13.3.1 Machine Learning Techniques Adopting Structural Features of URLs

A supervised ML approach presented by Ma et al. (2009) for identifying the URL automatically by adopting the lexical and host-based features of websites. They evaluated their approach with up to 30,000 instances and achieved promising results, with 90–95% accuracy and a low FPR. Tao et al. (2010) designed an approach for automatically identifying websites as malignant or benign by making the model learn through HTTP session information–based features, which was further subdivided into domain-based features and HTTP session header–based features. The accuracy of the approach was 92.5%, FPR was 0.1% and TPR rate was low. Aldwairi and Alsalman (2012) proposed a lightweight statistical self-learning system to classify the website by using lexical features of URLs, host-based features and some other special URL features. To improve the speed of classification, the authors applied an evolutionary genetic algorithm by expanding the training dataset automatically, hence reducing the training time. This approach used NB classifiers to identify the malignant web pages. The advantage of this methodology was low memory usage. A limitation to this approach is the consideration of a single classifier, which leads to less accuracy. A lightweight web page–classification approach (Darling et al. 2015) using only the lexical features of URLs (87 features that were further classified into 5 groups) and logistic regression, J48, NB and k-NN (k-nearest neighbor) classifiers. The average speed of this approach for processing and classifying the URL was 0.627 milliseconds and yielded 99% accuracy. The FPR value was 0.4%, but the FPR of our proposed approach is 0.01%. The F1 score of this approach was 98.7. Liu et al.

(2018) designed an approach by combining statistical analysis of URLs, ML techniques and character features of URLs to effectively classify malicious URLs. They used a statistical method to gather character distribution and structural features. To obtain the best performance, they chose six different classifiers, out of which RF classification achieved a precision of 99.7% and an FPR less than 0.4%. WU et al. (2018) identify malicious websites based on static features of URLs. All features are extracted statistically to reduce the time and constructed feature vectors. Then, they selected some reasonable ML classifiers, like NB, LR, SVM and DT. The proposed approach achieved high accuracy and efficiency. NB achieved a higher precision of 0.964, recall of 0.964, F1 score of 0.963 and mean accuracy of 0.964 with the shortest training time of 0.101s compared to other classification algorithms.

13.3.2 Machine Learning Techniques Adopting Structural and Web Page Content Features

Choi et al. (2011) developed an ML technique to detect malicious URLs and different types of threats URLs attempt to launch; hence, they can estimate the severity of the attack. In this approach, the lexical features link popularity features, web page content features and DNS for network features are used. The URL is classified as malicious or benign by SVM classifier with an accuracy of 98%, and both RAkEL and MLkNN were used to detect different types of attack with an accuracy of 93%. A holistic, lightweight approach (Eshete et al. 2012) called BINSPECT considered the URL features, page source–based features and social reputation–based features and implemented a supervised ML technique for identifying malicious web pages. BINSPECT is effective in identifying malicious web pages precisely with accuracy of 97% and a low false signal that was represented through large-scale evaluation of experiments with a very low performance overhead. They leveraged the static and dynamic analysis of web pages and lacked evolution-aware analysis. Eshete et al. (2013) developed an evolution-guided and learning-based approach called EINSPECT, which utilized the evolutionary approach of the genetic algorithm to identify and analyze malicious web pages. EINSPECT was executed and evaluated in a large-scale dataset and improved the precision using evolutionary searching and the model optimization technique of GA for the prediction of suspicious URLs. Wang et al. (2017) proposed a hybrid analysis (static, dynamic and hybrid) method for identification of malignant web pages. The approach considered features like HTML document–based features and Java Script–based features as well as some URL features. Their proposed approach achieved high performance. The precision of static analysis was 0.92, dynamic analysis was 0.86 and hybrid analysis was 0.95.

13.4 DISCUSSION

For suspicious URL identification, previously the blacklisting methodology was widely used. The blacklisting features have shown to be feeble in the zero hour of web page detection, as they can consume more time and may even take hours or days, and the blacklisting features are less effective than the URL lexical features used in ML techniques. Table 13.1 summarizes the limitations of various detection methods.

TABLE 13.1
Summary Table Lists the Limitation of Different Approaches for Detection of Malicious URLs.

Authors	Feature sets	Methodology	Limitations	Results
Ma et al. 2009	Lexical & host-based features	NB, SVM, LR	Low FPR	Accuracy: 95–99%
Tao et al. 2010	HTTP session information–based features	NB, DT, SVM	Characterize the web page exclusively based on HTTP session information–based features. Low TPR	FP: 0.1% TP: 92.2%
Aldwairi and Alsalman 2012	Lexical & host-based features	NB	Less precision value	Precision: 87%
Darling et al. 2015	Lexical features	LR, J48, NB and K-NN	FPR is more. Not succeeded to find the length of shortened URL.	Accuracy: 0.990 F1 score: 0.987 FP: 0.4%
Liu et al. 2018	Structural features	RF, J48, LR, Lib SVM, Multilayer Perceptron and NB	Discarded the HTTP session information–based features.	Precision: 0.997 FP < 0.4%
WU et al. 2018	Static features	NB, LR, SVM and DT	Limited to lexical features only	Accuracy: 0.964 Precision: 0.964 Recall: 0.964 F1 score: 0.963
Mohanty et al. 2020	Lexical and a few network features	NB, SVM, LR and RF	Limited number of feature sets	Accuracy: 95%

The research studies adopting page content features require time-intensive look-ups, which results in significant delays in a real-time system. Also, the limitation of page content–based classification leads to failure of audit instructions and search engines because the contents of malicious sites would appear to be normal and regular (Bai et al., 2016). URL-based classification is more reliable for identifying fraudulent sites than page content–based classification because the background information of a URL, like domain knowledge, host-based information and indexing information from search engines, are hard for hackers to manipulate. So, to address these challenges, in this chapter, we have presented a lightweight, learning-based approach using only the distinct structural features of URLs for classifying URLs as malicious or benign precisely and effectively.

13.5 PROPOSED ARCHITECTURE

This section describes the architecture of our framework for identifying fraudulent sites. To address the issue of page content–based classification, we designed a lightweight, learning-based classification approach based on structural features of URLs, which is depicted in Figure 13.5. The key steps associated with building this classification framework are URL dataset, data cleaning, relevant feature selection, data

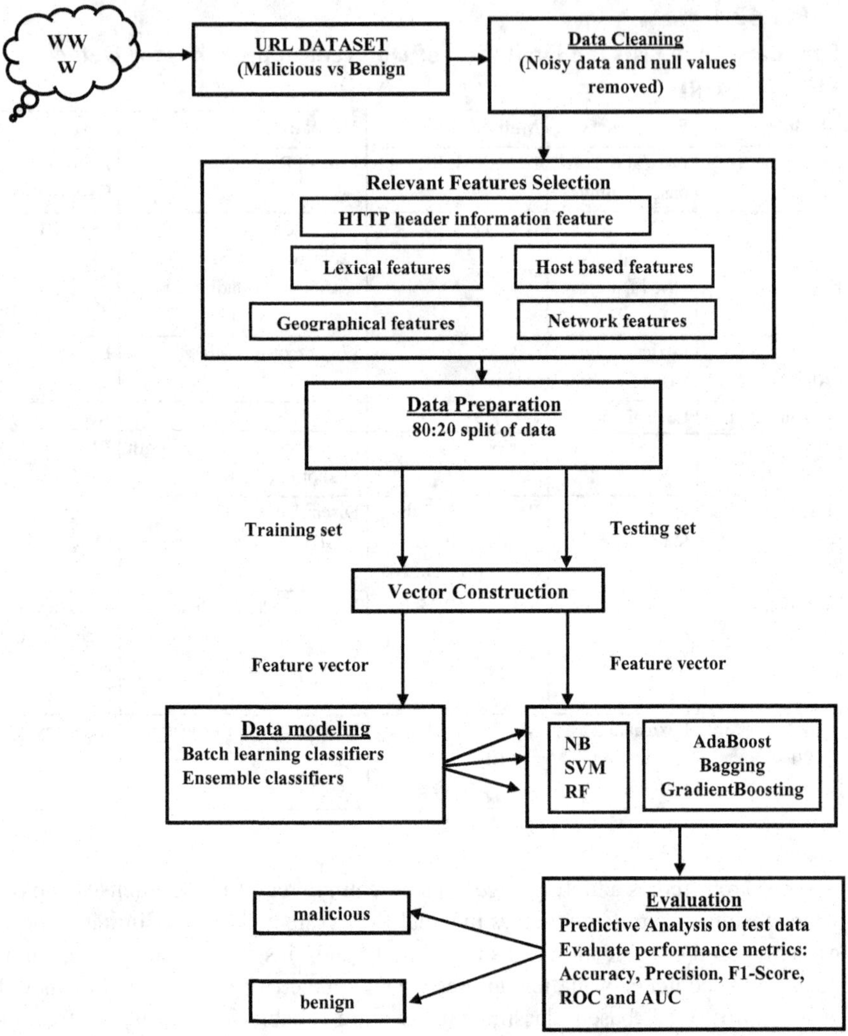

FIGURE 13.5 Proposed Framework for Identifying Suspicious URLs.

preparation, vector construction, data modeling and evaluation. In the subsequent sections, each phase is explained in detail.

13.5.1 URL Dataset and Data Cleaning

The selection of the dataset greatly influences the classification results. The dataset of our proposed approach is collected from internet sources (Urcuqui et al. 2017), where 185,181 malicious links and 345,000 benign URLs are implemented in a self-developed tool for creating the dataset. The dataset contains a total of 1,782 records and 21 attributes. Figure 13.6 shows a snapshot of the dataset.

```
In [475]:  df.head( )
Out[475]:
```

	URL	URL_LENGTH	NUMBER_SPECIAL_CHARACTERS	CHARSET	SERVER	CONTENT_LENGTH	WHOIS_COUNTRY	WHOIS_STATEPRO	WHOIS
0	M0_109	16	7	iso-8859-1	nginx	263.0	None	None	10-1(
1	B0_2314	16	6	UTF-8	Apache/2.4.10	15087.0	None	None	
2	B0_911	16	6	us-ascii	Microsoft-HTTPAPI/2.0	324.0	None	None	
3	B0_113	17	6	ISO-8859-1	nginx	162.0	US	AK	07-1(
4	B0_403	17	6	UTF-8	None	124140.0	US	TX	12-0

5 rows × 21 columns

```
In [476]:  df.shape
Out[476]: (1781, 21)
```

FIGURE 13.6 URL Dataset (Urcuqui et al. 2017).

The attribute server had 1 NULL value, conent_length had 812 NULL values and dns_query_time had 1 NULL value. We have cleaned the dataset by removing these NULL values of the attributes conent_length, server and dns_query_time in the python environment using Jupyter Notebook IDE. We labeled the dataset 0 for benign web pages and 1 for malicious web pages.

13.5.2 Relevant Feature Selection

This section presents the most relevant features applied in the field of malicious URL identification to distinguish between malicious and benign websites. Some URL features are more nontrivial and relevant than other features in contributing to identifying malicious and benign URLs. Hence, selecting these distinctive URL features plays a significant role in enhancing the accuracy of the model for malicious website detection. We have investigated and reviewed numerous popular URL features from the literature and categorized them into five types, including lexical features, HTTP header information–based features, host-based features, geographical features and network features depicted in Table 13.2.

Lexical features: Lexical features represent the structural component of the URL rather than the web page content to which it refers (Ma et al. 2009). These lexical features include the URL length and the number of special characters (., /, ?, =, -, _ etc.). The reason for using lexical features is because the malicious URL tends to look abnormal compared to the benign one; usually, the length of a malicious URL is longer than a benign one, and instead of a domain name, sometimes the IP address, which is rarely used in normal websites, is used by the intruder to confuse users (Begum et al., 2020). They may also use of lots of dots (.), underscores (_) and slashes (/) in the domain name, leading to an abnormal path length. So, focusing on these lexical characteristics, we can train a model easily in a shorter amount of time than going for the page content features as in the latter case. First, we have to download the page, then use its content for classification purposes. Hence, lexical features have the potential to improve classification accuracy.

HTTP header information feature: This includes HTTP header content length, the HTTP header server and the HTTP header charset. The majority of data communication over the internet occurs through the mechanism governed by http or https protocols (Kumar et al. 2018). So, the information regarding the http header contributes some kind of security to the web page. An intruder may set a negative value (malformed) to the content length field of their websites, so it must be noticed. Hence, HTTP header content length attribute acts as a discriminative feature for identifying malicious web pages.

Host-based features: These explain where the malicious web links are hosted; who the owner of that link is, how it is managed and what its registration date, updating date and expiration date are. The justification for using host-based features is that, unlike normal websites, suspicious websites are generally not hosted in reputable hosting centers. They are normally hosted in a computer of nonconventional web host or disreputable registrar.

Geographical features: These represent the country and state to which the IP address belongs.

Network features: These describe features like who is the domain, what the TCP port number is etc. Usually, malicious web pages exist on the Web for a very short time; hence, this TTL (time to live) feature is important for detecting fraudulent sites. Malicious websites tends to be hosted by less reputable service providers, so the DNS information provides some novel and discriminative features in identifying the malignant web pages. The features are represented in Table 13.2.

13.5.3 Data Preparation

After selecting the relevant features, we prepare the data for the classification models. The noisy data are cleaned by removing the NULL values contained in the value of the attributes of the dataset, and the unstructured data are formatted and then converted into a numerical vector. Then the dataset is divided into two parts in the ratio 80:20. Most (80%) of the dataset is used for training the classifiers and the rest (20%) is used for testing the predictive model.

13.5.4 Data Modeling with Machine Learning Classifiers

This section concisely describes the ML models we considered for URL classification. For detecting malicious URLs, different ML techniques are adopted, such as batch learning techniques, online learning techniques and representation algorithms (Sahoo et al. 2017). In this chapter, we have applied batch learning techniques and compared the results with EML techniques. Before implementing any technique, the URL is converted into some feature vectors, then the learning algorithms are applied to train the predictive model. In the batch learning algorithm, it ensures that the training data is available prior to the training task. Here, we have discussed some popular batch learning algorithms, like SVM, NB, RF, and some ensemble classifiers, including bagging, AdaBoost and gradient boosting algorithms for detecting malicious URLs. The different ML classifiers are described in detail in Section 13.2.

TABLE 13.2
Feature Sets.

A	Lexical Features	Description
1	URL length	Total length or size of the URL
2	No. of special characters	Total number of special characters (., /, ?, =, -, _ etc.) that appear in the URL string
B	**HTTP header information feature**	
3	HTTP header content length	Content length of entire http packet
4	HTTP header server	Provides information about the web page server (name, type and version etc.)
5	HTTP header charset	Indicates the encoding scheme of each URL (UTF-8) and says what character sets are acceptable for the response
C	**Host-based features**	
1	Whois regDate	Registration date of the website's server
2	Whois updatedDate	Last time the server was updated
D	**Geographical features**	
1	Whois Country	Country from where we got the server response
2	Whois State	State from where we got the server response
E	**Network-based features**	
1	Whois Domain	Domain of the website
2	TCP conversion Exchange	Number of packets exchanged between server and client by TCP protocol
3	Remote IPs	The total number of IPs connected to the client
4	PKT without DNS	Number of non-DNS packets
5	Source App_Pkts	Number of packets sent from the client to the server of the remote viewer
6	Remote App_Pkts	Number of packets obtained from the server
7	TTL or Duration	A value for DNS records associated with host name that specifies the duration time of the websites
8	Avg Local pkt_rate	Average local IP packets per second (packets sent over the duration)
9	Avg Remote pkt_rate	Average remote IP packets per second sent from the remote server to the client
10	App_pkts	Total number of IP packets created during the communication between the client and server.
11	DNS query_time	Number of DNS packets created during the communication between client and server.
1	Type	Represents the type of website analyzed; for suspicious websites, Type=1, for benign, Type=0

13.5.5 Performance Evaluation Measure

We have used the following metrics to monitor and measure the performance of the ML classifiers. We have considered malignant websites as positive examples and safe websites as negative examples.

***TP (true positive or sensitivity)*:** The ratio of malignant URLs predicted as malicious over all malicious websites.

***FP (false positive)*:** The ratio of safe websites predicted as malignant over all safe web pages.

***TN (true negative or specificity)*:** The number of websites predicted as non-malicious over all the safe websites.

***FN (false negative)*:** The number of malicious websites predicted as safe over all the malicious websites.

***Accuracy*:** The ratio of the number of correct predictions made with the total number of predictions made, denoted as:

$$\text{Accuracy} = \frac{\text{correct prediction}\left(\text{TP}+\text{TN}\right)}{Totalnumberofprediction\left(TP+FP+FN+TN\right)}$$

$$\boldsymbol{Precision\left(Exactness\right)} = \frac{\text{TP}}{TP+FP}$$

$$\boldsymbol{Recall\left(Completeness\right)} = \frac{\text{TP}}{TP+FN}$$

$$\boldsymbol{F1\text{-}Score} = \frac{2*\text{precision}*\text{recall}}{precision+recall}$$

***AUC (area under curve)*:** This can be defined as the area under ROC (receiver operating characteristics) curve. The classifier with the larger AUC can distinguish positive and negative classes perfectly (Liu et al. 2018).

13.6 EXPERIMENTS AND RESULTS

In this section, we assess the effectiveness of the models in classifying suspicious URLs. Here, we want to solve whether the set of features we have considered are effective to classify URLs appropriately and determine which classification model is best fit for the requirement. In our experiment, the proposed methodology is implemented and evaluated in the python environment using Jupyter Notebook (an ML platform) (Jupyter Organisation 2020), and the graph was plotted with the help of matplotlib. We have used some commonly used measures to study the efficiency and performance of the supervised ML techniques. The easiest way to measure the performance of a classification problem is the confusion matrix, which is a successful detection measure used for describing the performance of implemented classification models (Altay et al. 2019). It is a contingency table that enables the tabular visualization of the model prediction (predicted value) versus the ground truth level (actual value), and it is shown in Table 13.3.

TABLE 13.3
Confusion Matrix for the Problem of Suspicious URL Identification.

	Malicious Websites	Benign Websites
Classified as malicious	True Positive (TP)	False Positive (FP)
Classified as benign	False Negative (FN)	True Negative (TN)

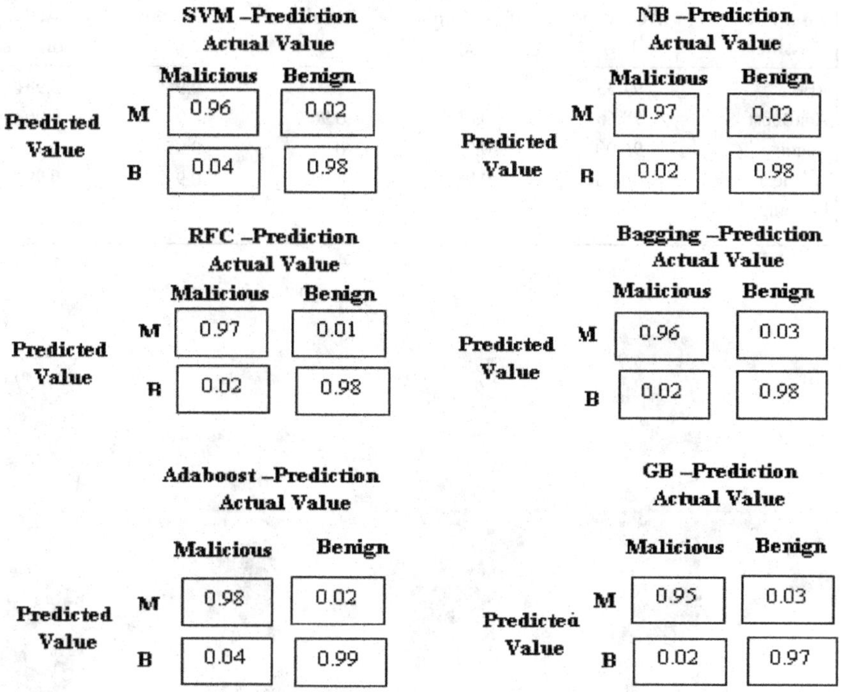

FIGURE 13.7 Confusion Matrix of Different ML Classifiers.

The confusion matrix has four sections, TP, FP, FN and TN, represented in Table 13.3. True positive denotes that a malicious web page is correctly identified as malicious. FP indicates that a benign web page is incorrectly identified as malignant. FN represents that a malignant web page is incorrectly labeled as benign. True negative means that a benign web site is correctly identified as benign. Malicious URLs are represented as ‘M’ and benign URLs as ‘B’ in Figure 13.7.

Figure 13.7 shows the confusion matrices of different ML classifiers in terms of TPR, FPR, TNR (true negative rate) and FNR (false negative rate). For better performance, the proposed method of malicious website detection should achieve high classification accuracy, TPR and TNR and must produce low FPR and FNR. We have evaluated the performance of both batch ML classifiers and ensemble classifiers with distinct feature sets and determined that the ensemble classifier AdaBoost algorithm

TABLE 13.4
Performance Evaluation of Different Classifiers.

Batch Learning Classifiers	Accuracy (%)	Precision (%)	Recall (%)	F1 score (%)	AUC score (%)
SVM	96.00	0.90	0.95	0.92	0.983
NB	97.22	0.50	0.75	0.60	0.978
Ensemble Classifiers	**Accuracy (%)**	**Precision (%)**	**Recall (%)**	**F1 score (%)**	**AUC score (%)**
RFC	97.33	0.92	1.00	**0.96**	0.981
AdaBoost	**98.66**	0.88	0.96	0.92	0.967
Bagging	96.00	0.93	0.96	0.95	0.980
Gradient boosting	94.66	**0.95**	0.92	0.94	**0.982**

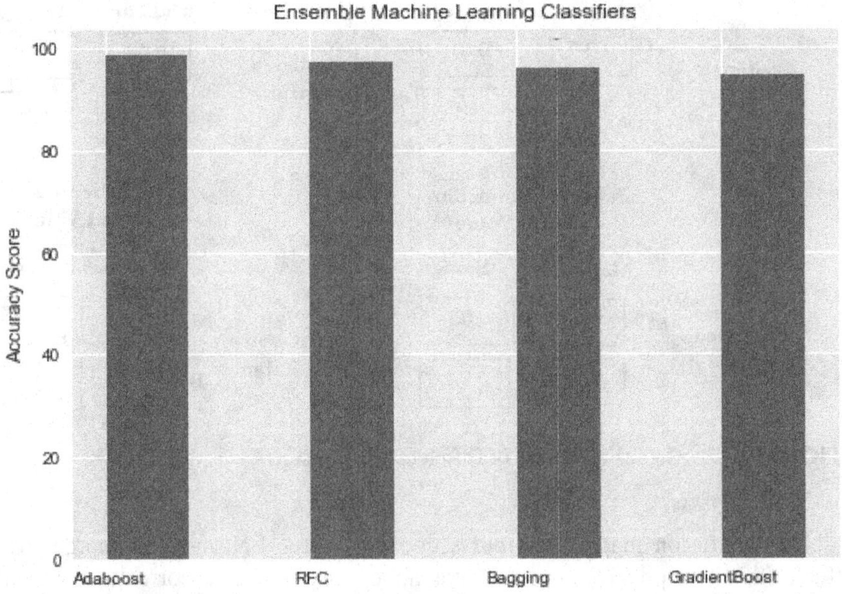

FIGURE 13.8 Classification Accuracy Obtained by ML Algorithms.

achieves the highest TPR at 0.98%, TNR at 0.99%, low FNR at 0.02%, as represented in Figure 13.7, and the highest accuracy at98.66%, depicted in Table 13.4.

Table 13.4 represents the values of different performance measures, like accuracy, precision, recall, F1 score and AUC. The Adaboost classifier of ensemble machine learning technique achieves the highest accuracy value of 98.66% among all the classifiers mentioned in Table 13.4.

Figure 13.8 shows the performance in terms of the classification accuracy of AdaBoost, RFC, bagging and gradient boosting. As mentioned, the highest classification accuracy is achieved by AdaBoost (98.66%), followed by RFC (97.33%), NB (97.22%), bagging (96.00%), SVM (96%) and gradient boosting (94.66%). GB produced the worst classification accuracy (94.66%), but precision (exactness) is higher, at 95%, than other estimators in detecting malicious URLs. Recall (completeness) value of RFC is highest (100%), and AdaBoost estimators achieve 96%. Gradient boosting achieves the highest precision (95%) and the highest AUC score (98.2%).

The ROC graph for the ML classifiers is used to validate the quality of the proposed method. The ROC graph plots the two parameters defined as true positives and false positives (Ali et al., 2020). TPR is represented on the X axis and FPR is represented on the Y axis at various classification threshold settings. The perfect classification model would produce a point in the upper left corner, or coordinate (0,1) of ROC, indicating 100% TPR and zero FPR. ROC curves of different ML classifiers, including AdaBoost, RFC, bagging and gradient boosting are plotted in Figure 13.9. As can be observed from Figure 13.9, AdaBoost has a good classification performance.

The mean accuracy of different ML classifiers is represented in Figure 13.10. The boosting classifiers show the lowest mean in comparison to batch ML classifiers because the boosting estimators uses the weak learners to predict the results. So, for this dataset, the mean accuracy of all the weak learners might be less, and we will consider it for our next extended work.

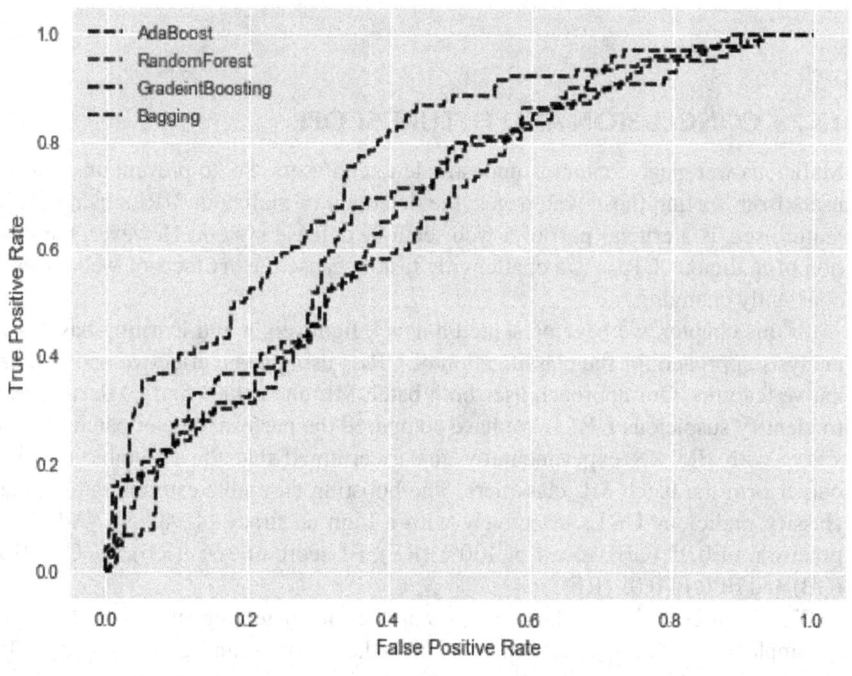

FIGURE 13.9 ROC Curve of Different ML Classifiers.

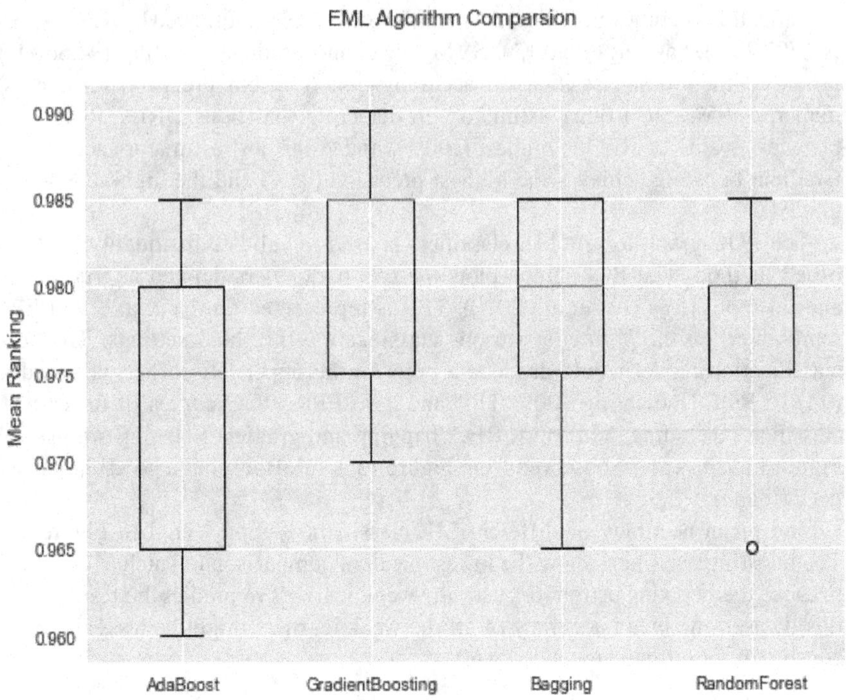

FIGURE 13.10 Mean Accuracy Ranking of ML Classifiers.

13.7 CONCLUSION AND FUTURE SCOPE

Malicious web page content is undesirable to end users. So, to prevent unsuspecting users from surfing those web pages, the detection of malicious URLs using distinct feature sets is a crucial part of a web security defense system. However, the detection of malicious URLs is a challenging task because the artifacts of web pages are constantly changing.

In this chapter, we have presented a novel, lightweight and learning-based static analysis approach for the classification of URLs using some effective and discriminative features. Our approach uses both batch ML algorithms and EML algorithms to identify suspicious URLs. We have compared the performance of batch ML classifiers with EMLCs experimentally and ascertained that the ensemble classifiers outperform the batch ML classifiers. The boosting ensemble estimators are able to classify malicious URLs effectively with a high accuracy of 98.66% (AdaBoost), precision of 0.95 (GB), recall of 100% (RF), F1 score of 96% (RF), AUC of 0.982 (GB) and FPR of 0.01 (RF).

The features and the ML techniques applied for detecting suspicious URLs can be employed in spam detection, web page classification and in identifying different attacks—web phishing attacks, XSS (Cross Site Scripting) attacks, drive-by-download attacks etc—.because the lexical features, host-based features and content-based features used for spamming, phishing, XSS attacks, drive-by-download attacks and web page classification are a subset of those features that are commonly used for suspicious URL detection.

In the future, we plan to include two or more datasets to optimize the ML model to improve the malicious URL detection performance and to enhance the feature sets by utilizing different feature selection techniques to lessen the training time and increase the detection accuracy of suspicious URLs. Also, we plan to detect some novel attacks like drive-by-download and XSS attacks by using the different URL-based features and adopting the ensemble classifiers.

REFERENCES

Aldwairi, M., & Alsalman, R. 2012. Malurls: A lightweight malicious website classification based on URL features. *Journal of Emerging Technologies in Web Intelligence*, *4*(2), 128–133.

Ali, W., & Malebary, S. 2020. Particle swarm optimization-based feature weighting for improving intelligent phishing website detection. *IEEE Access*, *8*, 116766–116780.

Altay, B., Dokeroglu, T., & Cosar, A. 2019. Context-sensitive and keyword density-based supervised machine learning techniques for malicious webpage detection. *Soft Computing*, *23*(12), 4177–4191.

Bai, J., & Wang, J. 2016. Improving malware detection using multi-view ensemble learning. *Security and Communication Networks*, *9*(17), 4227–4241.

Begum, A., & Badugu, S. 2020. A study of malicious URL detection using machine learning and heuristic approaches. In *Advances in Decision Sciences, Image Processing, Security and Computer Vision* (pp. 587–597). Springer, Cham.

Choi, H., Zhu, B. B., & Lee, H. 2011. Detecting malicious web links and identifying their attack types. *WebApps*, *11*(11), 218.

Darling, M., Heileman, G., Gressel, G., Ashok, A., & Poornachandran, P. 2015. A lexical approach for classifying malicious URLs. In *2015 International Conference on High Performance Computing & Simulation (HPCS)* (pp. 195–202). IEEE, Amsterdam, Netherlands.

Eshete, B., Villafiorita, A., & Weldemariam, K. 2012, September. Binspect: Holistic analysis and detection of malicious web pages. In *International Conference on Security and Privacy in Communication Systems* (pp. 149–166). Springer, Berlin, Heidelberg.

Eshete, B., Villafiorita, A., Weldemariam, K., & Zulkernine, M. 2013. Einspect: Evolution-guided analysis and detection of malicious web pages. In *2013 IEEE 37th Annual Computer Software and Applications Conference* (pp. 375–380). IEEE, Kyote, Japan.

The Genius Blog. Online from https://kindsonthegenius.com/blog/what-is-the-difference-between- classification-and-regression. Retrieved on September 2020.

IBM Developer. Online from https://developer.ibm.com/technologies/artificial-intelligence/articles/cc-supervised-learning-models. Retrieved on October 2020.

Javapoint Tutorial. Online from www.javatpoint.com/machine-learning-support-vector-machine-algorithm. Retrieved on October 2020.

Jupyter Organisation. Online from https://jupyter.org/. Retrieved on 15 May 2020.

Kazemian, H. B., & Ahmed, S. 2015. Comparisons of machine learning techniques for detecting malicious webpages. *Expert Systems with Applications*, *42*(3), 1166–1177.

Kumar, H., Gupta, P., & Mahapatra, R. P. 2018. Protocol based ensemble classifier for malicious URL detection. In *2018 3rd International Conference on Contemporary Computing and Informatics (IC3I)* (pp. 331–336). IEEE, Gurgaon, India.

Liu, C., Wang, L., Lang, B., & Zhou, Y. 2018. Finding effective classifier for malicious URL detection. In *Proceedings of the 2018 2nd International Conference on Management*

Engineering, Software Engineering and Service Sciences (pp. 240–244). ACM, New York, United States.

Ma, J., Saul, L. K., Savage, S., & Voelker, G. M. 2009. Beyond blacklists: Learning to detect malicious web sites from suspicious URLs. In *Proceedings of the 15th ACM SIGKDD International Conference on Knowledge Discovery and Data Mining* (pp. 1245–1254). ACM, Paris, France.

Mohammed, M., Muhammad Badruddin, K., Eihab, B., & Mohammed, B. 2016. *Machine Learning: Algorithms and Applications*. CRC Press, Boca Raton, FL.

Mohanty, S., Acharya, A. A., Sahu, L., & Mohapatra, S. K. 2020. Hazard identification and detection using machine learning approach. In *2020 4th International Conference on Intelligent Computing and Control Systems (ICICCS)* (pp. 1239–1244). IEEE, Madurai, India.

Sahoo, D., Liu, C., & Hoi, S. C. 2017. Malicious URL detection using machine learning: A survey. *arXiv preprint arXiv:1701.07179.*

Scikit Learn Organisation. Online from https://scikit-learn.org/stable/modules/ensemble.html. Retrieved on September 2020.

Tao, W., Shunzheng, Y., & Bailin, X. 2010. A novel framework for learning to detect malicious web pages. In *2010 International Forum on Information Technology and Applications* (Vol. 2, pp. 353–357). IEEE, Kunming, China.

Towards data science. Online from https://towardsdatascience.com/understanding-gradient-boosting-machines. Retrieved on October 2020.

TRAGHA, A. 2019. Machine learning for web page classification: A survey. *International Journal of Information Science and Technology*, *3*(5), 38–50.

Urcuqui, C., Navarro, A., Osorio, J., & García, M. 2017. Machine learning classifiers to detect malicious websites. In *SSN. CEUR Workshop Proceedings* (Vol. 1950, pp. 14–17).

Wang, R., Zhu, Y., Tan, J., & Zhou, B. 2017. Detection of malicious web pages based on hybrid analysis. *Journal of Information Security and Applications*, *35*, 68–74.

WU, C. M., Min, L. I., Li, Y. E., ZOU, X. C., & QIANG, B. H. 2018. Malicious website detection based on URLs static features. *DEStech Transactions on Computer Science and Engineering* (mso) (pp. 1–7), ISBN: 978-1-60595-542-1.

Yoo, S., Kim, S., Choudhary, A., Roy, O. P., & Tuithung, T. 2014. Two-phase malicious web page detection scheme using misuse and anomaly detection. *International Journal of Reliable Information and Assurance*, *2*(1), 1–9.

14 Finger Vein Authentication Using Convolutional Neural Networks and Feature Extraction

Prasannavenkatesan Theerthagiri and C. Gopala Krishnan

CONTENTS

14.1 INTRODUCTION

The biometric is a physiological and behavioral attribute that is unique for every individual. Physiological biometric refers to physical measurements of the human body, such as the face, fingerprint, finger vein, palm vein, hand geometry, DNA, retina, and iris. Behavioral biometric characteristics are the measure of trends in human behaviors, such as keystrokes, speech, signature, and gait, that are distinctly identifying and observable. Among several types of biometrics, veins are highly preferable because of their unique characteristics, such as liveness, user-friendliness, low intrusiveness, and hygiene. Hence, finger vein biometrics are paying more attention to the field of authentication.

A biometric security system is simply a pattern-recognition method that allows an individual to be identified using a feature vector associated with a measurable

DOI: 10.1201/9781003184140-14

physical or behavioral property that they acquire. Biometrics modalities are frequently one of a kind, quantifiable, and automatically verified, as well as permanent (Cherrat et al., 2017). To locate veins underneath the skin, modern camera technology must be used to display venous networks to identify them, and one method for doing so is the line tracking approach.

A range of biometric sources are combined in the multi-biometric recognition system. The key benefit of a multimodal system over a single biometric system is that the learning procedure is more secure and accurate (Unar et al., 2014). In this sense, studies of multimodal biometrics employing finger vein and facial photos have become increasingly popular and important in recent years (Kauba et al., 2019; Mehdi Cherrat et al., 2020).

In general, the finger vein authentication method comprises two steps, as indicated in Figure 14.1. The enrolling phase is where the finger vein template is stored in the database for the first time. The verification phase is the second step in the process, and it compares the input finger vein template to the database template.

The four steps of finger vein authentication are finger vein capture, preprocessing, feature extraction, and feature matching. Among these parts, feature extraction has a major contribution to authentication accuracy because the extracted features have a great influence on the subsequent matching phase. Several approaches are used to extract characteristics from finger vein images. Handcrafted feature extraction techniques are the traditional method that makes use of the filters designed prior to extracting the vein lines presented in the image. Poor-quality infrared light, light scattering while imaging finger tissues, ambient lighting conditions, cold weather, fat fingers, and poorly designed image capturing equipment, among other factors, can all have an influence on these approaches.

To address the issues with traditional feature extraction approaches, this paper offers a convolutional neural network–based finger vein authentication approach.

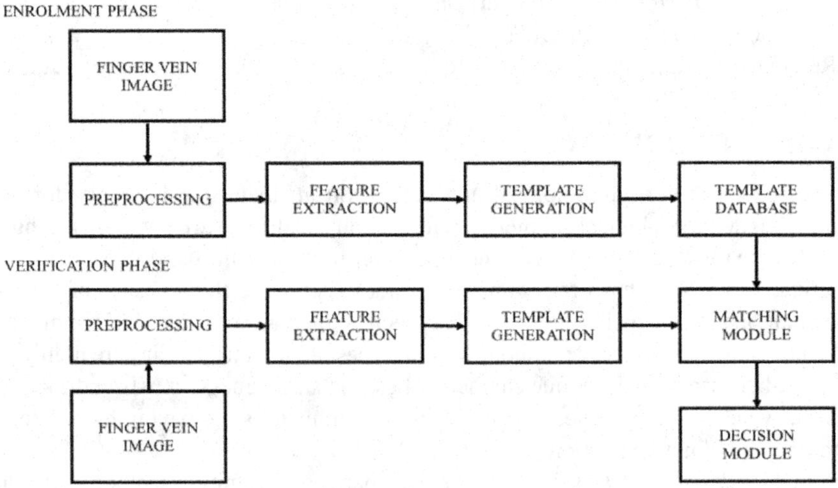

FIGURE 14.1 Finger Vein Authentication System.

This is a better way for feature extraction from finger vein images because it learns them during the training phase and categorizes them using the labels associated with the features.

14.2 FINGER VEIN AUTHENTICATION METHODS

Several researchers made their contribution to the field of finger vein authentication from the past decades.

Miura et al. (2004) suggested a repeated line tracking (RLT) approach with an equal error rate (EER) of 0.145%, for extracting finger vein pattern by iteratively tracking local lines beginning from multiple random points in the picture. The finger vein pattern extraction approach published by Miura et al. (2007) was based on the highest curvature points in the picture profile. The EER of this approach is 0.0009%, according to the findings. Cui and Yang (2011) combined finger vein and fingerprint pictures at the score level. This multimodal biometric authentication method has false accepted rate (FAR) 1.2% and false reject rate (FRR) of 0.75%. Modify finite radon transform (MFRAT) and GridPCA (principal component analysis)-based feature extraction method is proposed by Van et al. (2011). This method is implemented using the SDUMLA finger vein database and the obtained genuine acceptance rate is 95.67% with zero FAR.

With the advancement of AI in recent years, researchers have attempted to use machine learning methods to detect finger veins. Wu and Liu (2011) used PCA and a neuro-fuzzy system to conduct an experiment that generated high accuracy (ANFIS).

The methods described here are all handcrafted approaches for extracting features. In the recognition phase, these approaches attain a fast reaction speed (Yang et al., 2012). Despite these advantages, there are numerous evident flaws, including:

1. Handcrafted features are shallow features that only represent a small portion of the image and may contain irrelevant information.
2. Noise is sensitive to the handmade characteristics.
3. Handcrafted characteristics cannot be automatically recovered from the source images without human interaction, which is cumbersome and time consuming.

Deep learning algorithms have been used in the field of personal identification in recent years. The approaches for recognizing finger veins using convolutional neural networks (CNNs) are explained here.

The finger vein identification approach described by Liu et al. (2017) is based on altering the network topology and parameters of AlexNet. Using the SDUMLA-FV database, with three photos for training and three photos for testing for each class, this approach has an ERR of 0.80%. Hyung Gil Hong et al. (2017) proposed the CNN with 13 convolutional layers, which achieved an EER of about 3.906 for the SDUMLA database. The finger vein and finger-shaped multimodal finger vein authentication method is proposed by Kim et al. (2018). False rejection cases are highly occurring in this method due to finger position changes between input and enrolled image.

With the advancement of AI (artificial intelligence) in recent years, researchers have attempted to use machine learning methods to detect finger veins (Theerthagiri, 2020; Prasannavenkatesan, 2020). Zhang et al. (2019) suggested an adaptive Gabor CNN for finger vein detection to address the challenges of high complexity and huge parameters in CNNs. Boucherit et al. (2020) reported a merged CNN-based finger vein identification approach that obtains a recognition rate of roughly 99.48% for the SDUMLA database, and it employed the contrast limited adaptive histogram equalization (CLAHE) approach to combine the original and upgraded images.

14.3 PROPOSED WORK

14.3.1 Database Description

The proposed CNN is trained with the SDUMLA-HMT database (Fairuz et al., 2018, 2019). It includes 3,816 finger vein photographs from 106 people. On the left and right hands, six photos in gray level bmp format with a resolution of 320 x 240 pixels were obtained from the index, middle, and ring fingers. The input images are resized into 227 × 227 and feed as input to the CNN. In this paper, the index fingers of 50 classes were considered for experimentation. Sample images of index fingers in the database are shown in Figure 14.2.

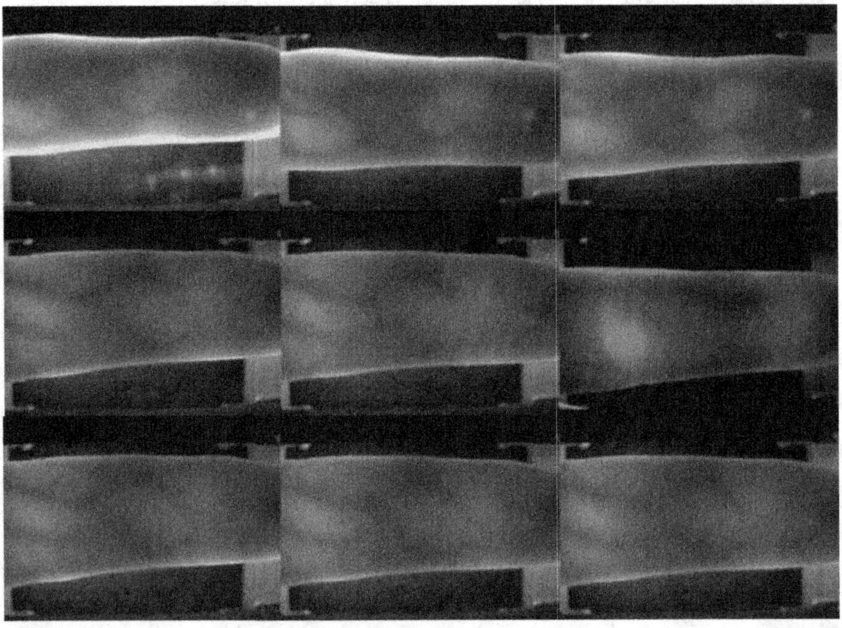

FIGURE 14.2 Sample Images of Index Finger in SDUMLA Database.

14.3.2 Learning Strategy

Each person's index finger is considered a different class. The experiment was conducted with the first four photographs of each finger image, and the remaining two photographs from each class were tested. The suggested CNN is trained utilizing the SGDM optimizer along stochastic gradient descent with momentum (Xie and Kumar, 2017, 2019). Experiments were run in MATLAB 2020a on a PC with 8 GB RAM, an NVIDIA GeForce 940MX graphics card, an i5–7500U CPU running at 2.5 GHz, with Windows 10.

14.3.3 Training Options and Configuration

Training options and configurations of the proposed CNN are mentioned in Table 14.1 and Table 14.2.

TABLE 14.1
Configuration of the Proposed Method.

Function		Layers	Filters & stride	Feature map
Input		Image input	-	227×227×3
Feature extraction	**Group 1**	Convolutional	Filters= 32; Size=3×3; Stride= [1 1]	227×227×32
		ReLU		227×227×32
		Batch normalization		227×227×32
		Average pooling	Stride= [2 2]	114×114×32
	Group 2	Convolutional	Filters= 64; Size=3×3; Stride= [1 1]	114×114×64
		ReLU		114×114×64
		Batch normalization		114×114×64
		Average pooling	Stride= [2 2]	52×52×64
	Group 3	Convolutional	Filters= 64; Size=3×3; Stride= [1 1]	57×57×64
		ReLU		57×57×64
		Batch normalization		57×57×64
		Average pooling	Stride= [2 2]	29×29×64
	Function	Layers	Filters & stride	Feature map
	Group 4	Convolutional	Filters= 128; Size=3×3; Stride= [1 1]	29×29×128
		ReLU		29×29×128
		Batch normalization		29×29×128
		Average pooling	Stride= [2 2]	15×15×128
	Group 5	Convolutional	Filters= 256; Size=3×3; Stride= [1 1]	15×15×256
		ReLU		15×15×256
		Batch normalization		15×15×256
		Average pooling	Stride= [2 2]	8×8×256
	Group 6	Convolutional	Filters= 256; Size=3×3; Stride= [1 1]	8×8×256
		ReLU		8×8×256
		Batch normalization		8×8×256
		Average pooling	Stride= [2 2]	4×4×256
Flatten		Fully connected	-	50×4096
Activation		Softmax	-	1×1×50

TABLE 14.2
Training Options Used in Experiment.

Options	Value
Maximum epochs	10
Mini batch size	20
Validation frequency	5
Decay	0.01
Momentum	0.9

14.3.4 Architecture of the Proposed CNN

CNN is a specific sort of neural network technique that uses learned characteristics to classify images. Figure 14.3 shows a block schematic of the proposed CNN-based authentication system.

The projected CNN's architecture is seen in Figure 14.4. The input images are preprocessing to make them suitable for feed into a CNN. The convolutional layer is the major building block to extract the features. It contains a set of filters, and each filter is convolved across the width and height of the input volume during the forward propagation. CNN convolves the input image with the weight matrix to extract the features from the input image without changing its spatial arrangement. At initial convolutional layers, CNN learns the basic shapes, and the specific features are in the deeper layers (Ruby et al., 2020; Theerthagiri et al., 2021).

The convolutional layer performs a bit-by-bit dot product between filter sized input x_m^l and $w_{n,m}^l$ filter, as mentioned in Equation 14.1.

$$y_n^l = \sum_{m}^{M^{l-1}} w_{n,m}^l * x_n^l + b_n^l \tag{14.1}$$

where l and m are the level and map indexes, respectively, and b_n^l is the bias of the n-th output map. The activation function for introducing non-linearity to the network is the ReLU. ReLU is the activation function to introduce the non-linearity to the network. ReLU returns zero for negative input and returns the same value for all the positive inputs, as depicted in Equation 14.2.

$$f(x) = \max(0, x) \tag{14.2}$$

ReLU not only helps to solve the problem of disappearing gradients, it also speeds up the training process. The pooling layer helps to reduce the dimension of each feature map and retains the important information in the input. Generally used, the pool and stride size are 2*2 and return the average value from each pool in case of average pooling. The mean and standard deviation of each input variable are calculated by the batch normalization for each layer and for all mini batches, then these statistics are used to perform the standardization. It can train the network faster, makes

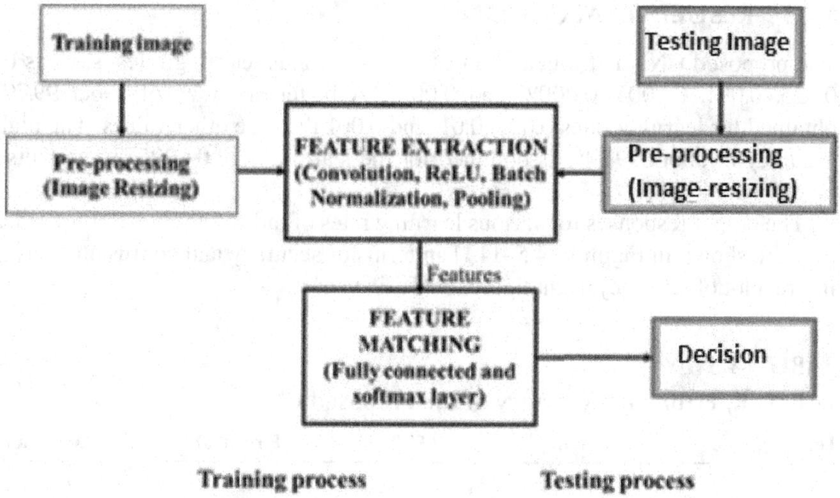

FIGURE 14.3 Block Diagram of CNN-Based Biometric Authentication.

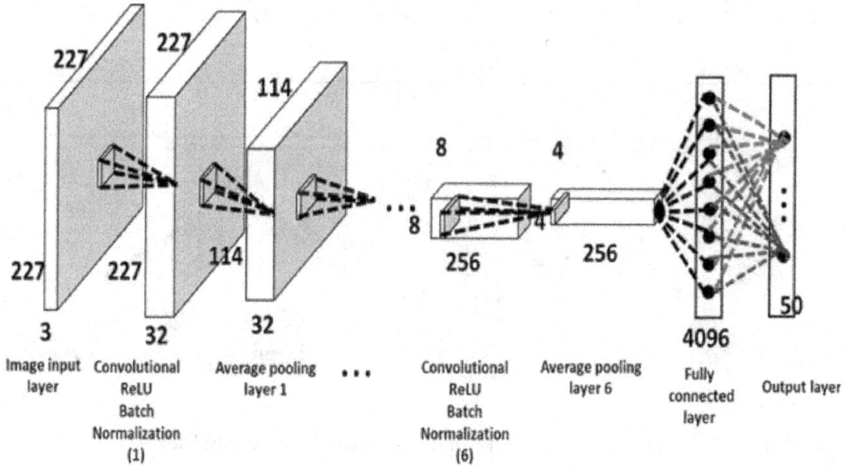

FIGURE 14.4 Architecture of Proposed CNN.

weights easier to initialize, and may give better results. The output of convolutional and pooling layers is used to extract the input image's high-level features. The fully connected layer translates the final average pooling layer's feature map into 1D data. Softmax is a loss function with an output probability of roughly 1 that is utilized in the classification part of CNN. The softmax function takes a vector of real-valued scores as input. These are turned into a vector of zero to one values that add up to one. Finally, using the probabilities supplied by the softmax function, the classification layer assigns each input to one of the classes and computes the loss.

14.4 RESULTS EVALUATION

The proposed CNN is trained and tested with various learning rates, such as 0.03, 0.025, 0.001, 0.0005, 0.0002, and 0.0001. A better accuracy of about 99.7% is obtained for learning rates 0.025, 0.01, and 0.001 than the other values. The highest accuracy of about 99.84% is obtained for the learning rate 0.0005. The results are given in Table 14.3.

The output responses for various learning rates of index fingers are observed, and they are shown in Figures 14.5–14.11 in terms of accuracy and confusion chart (left figure: plot of accuracy, right figure: confusion chart).

TABLE 14.3
FAR, FRR, Error, and Accuracy of the Proposed CNN.

Learning Rate	FAR	FRR	Error (%)	Accuracy (%)
0.03	0.00285	0.14	0.56	99.44
0.025	0.00147	0.07	0.29	99.71
0.01	0.00131	0.06	0.25	99.74
0.001	0.00136	0.05	0.27	99.73
0.0005	**0.00081**	**0.04**	**0.16**	**99.84**
0.0002	0.00450	0.16	0.87	99.12
0.0001	0.01553	0.23	0.92	99.08

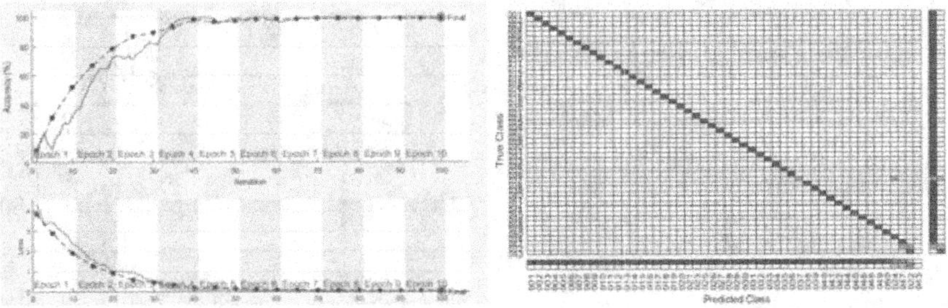

FIGURE 14.5 Plot of Accuracy and Confusion Chart for Learning Rate 0.01.

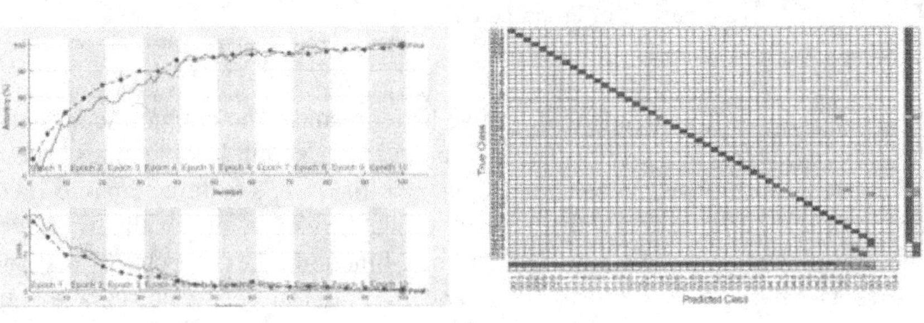

FIGURE 14.6 Plot of Accuracy and Confusion Chart for Learning Rate 0.025.

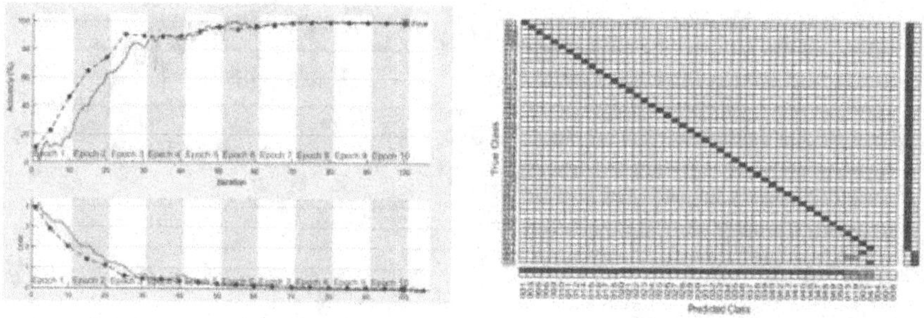

FIGURE 14.7 Plot of Accuracy and Confusion Chart for Learning Rate 0.03.

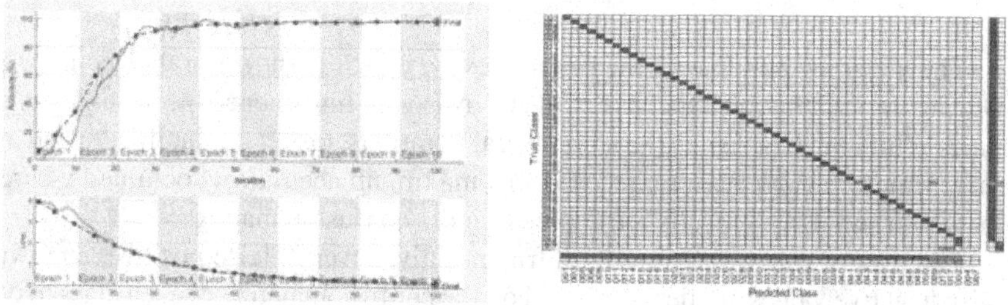

FIGURE 14.8 Plot of Accuracy and Confusion Chart for Learning Rate 0.001.

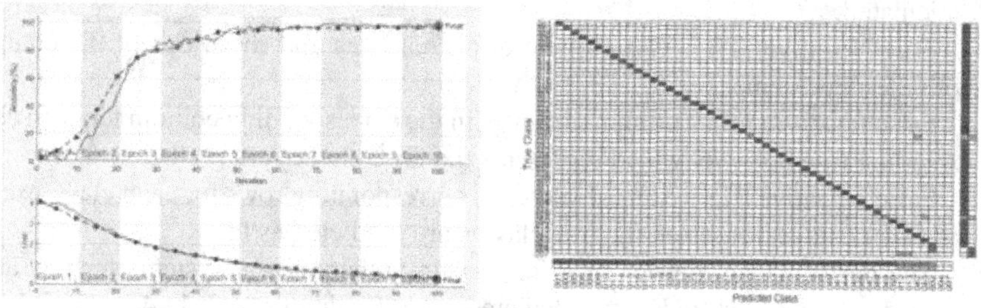

FIGURE 14.9 Plot of Accuracy and Confusion Chart for Learning Rate 0.0005.

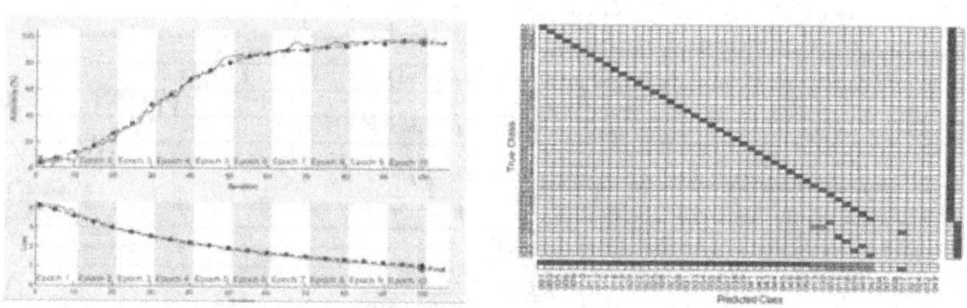

FIGURE 14.10 Plot of Accuracy and Confusion Chart for Learning Rate 0.0002.

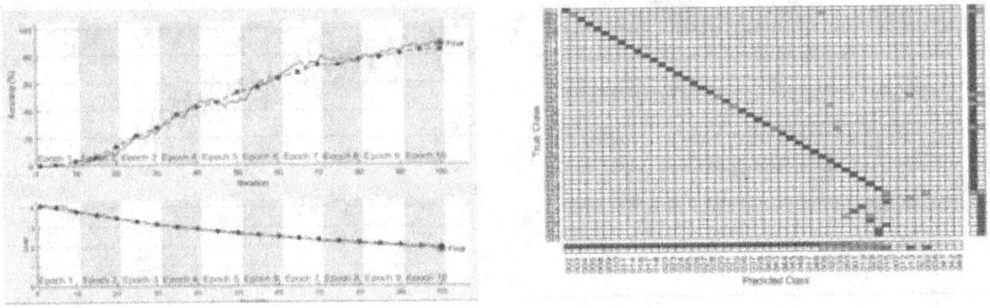

FIGURE 14.11 Plot of Accuracy and Confusion Chart for Learning Rate 0.0001.

The accuracy and loss of the CNN at every epoch is shown in the accuracy plot. During initial iterations, the loss of the CNN is very high, hence the accuracy is minimal. While the CNN goes through several iterations, loss is gradually reduced at the same time accuracy is increased. Finally, maximum accuracy is obtained where the loss is minimal. The results are depicted in the confusion matrix.

The overall true positive, overall true negative, overall false positive, and overall false negative values are derived based on the confusion matrix of each class to determine the biometric authentication system's performance metrics (Theerthagiri et al., 2021; Prasannavenkatesan, 2021).

True Positive (TP): The true classes' average of diagonal elements is accurately anticipated.

True Negative (TN): The average of all columns and rows, except the columns with the same name.

False Positive (FP): The sum of values in the corresponding column for each class, excluding true positive, which represents the false classes are predicted as true.

False Negative (FN): Sum of values in corresponding rows for each class, excluding true positive, represent the true classes are predicted as false.

The performance metrics of a finger vein authentication system, such as accuracy, error, FAR, and FRR, are calculated using Equations 14.3, 14.4, 14.5, and 14.6, respectively.

$$Accuracy = \frac{TP + TN}{TP + TN + FP + FN} \quad (14.3)$$

$$Error = \frac{FP + FN}{TP + TN + FP + FN} \quad (14.4)$$

$$FAR = \frac{FP}{FP + TN} \quad (14.5)$$

$$FRR = \frac{FN}{FN + TP} \quad (14.6)$$

Accuracy is defined as the number of right predictions divided by the total number of positives and negatives. The error rate is equal to the number of inaccurate predictions (FN + FP) divided by the total number of positives and negatives (Theerthagiri, 2019). The number of inaccurate positive predictions divided by the total number of negatives is known as FAR. By dividing the total number of positives by the number of erroneous negative estimations, the FRR is determined. The obtained results of the performance metrics are indicated in Table 14.3.

Learning rate 0.03 has the FAR, FRR, error (%), and accuracy (%) as 0.00285, 0.14, 0.56, and 99.44, respectively. For learning rate 0.025, the FAR, FRR, error (%), and accuracy (%) are 0.00147, 0.07, 0.29, and 99.71, respectively. Learning rate 0.01 has the FAR, FRR, error (%), and accuracy (%) as 0.00131, 0.06, 0.25, and 99.74, respectively. Learning rate 0.001 has the FAR, FRR, error (%), and accuracy (%) as 0.00136, 0.05, 0.27, and 99.79, respectively. For learning rate 0.0005, the FAR, FRR, error (%), and accuracy (%) are 0.00081, 0.04, 0.16, and 99.84, respectively. Learning rate 0.0002 has the FAR, FRR, error (%), and accuracy (%) as 0.00450, 0.16, 0.87, and 99.12, respectively. For learning rate 0.0001, the FAR, FRR, error (%), and accuracy (%) are 0.01553, 0.23, 0.92, and 99.08, respectively. Therefore, learning rate 0.0005 has highest learning rate and lowest error compared to the other learning rates.

14.4.1 Performance Comparison and Discussion

The comparison of the performance of the previous methods and the proposed method for the SDUMLA finger vein database is indicated in Table 14.4.

Traditional feature extraction strategies, like maximum curvature and repeated line tracking only get 86.01% and 87.11%, respectively, for the used learning strategy. The accuracy of the MFRTA and GridPCA-based feature extraction method is improved compared to the traditional methods. The proposed method, which utilizes six convolutional layers, achieves high accuracy of 99.84% compared to the previous methods. Hence, the proposed method can outperform the previous methods mentioned in the literature, and it provides a better solution to the finger vein authentication system. Therefore, the proposed method gives 4–14% improved accuracy for the finger vein authentication system as compared to existing methods.

TABLE 14.4
Comparison of Performance.

Paper	Feature Extraction Method	Accuracy
Miura et al. (2007)	Maximum Curvature	86.01%
Miura et al. (2004)	Repeated Line Tracking	87.11%
Boucherit et al. (2020)	Merged CNN	89.88%
Van et al. (2011)	Modify finite radon transform (MFRAT) and GridPCA	95.67%
Proposed method	CNN (6 convolutional layers)	99.84%

14.5 CONCLUSION AND FUTURE WORK

In this paper, a deep learning–based approach is proposed for identifying a person based on the vascular patterns in the finger. The CNN with six convolutional layers is designed, trained, and tested using the SDUMLA finger vein database. Adoption of the fewer convolutional layers helps to reduce the training time. The training and testing method uses a separate set of finger vein pictures from the same database. For training the CNN, several learning rates are utilized, and FAR, FRR, error, and accuracy are determined. The findings reveal that the proposed method's overall accuracy is 99.84% for learning rate 0.0005. The results are compared to those obtained using more traditional approaches. When compared to earlier techniques, the proposed CNN-based feature extraction methodology produces better results. The proposed technique clearly outperforms traditional approaches. As a result, the proposed framework for finger vein–based person authentication is effective.

Further, the authentication accuracy can be improved using the combination of features from multimodal biometric traits. The biometric features are stored in the database as a biometric template to use these in the verification phase. But these templates will be spoofed by the attackers. A multimodal biometric authentication system can be implemented to improve the authentication accuracy and template protection scheme to increase the security of the biometric template.

REFERENCES

Boucherit, I., Zmirli, M. O., Hentabli, H., & Rosdi, B. A. (2020). Finger vein identification using deeply-fused convolutional neural network. *Journal of King Saud University-Computer and Information Sciences* (In Press).

Cherrat, E. M., Alaoui, R., Bouzahir, H., & Jenkal, W. (2017, April). High density salt-and-pepper noise suppression using adaptive dual threshold decision based algorithm in fingerprint images. In *2017 Intelligent Systems and Computer Vision (ISCV)* (pp. 1–4). IEEE, Fez, Morocco.

Cui, F., & Yang, G. (2011). Score level fusion of fingerprint and finger vein recognition. *Journal of Computational Information Systems*, 7(16), 5723–5731.

Fairuz, S., Habaebi, M. H., & Elsheikh, E. M. A. (2019). Pre-trained based CNN model to identify finger vein. *Bulletin of Electrical Engineering and Informatics*, 8(3), 855–862.

Fairuz, S., Habaebi, M. H., Elsheikh, E. M. A., & Chebil, A. J. (2018, September). Convolutional neural network-based finger vein recognition using near infrared images. In *2018 7th International Conference on Computer and Communication Engineering (ICCCE)* (pp. 453–458). IEEE, Kuala Lumpur, Malaysia.

Hong, H. G., Lee, M. B., & Park, K. R. (2017). Convolutional neural network-based finger-vein recognition using NIR image sensors. *Sensors*, 17(6), 1297.

Kauba, C., Prommegger, B., Uhl, A., Busch, C., & Marcel, S. (2019). Openvein—an open-source modular multi-purpose finger-vein scanner design. In *Handbook of Vascular Biometrics* (pp. 77–112). Springer Science+ Business Media, Boston, MA.

Kim, W., Song, J. M., & Park, K. R. (2018). Multimodal biometric recognition based on convolutional neural network by the fusion of finger-vein and finger shape using near-infrared (NIR) camera sensor. *Sensors*, 18(7), 2296.

Liu, W., Li, W., Sun, L., Zhang, L., & Chen, P. (2017, June). Finger vein recognition based on deep learning. In *2017 12th IEEE Conference on Industrial Electronics and Applications (ICIEA)* (pp. 205–210). IEEE, Siem Reap, Cambodia.

Mehdi Cherrat, E., Alaoui, R., & Bouzahir, H. (2020). Convolutional neural networks approach for multimodal biometric identification system using the fusion of fingerprint, finger-vein and face images. *PeerJ Computer Science*, 6, e248.

Miura, N., Nagasaka, A., & Miyatake, T. (2004). Feature extraction of finger vein patterns based on iterative line tracking and its application to personal identification. *Systems and Computers in Japan*, 35(7), 61–71.

Miura, N., Nagasaka, A., & Miyatake, T. (2007). Extraction of finger-vein patterns using maximum curvature points in image profiles. *IEICE Transactions on Information and Systems*, 90(8), 1185–1194.

Prasannavenkatesan, T. (2020). Forecasting hyponatremia in hospitalized patients using multilayer perceptron and multivariate linear regression techniques. *arXiv preprint arXiv:2007.15554*.

Prasannavenkatesan, T. (2021). Probable forecasting of epidemic COVID-19 in using COCUDE model. *EAI Endorsed Transactions on Pervasive Health and Technology*, 7(26), e3.

Ruby, A. U., Prasannavenkatesan Theerthagiri, D. I., & Vamsidhar, Y. (2020). Binary cross entropy with deep learning technique for image classification. *International Journal*, 9(4).

The SDUMLA-HMT database. [Online]. Available: http://mla.sdu.edu.cn/sdumla-hmt.html.

Theerthagiri, P. (2020). FUCEM: Futuristic cooperation evaluation model using Markov process for evaluating node reliability and link stability in mobile ad hoc network. *Wireless Networks*, 26(6), 4173–4188.

Theerthagiri, P., Jeena Jacob, I., Usha Ruby, A., & Yendapalli, V. (2021). Prediction of COVID-19 possibilities using k-nearest neighbour classification algorithm. *International Journal of Current Research and Review*, 13(6), 156.

Theerthagiri, P., & Thangavelu, M. (2019). Futuristic speed prediction using auto-regression and neural networks for mobile ad hoc networks. *International Journal of Communication Systems*, 32(9), e3951.

Unar, J. A., Seng, W. C., & Abbasi, A. (2014). A review of biometric technology along with trends and prospects. *Pattern Recognition*, 47(8), 2673–2688.

Van, H. T., Tat, P. Q., & Le, T. H. (2011, October). Palmprint verification using GridPCA for Gabor features. In *Proceedings of the Second Symposium on Information and Communication Technology* (pp. 217–225). ACM, Hanoi Vietnam.

Wu, J. D., & Liu, C. T. (2011). Finger-vein pattern identification using principal component analysis and the neural network technique. *Expert Systems with Applications*, 38(5), 5423–5427.

Xie, C., & Kumar, A. (2017). Finger vein identification using convolutional neural network and supervised discrete hashing. In *Deep Learning for Biometrics* (pp. 109–132). Springer, Cham.

Xie, C., & Kumar, A. (2019). Finger vein identification using convolutional neural network and supervised discrete hashing. *Pattern Recognition Letters*, 119, 148–156.

Yang, G., Xi, X., & Yin, Y. (2012). Finger vein recognition based on (2D) 2 PCA and metric learning. *Journal of Biomedicine and Biotechnology*, 2012, 1–9.

Zhang, Y., Li, W., Zhang, L., & Lu, Y. (2019, May). Adaptive Gabor convolutional neural networks for finger-vein recognition. In *2019 International Conference on High Performance Big Data and Intelligent Systems (HPBD&IS)* (pp. 219–222). IEEE. Shenzhen, China.

15 Facial Expression Recognition in Real Time Using Swarm Intelligence and Deep Learning Model

Yogesh Kumar, Shashi Kant Verma, and Sandeep Sharma

CONTENTS

15.1 INTRODUCTION

In daily interpersonal communication, people prefer to understand facial expressions over the actual verbal interactions as the evaluation of emotional state through facial expressions can help to better analyze the actual status of a person's mind. Moreover, boring conversations can't easily come to an end, but the indication of boring facial expressions by the listening person can lead the speaker to finish the conversations quickly. This situation can only be considered efficient if the speaker demonstrates high emotional intelligence.

Although a human can express their emotions through numerous facial expressions, Ekman and Friesen (2003) has defined six standard expression classes: surprise, fear, anger, disgust, happiness, and sadness. The estimation of facial expressions requires humans to be aware of the possible expression classes along with higher emotional intelligence. In a similar manner, machines should also be

DOI: 10.1201/9781003184140-15

efficient enough to understand facial expressions with a deep knowledge of human emotions. The AFER system enables the human computing interaction with the understandability of a facial action coding system (Kumar et al. 2019; Sayette et al. 2001). There are a total of 64 action units in the facial action coding system based on the movement of facial muscles. For instance, the detection of the relative distance between the lower and upper lip for the different mouth positions as opened or closed is a feature for facial actions. The detection of only the facial muscle movement is not enough to determine the facial expressions; there is also a need to determine the change in texture as the local features. In the present work, the hybrid concept of the Gabor filter (GF) (Kumar et al. 2019) and local binary pattern (LBP) (Kumar et al. 2019; Kumar, Gupta et al. 2020) is used for the extraction of different facial features.

The overall process of autonomous facial expression recognition is composed of sub-modules of data collection for system training, face detection, features extraction, features selection, and classification of expressions. In this research work, the datasets of KDEF, JAFFE, RaFD databases, and some Indian expression images are utilized for the training of the proposed AFER system. The considered dataset images are passed through the Viola-Jones face detector to crop the face as the region of interest. Face recognition is required to process the facial components (Gupta et al. 2021). The selected face region is processed for the features extraction using hybrid GF and LBP. The extracted feature set is analyzed to select the significant features using swarm intelligence and quantum computing–based improved quantum-inspired gravitational search algorithm (IQIGSA). The IQIGSA serves as the optimization approach to select the appropriate feature set. The wide acceptability of swarm intelligence (SI) techniques in different applications (Goel et al. 2018; Gupta et al. 2011; Kumar 2017; Kumar 2019; Kaur et al. 2011; Singh et al. 2016) to optimize the solution set has enforced to adapt the SI technique for optimization. The auspicious features of fast convergence and lower computational cost are the reason for the GSA among the SI techniques. The GSA (Rashedi et al. 2009) is mass agents-based meta-heuristic SI technique that uses Newton's laws of motion and gravity. The feature-selection process of IQIGSA utilizes the attributes of quantum computing, GSA, and improved variant to handle the trapping of mass agents in local optima. The last module of the AFER system is the classification module to recognize facial expressions, which is performed using a hybrid deep convolutional recurrent neural network (DCRNN). The overall AFER system is tested for the expression recognition of humans ages 20–35. This real-time testing is performed on 24 male and 12 female candidates. Moreover, the multi-pose expressions are detected with evaluation in terms of different performance measures along with accuracy. The multi-pose expressions include the detection of expressions at 0 degrees (front pose) and ±30 degrees. To determine the effectiveness of the proposed AFER system, real-time experimentation is also conducted with the existing techniques.

The rest of the chapter is structured as follows: Section 15.2 describes the work related to facial expression recognition to determine the state-of-art research contributions in the field. Section 15.3 discusses the data collected from the different benchmark datasets for training the system. Section 15.4 elucidates the AFER system for the classification of expressions. Section 15.5 presents the results in terms of recall, precision, f-measure, accuracy, and recognition rate. Section 15.6 ends the chapter with the conclusion and future possibilities.

15.2 RELATED WORK

The work of facial expression recognition can be noticed with the possible experimentations on different datasets. A recent survey of FER on the deep learning techniques was conducted by Li and Deng (2020), and the authors determined a lack of training data leads to difficulty in recognition of facial expressions. The authors also mentioned that the pose invariant and occlusion robust are the lesser incorporated problems in facial expression recognition. Another extensive review was performed by Alexandre et al. (2020) for the recognition of expression from 3D facial images. The authors conducted the review in 2013–2018 with a focus on the classification models, preprocessing, and face recognition techniques. The authors indicated the incorporation of deep learning techniques to a higher extent due to the processing of huge dataset images.

Recent studies in the field of AFER are also discussed. Ding et al. (2017) used the Taylor feature pattern (TFP) for feature recognition and expression detection of the face. The TFP approach evaluated the features from the Taylor map using the Taylor theorem and LBP. The authors also proposed the double LBP approach for the detection of peak expression frames, which also helps to reduce the evaluation time. The Laplace transform of logarithm was also adapted to attain the robust feature set after the LBP operator. The work was efficient and can be extended for experimentation in an uncontrolled environment. Zhang et al. (2017) used the evolutional spatial-temporal network to improve the accuracy of expression recognition. The temporal features were adapted, and the method employed the PHRNN (part-based hierarchical bidirectional recurrent neural network) for the extraction of temporal sequence and MSCNN (multi-signal convolutional neural network) for the spatial features extraction. The work was significant to handle the higher error rate but inferior to handle the critical regions. Zeng et al. (2018) considered both the geometric and appearance-based features for the training of deep sparse auto-encoder to recognize the expressions. The authors conducted the evaluation for the seven and eight classes of expressions. The work for the classification of seven expression classes was notable but got confused when distinguishing between contempt and sadness in the case of the eight expression category.

Further, Yang et al. (2018) presented an improved random forest algorithm for the classification of expressions. The method considered the fused feature set of deep geometric features along with grayscale images. The features unable to identify with geometric changes are extracted with local binary patterns. The method was significant but failed to handle the drastic head deflection and heavy occlusion. Tsai and Chang (2018) used the support vector machine (SVM) along with the consideration of multiple features for expression recognition. The authors initially developed a face-detection model using a self-quotient image filter that overcame the shade light and insufficient light. The features were extracted using the simultaneous approach of Gabor filter, discrete cosine transform, and angular radial transform. The evaluation was better than state-of-art techniques. Sun et al. (2018) focused on three regions of the mouth region, left eye, and right eye. A separate convolutional neural network (CNN) was trained for each of the regions. The overall results were evaluated by incorporating the results from each of the networks. Sadeghi and Raie (2019) presented the local metric learning for expression recognition. The authors indicated the improvement of the method for the uncontrolled environment by training the learnable deep histogram features. Ye et al. (2019) proposed the region-based convolutional fusion network for the recognition of facial expressions. The method was effective for the different datasets of Oulu-CASIA, CK+, and KDEF. Kumar et al. (2020a) used the combination of principal component analysis (PCA) as the feature selector and deep CNN as the classifier. Further, the authors (Kumar et al. 2020b) extended the work to improve the recognition rate by changing the feature selector of PCA to QIBGSA (quantum-inspired binary gravitational search algorithm). The recognition rate with the QIBGSA was higher than the PCA-based method. Recently, Zheng et al. (2020) tested discriminative deep multitask learning for expression recognition. The techniques used the Siamese network and determined it to be constructive for the small learning samples as well.

On the basis of discussed research contributions, it can be analyzed that the researchers primarily focused on the classification of expressions with a front pose from static images. For real-time expression recognition, multi-pose recognition is needed to consider the ability of the model to recognize the features with moving faces as well. Moreover, the feature components also need more optimistic techniques to determine the significant features. Therefore, the present work considers the IQIGSA approach for feature selection and DCRNN for the classification of expressions in real time.

15.3 DATA COLLECTION

The proposed AFER system is trained by considering the benchmark datasets of KDEF, JAFFE, and RaFD. All the datasets are image-based datasets with the availability of six Ekman emotion classes and one neutral class. Along with these datasets, the proposed system is also trained with some Indian expression images.

KDEF dataset (Lundqvist et al. 1998) is a composition of facial expressions performed by 70 different male and female models for the variation of poses at −90, −45, 0, +45, and +90 degrees. In this research work, the real-time evaluation is conducted for front pose and 30 degrees yaw angle in left and right. Therefore, the KDEF dataset images are considered for the −45, 0, and +45 degrees, which constitute the 2,940 images for training the proposed system.

The RaFD dataset (Langner et al. 2010) is a composition of 67 models showing 8 expression classes. In this research work, seven emotion classes are considered (contempt was left out). Similar to the KDEF dataset, the RaFD database is also available with the five yaw angle poses. Among these poses, −45, 0, and +45 degree poses with seven expressions are considered, which make a total count of 4,221 images, as each pose is available with left, right, and front gaze directions without any movement of the head.

The JAFFE dataset (Lyons et al. 1998) is available only with front pose images for the 7 expression classes performed by 10 female models, which makes a total of 213 expression images.

Moreover, some Indian expression images (Tanveer 2020; Sobia and Abudhahir 2017) are also considered as per their availability. These expressions are performed by a researcher for the front pose with the expression classes of happiness, sadness, disgust, neutrality, and anger. There are a total of 50 images available, including 7 neutral, 9 sad, 10 angry, 11 disgusted, and 13 happy expressions.

15.4 PROPOSED AFER SYSTEM

The proposed AFER system is discussed with the steps of input data images for training, face detection, features extraction, features selection, and classification of expressions. This process is also depicted in Figure 15.1. These steps are discussed as follows.

Step 1: The initial step of the proposed AFER system is the consideration of dataset images from the data collection.

Step 2: The second step is the detection of the face from the considered images. The face region is detected with the Haar-like features using the Viola-Jones algorithm (Kumar et al. 2020a). It discards the other regions after selecting the face as the region of interest. The evaluation of Haar features is described in Eq. (15.1).

$$H(x) = \begin{cases} 1, if \sum_{k=1}^{K} \alpha_k h_k(x) \geq \varnothing \\ 0, \quad otherwise \end{cases} \tag{15.1}$$

Where the presence and absence of the face in the image is determined by a value of 1 and 0, respectively. The notations $\varnothing$ and $h_k(x)$ are the decision threshold and weak classifier at weight α_k, respectively.

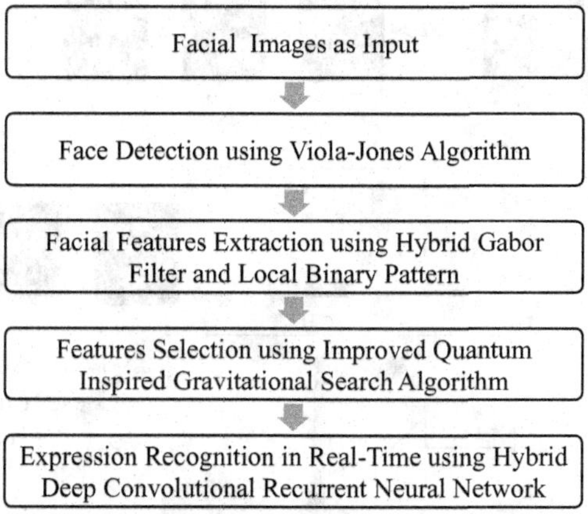

FIGURE 15.1 Process of Proposed Autonomous Facial Expression Recognition System in Real Time.

The attentional cascade structure in the Viola-Jones algorithm extracts the face region by discarding the other regions as negative features.

Step 3: The detected face regions are considered for the features extraction. In this work, the hybrid GF and LBP approach is employed for the features extraction (Kumar et al. 2020a). The features are extracted by evaluating the similarity of the divided image into the 64 sub-blocks. The pixel value is calculated for pixel position (P_p) and center pixel (P_c) using Eq. (15.2).

$$S\left(P_p - P_c\right) = \begin{cases} 1, if\ P_p \geq P_c \\ 0,\ otherwise \end{cases} \tag{15.2}$$

The discriminative local features are extracted with the Gabor filter. The hybrid GF with LBP operator determines the improved values of special histograms and comprehensively robust features. The evaluation of histograms for the hybrid feature extractor with the availability of 64 blocks and 40 filters is evaluated in Eq. (15.3).

$$Hist\left(Hybrid\,GF\ \&LBP\right) = \left(H_{0,0,0,}\ \ldots,\ H_{u,v,i},\ \ldots\ H_{1,7,63}\right) \tag{15.3}$$

The comprehensive histogram is determined with the concatenation of the histograms evaluated for all 64 blocks. The feature extraction process is also described in Figure 15.2 by considering an image of the KDEF dataset. The face is extracted from the image using the Viola-Jones algorithm.

Step 4: The extracted features are passed through the feature selector to determine the significant features. Here, IQIGSA (improved quantum inspired

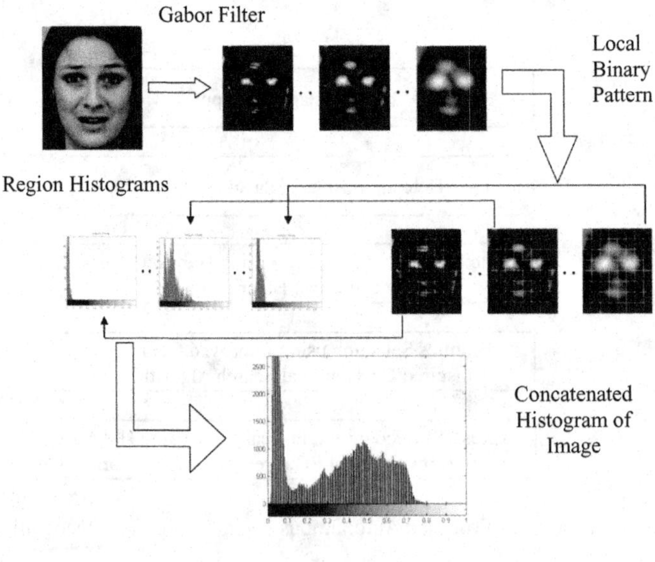

FIGURE 15.2 Hybrid GF and LBP for Feature Extraction.

gravitational search algorithm) (Moghadam and Nezamabadi-Pour 2012) is adapted for the feature selection (Moghadam et al. 2012) to handle the local trapping of mass agents and retaining the stochastic attributes. The IQIGSA modifies the $XMbest_i$ for the mass agents as illustrated in Eq. (15.4).

$$XMbest_i = \frac{\sum_{p=1}^{K} \frac{M_i}{distance_{i,p}} random_p . Xkbest_p}{\sum_{q=1}^{K} \frac{M_i}{distance_{i,q}} random_q} \tag{15.4}$$

Where $distance_{i,p}$ is determined based on the difference of the current position of mass agents to the best position of the mass agent (*Xkbest*) with a randomized vector of $random$.

Further, the fitness of the mass agents is determined for the iterations (*itr*) as described in Eq. (15.5).

$$fit_i^{itr} = \frac{DSE_i^{itr}}{DDE_i^{itr}} \tag{15.5}$$

Where the DSE_i^{itr} and DDE_i^{itr} refer to the Euclidean distance of features of the same expression and different expression, respectively.

The final selection of features is performed based on the final position of the mass agents as the optimized one. The binarization criterion is considered with the probability distribution of the quantum state. The outcome of values 1 and 0 refer to the selection of considered features and deny the feature, respectively.

Step 5: The final recognition of the expressions is executed using the hybrid DCRNN (Jain et al. 2018). The CNN optimizes the parameters for the feature vectors and recurrent neural network (RNN) is responsible for combining the sequential information.

The CNN considers the convolution, pooling, and fully connected layer for the mapping of the feature attributes. For each convolution layer, the rectified linear function is considered as the activation function. Moreover, the feature map is determined for the convolution layers between the weight vector and local patch. The dimensionality of the feature map is reduced in the pooling layer, and the local activation field is yield to the maximum activation. The fully connected layer is responsible for the final output in the form of neurons, which is proportional to the input feature vectors. The network training is formulated as depicted in Eq. (15.6).

$$\Delta w_{ij}(t) = -\eta \frac{\partial E_p(t)}{\partial w_{ij}(t)} \tag{15.6}$$

Where the rate of learning is signified as η. The value of E_p is the network error for the pth pattern.

The output neuron information is passed to the RNN that considers it as input to generate the sequential outcome. The RNN model employs the Elman network, which is the composition of input, hidden, and output layers. By considering the activation function of σ, the hidden and output layer vectors are determined as depicted in Eq. (15.7) and Eq. (15.8), respectively.

$$h_t = \sigma_t \left(W_h x_t + W_{rec} h_{t-1} \right) \tag{15.7}$$

$$y_t = \sigma_y \left(W_{out} h_t \right) \tag{15.8}$$

Where W_h, W_{rec}, and W_{out} are the weight of the hidden, recurrent, and output matrices, respectively. The final results are assigned for the different expression classes by considering the outcomes of the output layer in a serial manner.

15.5 RESULTS AND DISCUSSION

The result outcomes of the proposed AFER system are evaluated for the measures of recall, precision, f-measure, recognition rate, and accuracy. The learning of the system is executed using the KDEF, JAFFE, RaFD, and some Indian expression images.

The real-time testing is conducted for the front pose with a 0-degree yaw angle, and the side pose with the movement of the face at 30 degrees right and left is entertained. The movement of the face is considered fixed for both the yaw angular movement at 30 degrees. The experimentation is conducted on a window-based software system with simulation software of MATLAB. The experimentation is performed by considering the same background for all 36 candidates (24 male and 12 female) and the wearing of restricting objects, such as beards, glasses, nose jewelry, and makeup are kept away during expression recognition. Figure 15.3 depicts some of the outcomes as the evaluated expressions in real time.

15.5.1 Results with Proposed AFER System

Each expression is performed once by each candidate at angles of –30 degrees, 0 degrees, and +30 degrees. The respective results for each expression are noted using the proposed AFER system. The confusion matrix results for the 36 candidates with front pose and side pose are illustrated in Table 15.1 and Table 15.2, respectively. As the side pose expressions are determined at –30 and +30 degrees, this makes the total of 72 cases (36 cases for each of the –30 degrees and +30 degrees) for the side pose expression recognition. These confusion matrices are utilized for the evaluation of performance in terms of precision, recall/recognition rate, f-measure, and accuracy. The performance assessment results for the proposed AFER system are described in Table 15.3.

The confusion matrix result indicated in Table 15.1 illustrates that the angry, happy, and neutral expressions are determined with full accuracy for the front-posed faces. The other expression classes are also efficiently determined. The major confusion related to the fear, disgust, and sadness expression classes are observed. In Table 15.2, the expressions got more confusing due to the sideward movement of the face, which leads to lesser clarity of feature attributes for determining the expressions. There is the least confusion in the happiness expression class. The anger and

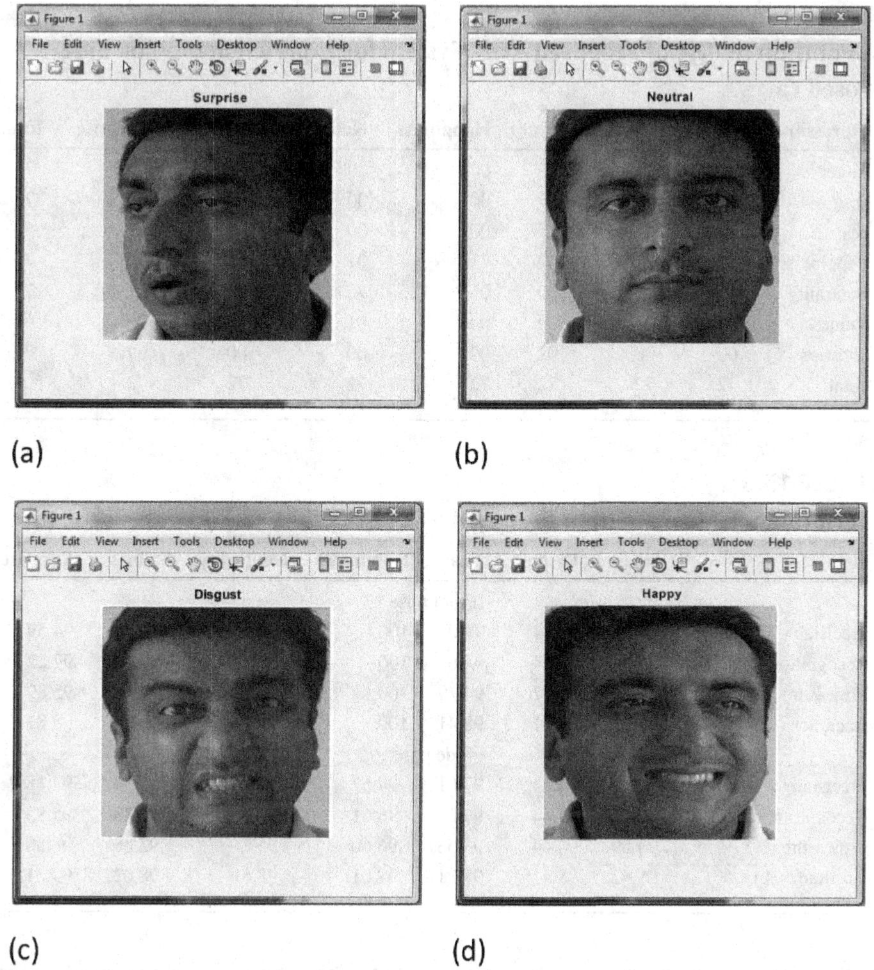

FIGURE 15.3 Some Expression Outcomes of a Candidate in Real-Time Using Proposed AFER System: (a) Surprise, (b) Neutrality, (c) Disgust, (d) Happiness.

TABLE 15.1
Confusion Matrix of Proposed AFER System for Experimentation on Front-Posed Faces.

Expressions	Anger	Disgust	Fear	Happiness	Neutrality	Sadness	Surprise	Total
Anger	36	01	00	00	00	00	01	38
Disgust	00	34	01	00	00	00	00	35
Fear	00	01	33	00	00	00	00	34
Happiness	00	00	00	36	00	00	00	36
Neutrality	00	00	00	00	36	02	00	38
Sadness	00	00	00	00	00	34	00	34
Surprise	00	00	02	00	00	00	35	37
Total	36	36	36	36	36	36	36	252

TABLE 15.2
Confusion Matrix of Proposed AFER System for Experimentation on Side-Posed Faces.

Expressions	Anger	Disgust	Fear	Happiness	Neutrality	Sadness	Surprise	Total
Anger	70	01	02	00	00	02	01	76
Disgust	01	68	03	00	00	00	00	72
Fear	00	01	65	00	00	00	01	67
Happiness	00	01	00	71	01	02	00	75
Neutrality	00	00	00	01	70	03	01	75
Sadness	01	01	00	00	01	65	00	68
Surprise	00	00	02	00	00	00	69	71
Total	72	72	72	72	72	72	72	504

TABLE 15.3
Performance Assessment of Proposed AFER System.

	Anger	Disgust	Fear	Happiness	Neutrality	Sadness	Surprise
			Front pose				
Precision (%)	94.74	97.14	97.06	100	94.74	100	94.59
Recognition rate (%)	100	94.44	91.67	100	100	94.44	97.22
F-measure (%)	97.30	95.77	94.29	100	97.30	97.14	95.89
Accuracy (%)	99.21	98.81	98.41	100	99.21	99.21	98.81
			Side pose				
Precision (%)	92.11	94.44	97.01	94.67	93.33	95.59	97.18
Recognition rate (%)	97.22	94.44	90.28	98.61	97.22	90.28	95.83
F-measure (%)	94.59	94.44	93.53	96.60	95.24	92.86	96.50
Accuracy (%)	98.41	98.41	98.21	99.01	98.61	98.02	99.01

neutrality classes are also significantly correct. The major confusion of side-pose expressions is related to the fear and sadness expression classes.

The results calculated in Table 15.3 indicate that the front-pose expressions are evaluated with a higher recognition rate, with an average value of 96.83%, than the side-pose expressions, which achieved 94.84% recognition. In terms of accuracy, the proposed AFER system outperformed, with an average value of 99.09% for front poses and 98.53% for side poses.

15.5.2 Results with Existing AFER Systems

To determine the effectiveness of the proposed AFER system, the experimentation for the same 36 candidates is also conducted with existing techniques. The existing AFER systems proposed by Kumar et al. (2020a, 2020b) are considered, and the same experiments are conducted in real time. Kumar et al. (2020a) used the combination of PCA and DCNN for feature selection and expression classification, respectively. Further, Kumar et al. (2020b) used the combination of QIBGSA and DCNN for feature selection and expression classification, respectively. Apart from these two experimentation techniques, the third experiment is also conducted with the combination of IQIGSA for feature selection and DCNN for expression classification. The third experiment is conducted to determine the importance of the hybrid DCRNN approach used in the proposed AFER system.

The first experiment for the comparative analysis is performed by considering the techniques of PCA+DCNN that Kumar et al. (2020a) used. The confusion matrix results using PCA+DCNN are determined in Tables 15.4 and 15.5 for the front- and side-posed faces, respectively. The respective performance assessment is described in Table 15.6.

The second experiment for the comparative analysis is conducted for the existing techniques of QIBGSA+DCNN that Kumar et al. (2020b) used. The confusion matrix results using QIBGSA+DCNN are described in Tables 15.7 and 15.8 for the front- and side-posed faces, respectively. The respective performance assessment for the QIBGSA+DCNN is described in Table 15.9.

The third experiment is performed by considering the combination of IQIGSA and DCNN instead of DCRNN used by the proposed system. This experiment is

TABLE 15.4

Confusion Matrix of PCA+DCNN for Experimentation on Front-Posed Faces.

Expressions	Anger	Disgust	Fear	Happiness	Neutrality	Sadness	Surprise	Total
Anger	31	02	01	00	00	02	01	37
Disgust	02	30	02	01	01	01	00	37
Fear	01	03	29	00	00	02	03	38
Happiness	00	00	00	32	00	00	01	33
Neutrality	01	00	00	02	32	03	00	38
Sadness	01	01	00	00	03	28	00	33
Surprise	00	00	04	01	00	00	31	36
Total	36	36	36	36	36	36	36	252

TABLE 15.5

Confusion Matrix of PCA+DCNN System for Experimentation on Side-Posed Faces.

Expressions	Anger	Disgust	Fear	Happiness	Neutrality	Sadness	Surprise	Total
Anger	62	02	03	00	01	04	02	74
Disgust	04	59	05	01	03	01	01	74
Fear	01	05	57	01	01	03	05	73
Happiness	00	01	00	65	02	02	01	71
Neutrality	03	04	02	03	60	07	02	81
Sadness	02	01	00	00	05	54	00	62
Surprise	00	00	05	02	00	01	61	69
Total	72	72	72	72	72	72	72	504

TABLE 15.6
Performance Assessment of PCA+DCNN System.

	Anger	Disgust	Fear	Happiness	Neutrality	Sadness	Surprise
			Front pose				
Precision (%)	83.78	81.08	76.32	96.97	84.21	84.85	86.11
Recognition rate (%)	86.11	83.33	80.56	88.89	88.89	77.78	86.11
F-measure (%)	84.93	82.19	78.38	92.75	86.49	81.16	86.11
Accuracy (%)	95.63	94.84	93.65	97.93	96.03	94.84	96.03
			Side pose				
Precision (%)	83.78	79.73	78.08	91.55	74.07	87.10	88.41
Recognition rate (%)	86.11	81.94	79.17	90.28	83.33	75	84.72
F-measure (%)	84.93	80.82	78.62	90.91	78.43	80.60	86.52
Accuracy (%)	95.63	94.44	93.85	97.42	93.45	94.84	96.23

TABLE 15.7
Confusion Matrix of QIBGSA+DCNN for Experimentation on Front-Posed Faces.

Expressions	Anger	Disgust	Fear	Happiness	Neutrality	Sadness	Surprise	Total
Anger	34	02	01	00	00	01	01	39
Disgust	01	32	01	00	01	00	00	35
Fear	00	01	31	00	00	01	00	33
Happiness	00	00	00	35	00	00	01	36
Neutrality	01	00	00	00	33	02	00	36
Sadness	00	01	00	00	02	32	00	35
Surprise	00	00	03	01	00	00	34	38
Total	36	36	36	36	36	36	36	252

TABLE 15.8
Confusion Matrix of QIBGSA+DCNN System for Experimentation on Side-Posed Faces.

Expressions	Anger	Disgust	Fear	Happiness	Neutrality	Sadness	Surprise	Total
Anger	67	01	03	00	01	02	01	75
Disgust	02	64	05	01	01	00	00	73
Fear	01	02	60	00	00	03	02	68
Happiness	00	01	00	68	02	02	01	74
Neutrality	01	03	02	02	67	04	02	81
Sadness	01	01	00	00	01	61	00	64
Surprise	00	00	02	01	00	00	66	69
Total	72	72	72	72	72	72	72	504

performed to analyze the effect of the DCNN classifier as compared to the DCRNN classifier. The confusion matrix results using IQIGSA+DCNN are described in Tables 15.10 and 15.11. The performance assessment for the same is described in Table 15.12.

15.5.3 Comparative Analysis

The comparison of the proposed AFER system is performed with the existing techniques of PCA+DCNN, QIBGSA+DCNN, and IQIGSA+DCNN. The expression-wise performance assessment for the proposed AFER system is depicted in Table 15.3, and the results of the PCA+DCNN, QIBGSA+DCNN, and IQIGSA+DCNN techniques are illustrated in Tables 15.6, 15.9, and 15.12, respectively. The overall average result values and comparative analysis are presented in Table 15.13.

TABLE 15.9
Performance Assessment of QIBGSA+DCNN System.

	Anger	Disgust	Fear	Happiness	Neutrality	Sadness	Surprise
			Front pose				
Precision (%)	87.18	91.43	93.94	97.22	91.67	91.43	89.47
Recognition rate (%)	94.44	88.89	86.11	97.22	91.67	88.89	94.44
F-measure (%)	90.67	90.14	89.86	97.22	91.67	90.14	91.89
Accuracy (%)	97.22	97.22	97.22	99.21	97.62	97.22	97.62
			Side pose				
Precision (%)	89.33	87.67	88.24	91.89	82.72	95.31	95.65
Recognition rate (%)	93.06	88.89	83.33	94.44	93.06	84.72	91.67
F-measure (%)	91.16	88.28	85.71	93.15	87.58	89.71	93.62
Accuracy (%)	97.42	96.63	96.03	98.02	96.23	97.22	98.21

TABLE 15.10
Confusion Matrix of IQIGSA+DCNN for Experimentation on Front-Posed Faces.

Expressions	Anger	Disgust	Fear	Happiness	Neutrality	Sadness	Surprise	Total
Anger	34	02	01	00	00	01	01	39
Disgust	01	33	01	00	01	00	00	36
Fear	00	01	32	00	00	00	00	33
Happiness	00	00	00	35	00	00	00	35
Neutrality	01	00	00	00	35	02	00	38
Sadness	00	00	00	00	00	33	00	33
Surprise	00	00	02	01	00	00	35	38
Total	36	36	36	36	36	36	36	252

TABLE 15.11
Confusion Matrix of IQIGSA+DCNN System for Experimentation on Side-Posed Faces.

Expressions	Anger	Disgust	Fear	Happiness	Neutrality	Sadness	Surprise	Total
Anger	69	01	02	00	00	02	01	75
Disgust	01	66	03	01	01	00	00	72
Fear	01	02	63	00	00	01	01	68
Happiness	00	01	00	70	02	02	00	75
Neutrality	00	01	02	01	68	03	02	77
Sadness	01	01	00	00	01	64	00	67
Surprise	00	00	02	00	00	00	68	70
Total	72	72	72	72	72	72	72	504

TABLE 15.12
Performance Assessment of IQIGSA+DCNN System.

	Anger	Disgust	Fear	Happiness	Neutrality	Sadness	Surprise
			Front pose				
Precision (%)	87.18	91.67	96.97	100	92.11	100	92.11
Recognition rate (%)	94.44	91.67	88.89	97.22	97.22	91.67	97.22
F-measure (%)	90.67	91.67	92.75	98.59	94.59	95.65	94.59
Accuracy (%)	97.22	97.62	98.02	99.60	98.41	98.81	98.41
			Side pose				
Precision (%)	92	91.67	92.65	93.33	88.31	95.52	97.14
Recognition rate (%)	95.83	91.67	87.5	97.22	94.44	88.89	94.44
F-measure (%)	93.88	91.67	90	95.24	91.28	92.09	95.77
Accuracy (%)	98.21	97.62	97.22	98.61	97.42	97.82	98.81

TABLE 15.13
Comparative Analysis of Proposed AFER System with Other Systems.

	PCA+DCNN	QIBGSA+DCNN	IQIGSA+DCNN	Proposed System (IQIGSA+DCRNN)
		Front pose		
Precision (%)	84.76	91.76	94.29	96.90
Recognition rate (%)	84.52	91.67	94.05	96.83
F-measure (%)	84.57	91.65	94.07	96.81
Accuracy (%)	95.57	97.62	98.30	99.09
		Side pose		
Precision (%)	83.25	90.12	92.95	94.91
Recognition rate (%)	82.94	89.88	92.86	94.84
F-measure (%)	82.98	89.89	92.85	94.82
Accuracy (%)	95.12	97.11	97.96	98.53

TABLE 15.14
Average Computation Time for Facial-Expression Recognition.

AFER system	Computation time (ms)
PCA+DCNN	168.12±47.07
QIBGSA+DCNN	116.48±14.52
IQIGSA+DCNN	109.85±22.61
IQIGSA+DCRNN (proposed system)	79.13±18.34

The results depicted in Table 15.13 indicate that the proposed AFER system outperformed the others with a higher recognition rate of 96.83% for the front poses and 94.84% for the side poses. Another aspect of facial-expression recognition is that the recognition rate of the side-pose expressions lacks than the front-pose expressions, as the sideward movement of the face hides some of the action units and feature components of the face that help in identifying expressions.

It can also be noticed that the recognition rate for all the variants of the QIGSA is quite higher than the standard feature-selection method of PCA. The advanced QIGSA concept is not only best in terms of reducing the feature dimensionality, but it also takes less time for facial-expression recognition. Although the training time for these systems is higher due to the advanced concept of quantum computing, iteration criteria in gravitational search algorithm, and deep learning models, the computation time to determine the facial expression in real time as the testing criteria is much less. The comparison of the proposed AFER system with other systems based on computation time for expression recognition is described in Table 15.14. The computation time is described with the notation X±Y. Here, X indicates the time in milliseconds (ms) and Y indicates the standard deviation.

The computation time for the expression recognition in the real-time scenario is less than a second. The proposed system takes only 79.13 ± 18.34 ms to determine the expression. The other systems have also lesser time but still higher than the proposed AFER system.

15.6 CONCLUSION

Facial-expression recognition application has attracted researchers due to the increasing adaptability of AI and human–computer interaction applications in real life. These applications are helpful for Society 5.0 by determining human intentions in real time, and possible steps can be taken with respect to that. In this research work, an autonomous facial-expression recognition system is proposed, and real-time evaluations are presented. The proposed AFER system is a pool of different techniques for face recognition, feature extraction, feature selection, and classification. The preceding steps are fulfilled using the Viola-Jones algorithm, hybrid GF and LBP, IQIGSA, and DCRNN techniques. The system evaluation is conducted in real time to recognize the expressions of 36 candidates ages 20–35. The average recognition rate of candidates from the front is 96.83%, and the rate for side poses is 94.84%. The comparison conducted with the other techniques indicates that the proposed system outperformed the other techniques analyzed. In the future, the work can be extended to the recognition of facial expressions of people belonging to different cultures for the analysis of the effect of culture on expressions.

REFERENCES

Alexandre, G. R., Soares, J. M., Thé, G. A. P. 2020. Systematic review of 3D facial expression recognition methods. *Pattern Recognition*, 100, pp. 107108(1–16).

Ding, Y., Zhao, Q., Li, B., Yuan, X. 2017. Facial expression recognition from image sequence based on LBP and Taylor expansion. *IEEE Access*, 5, pp. 19409–19419.

Ekman, P., Friesen, W. V. 2003. *Unmasking the Face: A Guide to Recognizing Emotions from Facial Clues*, Vol. 10. Prentice-Hall, Englewood Cliffs, NJ.

Goel, S., Khurana, G., Panchal, V. K. 2018. Remote sensing image classification for Jabalpur region using swarm classifiers. *International Journal of Artificial Intelligence and Soft Computing*, 6, 4, pp. 326–347.

Gupta, S., Bhuchar, K., Sandhu, P. S. 2011. Implementing color image segmentation using biogeography based optimization. In *2011 International Conference on Software and Computer Applications IPCSIT*, IACSIT Press, Singapore, pp. 167–170.

Gupta, S., Thakur, K., Kumar, M. 2021.2D-human face recognition using SIFT and SURF descriptors of face's feature regions. *The Visual Computer*, 37, pp. 447–456.

Jain, N., Kumar, S., Kumar, A., Shamsolmoali, P., Zareapoor, M. 2018. Hybrid deep neural networks for face emotion recognition. *Pattern Recognition Letters*, 115, pp. 101–106.

Kaur, R., Girdhar, A., Gupta, S. 2011. Color image quantization based on bacteria foraging optimization. *International Journal of Computer Applications*, 25, 7, pp. 33–42.

Kumar, M., Gupta, S., Mohan, N. 2020. A computational approach for printed document forensics using SURF and ORB features. *Soft Computing*, 24, 17, pp. 13197–13208.

Kumar, S., Jha, R. K. 2017. Weak signal detection from noisy signal using stochastic resonance with particle swarm optimization technique. In *2017 International Conference on Noise and Fluctuations (ICNF)*. IEEE, Vilnius, Lithuania, pp. 1–4.

Kumar, S., Jha, R. K. 2019. Noise-induced resonance and particle swarm optimization-based weak signal detection. *Circuits, Systems, and Signal Processing*, 38, 6, pp. 2677–2702.

Kumar, Y., Verma, S. K., Sharma, S. 2019. Appearance based feature extraction and selection methods for facial expression recognition. In *Proceedings of the International Conference on Sustainable Computing in Science, Technology and Management (SUSCOM)*. Amity University Rajasthan, Jaipur-India, pp. 958–966.

Kumar, Y., Verma, S. K., Sharma, S. 2020a. Multi-pose facial expression recognition using appearance-based facial features. *International Journal of Intelligent Information and Database Systems*, 13, 2–4, pp. 172–190.

Kumar, Y., Verma, S. K., Sharma, S. 2020b. Quantum-inspired binary gravitational search algorithm to recognize the facial expressions. *International Journal of Modern Physics C*, 31, 10, pp. 2050138(1–24).

Langner, O., Dotsch, R., Bijlstra, G., Wigboldus, D. H., Hawk, S. T., Van Knippenberg, A. D. 2010. Presentation and validation of the Radboud faces database. *Cognition and Emotion*, 24, 8, pp. 1377–1388.

Li, S., Deng, W. 2020. Deep facial expression recognition: A survey. *IEEE Transactions on Affective Computing*. DOI: 10.1109/TAFFC.2020.2981446.

Lundqvist, D., Flykt, A., Öhman, A. 1998. The Karolinska directed emotional faces (KDEF). *CD ROM from Department of Clinical Neuroscience, Psychology Section, Karolinska Institutet*, 91, 630.

Lyons, M., Akamatsu, S., Kamachi, M., Gyoba, J. 1998. Coding facial expressions with Gabor wavelets. In *Proceedings Third IEEE International Conference on Automatic Face and Gesture Recognition*. IEEE, Nara, Japan, pp. 200–205.

Moghadam, M. S., Nezamabadi-Pour, H. 2012. An improved quantum behaved gravitational search algorithm. In *20th Iranian Conference on Electrical Engineering (ICEE2012)*. IEEE, Tehran, Iran, pp. 711–715.

Moghadam, M. S., Nezamabadi-Pour, H., Farsangi, M. M. 2012. A quantum behaved gravitational search algorithm. *Intelligent Information Management*, 4, 6, pp. 390–395.

Rashedi, E., Nezamabadi-Pour, H., Saryazdi, S. 2009. GSA: A gravitational search algorithm. *Information Sciences*, 179, 13, pp. 2232–2248.

Sadeghi, H., Raie, A. A. 2019. Histogram distance metric learning for facial expression recognition. *Journal of Visual Communication and Image Representation*, 62, pp. 152–165.

Sayette, M. A., Cohn, J. F., Wertz, J. M., Perrott, M. A., Parrott, D. J. 2001. A psychometric evaluation of the facial action coding system for assessing spontaneous expression. *Journal of Nonverbal Behavior*, 25, 3, pp. 167–185.

Singh, V., Kumar, G., Arora, G. 2016. Analytical evaluation for the enhancement of satellite images using swarm intelligence techniques. In *2016 3rd International Conference on Computing for Sustainable Global Development (INDIACom)*. IEEE, New Delhi, India, pp. 2401–2405.

Sobia, M. C., Abudhahir, A. 2017. Facial expression recognition using a hybrid kernel based extreme learning machine. *Journal of Computational and Theoretical Nanoscience*, 14, 6, pp. 2894–2904.

Sun, A., Li, Y., Huang, Y. M., Li, Q., Lu, G. 2018. Facial expression recognition using optimized active regions. *Human-centric Computing and Information Sciences*, 8, 1, pp. 1–24.

Tanveer, I. 2020. Eigenface based facial expression classification. Retrieved January 2, 2021, from www.mathworks.com/matlabcentral/fileexchange/33325-eigenface-based-facial-expression-classification.

Tsai, H. H., Chang, Y. C. 2018. Facial expression recognition using a combination of multiple facial features and support vector machine. *Soft Computing*, 22, 13, pp. 4389–4405.

Yang, B., Cao, J. M., Jiang, D. P., Lv, J. D. 2018. Facial expression recognition based on dual-feature fusion and improved random forest classifier. *Multimedia Tools and Applications*, 77, 16, pp. 20477–20499.

Ye, Y., Zhang, X., Lin, Y., Wang, H. 2019. Facial expression recognition via region-based convolutional fusion network. *Journal of Visual Communication and Image Representation*, 62, pp. 1–11.

Zeng, N., Zhang, H., Song, B., Liu, W., Li, Y., Dobaie, A. M. 2018. Facial expression recognition via learning deep sparse autoencoders. *Neurocomputing*, 273, pp. 643–649.

Zhang, K., Huang, Y., Du, Y., Wang, L. 2017. Facial expression recognition based on deep evolutional spatial-temporal networks. *IEEE Transactions on Image Processing*, 26, 9, pp. 4193–4203.

Zheng, H., Wang, R., Ji, W., Zong, M., Wong, W. K., Lai, Z., Lv, H. 2020. Discriminative deep multi-task learning for facial expression recognition. *Information Sciences*, 533, pp. 60–71.

16 Adversarial Attacks and Defenses against Deep Learning in Cybersecurity

Dr. B Gomathi and Ms. J Uma

CONTENTS

DOI: 10.1201/9781003184140-16

16.1 INTRODUCTION

16.1.1 Deep Learning

Adversarial attacks and defenses on cyber-physical systems is basically an AI (artificial intelligence) technique that mimics the human mind, i.e., the process of human thinking. Besides the emerging field of machine learning (ML), deep learning (DL) techniques deal with huge amounts of big data during the training phase. DL uses more hidden layers in its structure, as shown in Figure 16.1, to analyze the input that makes these DL techniques to produce the output more accurately when compared to ML.

Among the diversified learning models, DL learning models are classified into two types: supervised learning and unsupervised learning. Convolutional neural networks (CNNs), recurrent neural networks (RNNs) and deep belief networks (DBN) fall under supervised learning. Restricted Boltzmann machines (RBMs), generative adversarial networks (GANs) come under unsupervised learning. One unique feature of DL is that any DL model can directly learn the features of the raw data, such as images and text, without the preprocessing the data with any model. Thus, DL gives more accurate outputs than earlier AI methods (Liu and Lang 2019).

16.1.2 Deep Learning in Cybersecurity Applications

Cybersecurity is an emerging sector in the field of DL. Through Cybersecurity, we can protect the systems connected through a network, such as hardware, software and mainly the data and information from the internet. The Word "cyber" refers

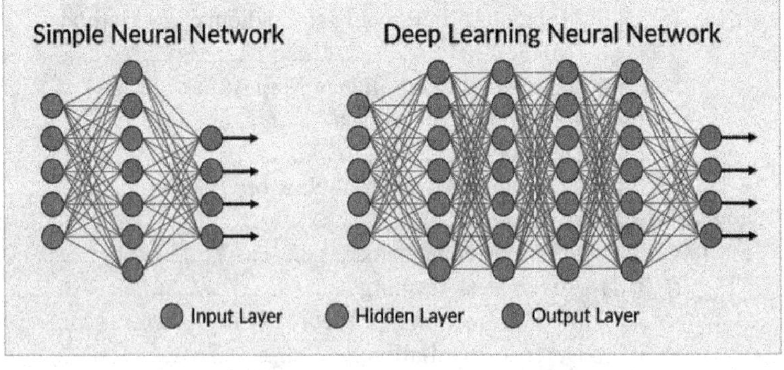

FIGURE 16.1 Deep Neural Network Architecture.

to the technique that is embedded into the component to protect it. "Security" is related to the component for which the security is needed. Cyberattacks are basically divided into two types: web-based and system-based. Let's recall few attacks based on these two types.

Web-based attacks: Attacks or threats happening on websites or in web applications, like DNS spoofing, session hijacking, injection attacks, phishing, denial of service, URL interpretation, dictionary attacks etc.

System-based attacks: Hardware-level threats and attacks to the set of machines, their components, or inside the networks. Such attacks are namely viruses, worms, Trojan horses, backdoors and bots.

AI techniques should be used in the right way to detect wide range of threats, which may include new threats as well as branded ones like IDPS, phishing, spam detection etc., and applying DL methods to cybersecurity applications is a state-of-the-art technique that improves the accuracy of the model. The main goal of this chapter is to analyze various cyberattacks and prevention methods using DL techniques. The cybersecurity applications process involves huge hash values to detect the cybercrimes. To process the huge amount of big data involved in cybercrimes, DL algorithms and techniques will produce better solutions than shallow algorithms and ML algorithms.

Common threats to cybersecurity are malicious software intrusion, data breeches, phishing, SQL injection, denial of service etc. Besides these threats, adversarial attacks are one of the most rigorously emerging in the field of cyberthreats.

16.1.3 Applications of Deep Learning in Cybersecurity

Here are some listed applications of DL in cybersecurity and an analysis of how DL can be used to improve security.

16.1.3.1 Intrusion Detection and Prevention Systems

Intruders create newer attacks every day. To prevent the intruders from misusing data, the detection system must be continually strengthened. Recently, many DL approaches have been used to train the network big data with many distinctive learning methods with high accuracy. Techniques such as artificial neural networks (ANNs), support vector machines (SVMs), rule-based systems and fuzzy logic aims to prevent intrusions with DL algorithms (Karatas, Demir, and Sahingoz 2019).

16.1.3.2 Phishing Cybercrime

Phishing is one of the major cybercrimes among Internet users. These kinds of criminals will direct the user to malicious websites to crack their login credentials. Malicious web links or enticing messages will be sent to users' email addresses or messaging apps and lead users to access unauthorized websites. This will result in users' sensitive information being downloaded to a malicious database. The users are psychologically convinced through such emails or messages, thereby they may enter their personal details, such as email address, password, bank account details, credit card number etc. Highly intelligent agents are acting under these sets of crimes that steal user credentials and spread malware among internet users (Thomas, Vijayaraghavan, and Emmanuel n.d.).

16.1.3.3 Spam Detection in Social Area Network

Huge adversaries and cybercrimes are evolving in the field of social engineering attacks. Spam is a type of junk mail sent to an email address. Spam does not only include unsolicited commercial emails; sometimes, they may be fraudulent messages. These dangerous types of spam may clog the information stored and damage the network as well. The filtering of spam can be done based on the textual information of the emails. Many AI algorithms, like Naïve Bayes (NB), term frequency-inverse document frequency (TF-IDF) and SVM, will boost up the filtration of spam mail and prevent fraudulent messages. By using advanced techniques, like deep neural networks (DNNs) and case-based reasoning fuzzy logic systems, these kinds of crimes are being prevented to some extent (Lansley et al. 2019). As an advanced technique, the suspected emails can be analyzed based upon feature vectors, such as attachments, mail size, IP address as well as the address of the recipient and sender of the email. To have such deep analysis of detecting cybercrime, advanced SVM and DNN methods are used in cybersecurity applications.

16.1.3.4 Denial of Service Attack

Denial of service (DoS) is an attack that disrupts the normal traffic of a targeted server or a network by overpowering the targeted network or server with a flood of internet traffic. If there are multiple such attacks in a distributed manner, they are known as distributed denial of service (DDoS) attacks. This DoS attack creates numerous malicious packets to the targeted network. They are usually detected in one of two ways: signature-based IDPS and anomaly-based IDPS. Signature-based detection relies on predefined network traffic to detect DoS threats, whereas anomaly-based detection analyzes the components of the network, such as packet header information, packet size etc. to detect various intrusions in a network. ML algorithms, such as NB, k-nearest neighbor (k-NN) and k-means algorithms, will work for DoS and DDoS detection and prevention.

16.2 ADVERSARIAL ATTACKS

Nowadays, intrusion detection systems (IDS) are incorporated with ML models to improve efficiency in identifying different kinds of cyberattacks. This is called adversarial machine learning. The main objective of the adversary is to deceive the prediction process of a trained model by producing an adversarial sample. Faulty input data that fools the trained model is called an adversarial sample (Dong et al. 2017). In an adversarial sample, the trained model comes over the decision boundary. The adversarial approach shifts the input samples towards a linear approximation of the decision boundary by orthogonally projecting them onto the boundary, resulting in an adversarial sample. Since the adversarial samples and input of the trained model are almost identical, it is difficult to determine that the trained model misclassified the data. When adversarial samples are mixed with input data, it is hard to identify them. When adversarial samples are crafted in high-dimensional space, slight changes to a few dimensions of actual input samples are not enough to modify the prediction results of the ML model. Instead, small modifications in all the dimensions of the actual sample input will result in effective modifications to the prediction result.

16.2.1 Adversarial Capabilities

Adversarial capabilities represent the portion of knowledge about the target model that can be gained and utilized by the adversaries. Noticeably, the adversaries are stronger than others by accessing and gaining more knowledge about the ML model. We can categorize the adversarial capabilities based on the usage time, such as the training and testing stages.

16.2.1.1 Training Stage Capabilities

Attacks at the training time try to manipulate the functionality of a target model by modifying the actual training data set. When all or partial training data are accessed in the training stage by any attack, it is called a weakest attack. Adversarial capability attacks can be categorized into three types:

1. Data injection: Adversaries don't have privilege to manipulate the training data and ML model, but they can append new data in the training set. Hence, the adversarial samples are injected into the training set by adversaries to change the behavior of the target ML model.
2. Data modification: Adversaries can manipulate the training set, but it is not possible to access the ML algorithm. Hence, adversaries can corrupt the training data before it is used to train the ML model.
3. Logic corruption: Adversaries can meddle with the behavior of the ML model.

16.2.1.2 Testing Stage Capabilities

In the testing stage, adversaries force the ML model to misclassify the actual sample input. It is based on the knowledge acquired by adversaries about the target model. There are different ways to attack a faulty trained model. These attacks lead the model to carry out malicious attacks that are against the actual task performed by the model. There are some attacks that disturb the main features of the trained model to confuse the model's decision-making capabilities. This is called a white box attack. Here, the adversary will carefully generate adversarial samples by making use of the whole information about the trained model. The adversary can grasp the weaknesses of the trained model in a clear way. In other ways, attackers confuse the trained model without any knowledge of the features of the model. This is called a black box attack. Here, the adversary sends the input data to the trained model and asks for the outcome of the trained model. It helps to identify the input–outcome relationship in the model that leads to finding the shortcomings of the trained model. Since the parameters are unidentified in the model, black box attacks are utilized in more applications than white box attacks.

Adversarial attacks make the trained model more vulnerable and make the system unacceptable for the specific application. One famous instance of adversarial attacks is email spam filtering. Here, the adversary attempts to manipulate a few important words or symbols in an email message. Hence, it can masquerade non-spam email as spam and spam email as non-spam. These two miscategorizations reduce the integrity of email spam filtering by stopping up authorized emails and bringing down the accuracy of the spam filter system.

Adversarial attacks can be classified depending on the

- Complexity: The results of adversarial attacks can vary from slightly bringing down the accuracy of the prediction of the trained model to classify all the input data in the wrong way.
- Knowledge: Based on the knowledge about the trained model acquired by attackers, it can be classified as a white box attack or black box attack.

16.2.2 Features of Adversarial Samples

The adversarial samples have the following features:

1. Transferability: When adversarial samples are crafted for an ML model M_1 and adversaries gained knowledge on model M_2, which provides same prediction results as M_1, then it is not necessary for the adversary to have knowledge of the M_1 model.
2. Adversarial instability:After applying perturbation on an actual sample to generate an Adversarial Sample, Adversarial Samples may lose their property. Hence, it is necessary to modify the crafted sample to classify under the targeted class.
3. Regularization effect: The regularization method can identify defects of models and enhance the robustness of samples, but it is expensive.

16.2.3 Adversarial Goals

The main goal of adversaries is to create uncertainties in ML models. Goals can be further partitioned into four types:

1. Confidence reduction: Adversaries try to reduce the confidential level of prediction for an ML model.
2. Misclassification: The adversaries try to generate crafted samples that make the model to classify the input sample under a class other than original one.
3. Targeted misclassification: Adversarial samples make the ML model classify the input sample under the target class.
4. Source/target misclassification: The adversaries map and classify the specific input under a specific target class.

16.2.4 Techniques to Generate Adversarial Attacks

ML models (Papernot et al. 2015) are attacked by various techniques that would turn the ML model to perform irrelevant tasks. The techniques can vary based on sample generation speed, performance and complication. Earlier input data are perturbed manually to generate adversarial samples. Automatic techniques are faster than manual perturbation in producing and assessing samples. In addition, adversarial samples generated for one model can affect other models that provide the same functionality. Furthermore, white box attacks, which require implementation details of the training model, are not always possible. In such instances, black box attacks are selected for adversarial attacks to generate adversarial samples. This portion of the chapter can provide techniques that are capable of masquerading the ML models.

16.2.4.1 Fast Gradient Sign Methods (FGSM)

Goodfellow et al. (2020) introduced an effective white box untargeted attack to produce adversarial samples of the ML model. The gradient descent optimization technique is utilized for one step gradient update along with the direction of improvement in adversarial loss. The actual input sample p is updated by diminishing or magnifying the magnitude of perturbation ε to each feature of p if p∈P, where P is the actual sample data set. If the input is an image, the individual pixel value of the image is updated. The adversarial sample is generated by using the following equation:

$$p' = p + \varepsilon.sign(\Delta_p.C(x, p, q)) \tag{16.1}$$

Where p is a sample data from an actual data set, q is the label of p and x and Δ_p are the parameters and gradient value of the specific ML model, respectively. To train the ML model, cost function C(x,p,q) is utilized and can be used to obtain perturbation to generate the adversarial sample. Here, p' is the adversarial sample from actual sample p. The magnitude of loss decides the sign of the gradient. In image classification, the intensity of the pixels is decided. However, random perturbation improves the achievement and diversity of the adversarial sample in the FCGM (Carlini and Wagner 2017).

16.2.4.2 Basic Iterative Method (BIM)

Alexey Kurakin et al. (2017) proposed BIM to enhance the accomplishment of FGSM by introducing the step size for several loop repetitions. When an adversarial sample is generated by FGSM, the Euclidean distance between the actual sample and the adversarial sample is too lengthy and makes the adversarial sample invalid by categorizing it as a dead zone. Instead of injecting noise with each parameter separately, as in FGSM, BIM is repeatedly injected many times with a small step size. In addition, the following equations help to clip the adversarial sample by moving it into a valid range within multiple iterations:

$$p_0' = p \tag{16.2}$$

$$p_{n+1}' = Clip_{p,\varepsilon}\left[p_n' + \alpha.sign\left(\Delta_p.C\left(x, p_n', q\right)\right)\right] \tag{16.3}$$

Let p be an actual sample data, q be the class label for actual data, $C\left(x, p_n', q\right)$ be the cost function of the L model for the specified sample data p and q, α decides the size of the perturbation at each loop step. $Clip_{p,\varepsilon}$ decides the clipping portion of the individual data in p that is utilized to have an effect on the prediction accuracy of the ML model. Clipping ensures that interim outcomes of each iteration are cropped such that adversarial samples are within the neighborhood of actual samples. This method generally does not rely on the approximation of the model and produces additional harmful adversarial samples when this algorithm runs for more iterations. A BIM attack is proven to be more efficient than an FGSM attack on ImageNet models. But BIM attacks only made an effort to magnify the loss value of classification. However, incorrect class labels chosen by model are not identified clearly.

16.2.4.3 Momentum Iterative Fast Gradient Sign Method (MI-FGSM)

Momentum iterative fast gradient sign method (MI-FGSM) is a kind of white box attack that helps to steady the optimization. Usually, iterative techniques compute gradients at each looping process and that helps to update the adversarial sample; hence, it can delude white box attacks. This does not, however, work for black box attacks. MI-FGSM is controlled in practical attacks. In various optimization techniques, the momentum can help to provide quicker convergence and get away from local extremes by adding gradients of loss function with every loop process. The momentum term makes the direction of update unchangeable. Hence, the momentum term is included in MI-FGSM to produce the adversarial samples as well as to avoid poor local convergence during the looping process. MI-FGSM introduces noise by using the following equations:

$$v_{t+1} = \mu v_t + \frac{\nabla_p C\left(x, p_t^{'}, q\right)}{\left\| \nabla_p C\left(x, p_t^{'}, q\right) \right\|_1} \tag{16.4}$$

$$p_{t+1}^{'} = p_t + \mu.\ \text{sign}(v_{t+1}) \tag{16.5}$$

Let us assume that v_{t+1} collects velocity values in the direction of gradient, μ is the decay factor of the momentum term and v_t is the accumulated gradient at iteration t. It enhances the chances of attaining a global minimum by using the momentum of physics that helps to overcome the local minimum. This physical momentum causes faster convergence and over-acceleration. Hence, Nesterov momentum is introduced to avoid moving too fast to modify the gradient value.

16.2.4.4 Jacobian-Based Saliency Map Attack (JSMA)

Papernot proposed the JSMA technique, which is a white box attack for targeted misclassification. Let us assume that the input is p, the ML model is M and the outcome of the class q is represented as $M_q(p)$. To achieve the target class s, the probability of $M_s(p)$ will be magnified while probabilities of $M_q(p)$ of the remaining classes q≠s will be diminished by the given changes in input feature i. Using the saliency map helps to achieve the aforementioned goal as shown here:

$$JS(p,s)[i] = \begin{cases} 0, if\ \frac{\partial F_s}{\partial p_i}(p) < 0\ or \sum_{q \neq s} \frac{\partial F_q}{\partial p_i}(p) > 0 \\ \frac{\partial F_s}{\partial p_i}(p) \left| \sum_{q \neq s} \frac{\partial F_q}{\partial p_i}(p) \right|,\ otherwise \end{cases} \tag{16.6}$$

Jacobian matrix is utilized to assess the sensitivity of a model with each input feature that makes important modifications in the output of the ML model. Instead of changing the whole image at once, it changes small sets of pixels at a time. This will be iterated until the ML model misclassifies the input or maximum counts of the pixels are modified. Since the forward propagation is utilized to compute salient points in JSMA, it makes the computation process as simple as possible. But it takes more time to do the computation.

16.2.4.5 Carlini-Wagner Attack

Delineating adversarial attacks to generate adversarial samples is developed as an optimization problem. The main objective of optimization problems is to identify small perturbations for adversarial samples that are classified under different labels by the ML model. Compared to FGSM attacks, Carlini-Wagner attacks require very little perturbation on an input data set for adversarial attacks. The objective function for the optimization problem to generate an adversarial sample is represented as follows:

$$\text{Minimize } d' - d_p{}' + c.g(d') \tag{16.7}$$

Where d is the actual input data, d’ is the adversarial sample, c is the constant value that helps to merge different functions in minimization function and g is the loss function as given here:

$$g(d') = Max\left(Max\left\{Z(d')_i : i \neq s\right\} - Z(d')_s, -f\right) \tag{16.8}$$

Let us assume that, as the target classification class, Z represents the softmax function. $Z(d')_s$ denotes the probability of an adversarial sample that belongs to the target class. $Z(d')_i$ represents the probability of the best prediction of the adversarial sample among non-target classes. F decides the confidence level of the adversarial sample. Adversarial samples have high confidence when f value is improved. If f=0, then crafted samples have less confidence and indicates that both the actual and adversarial samples are categorized under the same class.

16.2.4.6 Deepfool Attack

Deepfool (Moosavi-Dezfooli, Fawzi, and Frossard 2015) is an untargeted white box attack. Deepfool can identify the nearest distance from the actual sample to the decision boundaries of the crafted sample. It can generate small adversarial perturbation for the given actual sample using a repetition method. Deepfool finds decision boundaries of the classifier based on the region of the actual sample data. At each iteration, Deepfool provides a minor perturbation for the actual data, leading in an adversarial image generated by linearizing the outline of the region where the actual data is located.

16.2.4.7 Generative Adversarial Network (GAN)

In GANs, the distribution of an input data set is learned to produce an adversarial samples (Figure 16.2). Furthermore, DNNs are utilized to implement various applications, like image processing, self-driving cars, natural-language processing, facial recognition etc. Nevertheless, DNNs are more vulnerable and deviate from the expected outcome by making adversarial attacks on the training model. For example, GAN generates images from noise that creates enormous threats to image-processing systems. Massive research projects are going on to produce adversarial samples against DNNs.

Goodfellow et al. (2020) proposed GANs that generate crafted samples identical to actual training data sets. The GAN framework has two different neural networks: generators (G) and discriminators (D). First, the generator is initialized with random noise. The generator G is a deep CNN model that keenly generates samples that are anticipated to be similar to the actual data sample set. However, the training is provided to G that makes D mistake the actual sample for the crafted sample with high probability. Network D is to examine an input sample to check whether the input samples are actual or crafted. This race drives both the models to enhance their accuracy. This process ends when D is not able to differentiate between actual input samples and samples produced by G. Here, G makes D to be a fool, and D tries to avoid being fooled. The function V (G, D) is proposed by Goodfellow as shown here:

$$Min_G Max_D F(D,G) = E_{d \sim p_{data}(x)}\left[\log D(d)\right] + E_{t \sim p_t(t)}\left[\log\left(1 - D\left(G(t)\right)\right)\right] \quad (16.9)$$

Let us take $p_{data}(x)$ as the generator's distribution and $p_t(t)$ as a prior on input noise variables. The objective is to train D to maximize the probability of assigning the correct label to the training and sample examples while simultaneously training G to minimize it.

The goal of the discriminator D is to generate a value near 1 for an input sample from the actual training set and values close to 0 for samples from the generator. Here, crafted samples cannot be different from the actual input sample under the original class. GAN can misguide the target ML model by producing output results

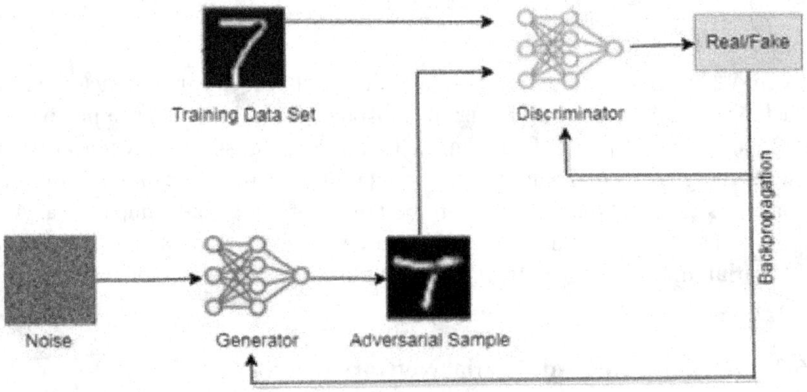

FIGURE 16.2 Framework of Generative Adversarial Network.

that are not always the actual outcome. To highlight more about GAN, the binary malware classification problem is considered. The steps are as follows:

1. Input random noise to generator G. Feed some random noise to G. Let us take (p,q) be an input-label set that changes to (p',q'), where p' is the adversarial sample and q' is a fake label.
2. Input the actual sample and adversarial sample alternatively to the discriminator D.
3. D outputs a probability that ranges from 0 to 1.
4. Both G and D have feedback loops. G is a feedback loop with D, while D is a feedback loop with the actual set of data from a training feature set.
5. As a binary malware classifier, it categorizes the inputs as malware or goodware.
6. The loss of D is computed by adding the loss of the neural network when both the actual and crafted samples are fed as input. G computes its noise separately based on the objective function.
7. An optimization algorithm is applied, and the previous steps are repeated for a certain number of iterations.

In the context of cybersecurity applications for the GAN, one can clearly observe two distinct paths. The first involves research where the ability to detect adversarial attacks is strengthened using GANs. The second path involves research that makes use of the GAN to create adversarial models or adversarial data.

16.3 DEFENSE MECHANISM

DNN approaches have made remarkable advancements against adversarial examples. The characteristics of defending against the adversarial examples is said to be "robustness against adversarial samples". It is also proven in many research articles that adversarial perturbations are able to produce vulnerable results against the adversarial attacks. The reason behind the adversarial attacks is still a nightmare for experts in the field of cybercrimes (Li et al. 2020). Therefore, building a better defense model to prevent and at the same time to defend against the attack is a challenging and tedious job for researchers. The defense mechanism can be divided into three types based on the knowledge of the attack: zero knowledge adversary, perfect knowledge adversary and partial knowledge adversary.

Zero knowledge adversaries are not aware of the history of information about the mechanism behind the attacks. Perfect knowledge adversaries have wide knowledge about the parameters involved during the attack and the mechanisms behind the attack history, and partial knowledge adversaries fall under both white and black box attacks. It is not aware of the inner schema of the attack that happened. Thus, zero knowledge adversaries are said to be a black box defense, and perfect knowledge is known as a white box defense.

Among these types, zero knowledge adversaries show more efficacy than the other two types. It shows a robust detection rate against a Modified National Institute of Standards and Technology (MNIST) training data set for about a 70% detection rate and a 40% false positive rate, but the other two methods show very poor rates of detection (Short, La Pay, and Gandhi 2019).

Defense mechanisms (Yu et al. 2020) are a challenging task since we have two different approaches to defend against adversarial examples: i) training DNNs based on the adversarial examples and defensive distillation and ii) detecting adversarial examples or eliminating adversarial noise after DNNs are built.

Hence, designing a defense mechanism is a bottleneck task for any application. The following are reasons for the complexity in designing the defense mechanism discussed by Yu et al. (2020) in their article "The Defense of Adversarial Example with Conditional Generative Adversarial Networks".

1. Huge data is required to frame an efficient defense mechanism, which leads to high computational power.
2. The defense mechanism will be effective based on the type of attack.
3. Since the noisy data and clear data have only minute divergence, it is difficult for the defense technique to detect the attack.

16.3.1 Various Defense Mechanisms

In this chapter, we give a wide summary about various defense mechanisms and their framework.

16.3.1.1 Adversarial Training

Adversarial training is one of the most common methods to defend against attacks. This approach is mainly used to train CNN data sets based on the knowledge of the adversarial attack. This idea was first introduced by Szegedy et al. (2015) to build a robust classifier to include adversarial information in the training process, which we refer to as adversarial training. The adversarial noise from those models is fed into the training data set of the defense model. These training data sets are amplified during the adversarial training to detect the attacks efficiently. These types of defense models are more robust against black box attacks and achieve remarkable computational efficiency using FGSM (Khalid et al. 2019). At the same time, this method of defense can be applied only to trained adversarial examples. It does not provide efficient results among the new attacks apart from the trained data set.

16.3.1.2 Input Validation and Preprocessing

This is one of the primary defense mechanisms against adversarial attacks. Perhaps validating the input is mandatory before it is given to any model; after this validation, it is safer to preprocess the input image/data to get rid of adversarial perturbations. This preprocessing follows two steps: i) normalization with compression and ii) noise reduction. During normalization with compression in DNNs, all the images will be converted to the same size, irrespective of the original sizes of the images from the data set. Noise reduction will be performed with the help of Gaussian transformation to make the images smoother than the originals. By the aforementioned preprocessing steps, the precision of the image is reduced so that the chance for adversarial attacks is minimized. As discussed, the precision of the image can be determined by its spatial resolution and pixel density (Samangouei, Kabkab, and Chellappa 2018).

16.3.1.3 Defense Distillation

This is a method that follows the training procedure to generate a new model whose gradients are smaller than the original defense model. It smooths the decision surface of the adversarial images in two phases using the variant of the distillation and reduces the amplitude of the gradient around the input points to reduce the attack rate. It trains one model to predict the output of the other model, which is of same size and architecture. This initial model is trained with the property of "hard labels", which achieves maximum accuracy.

For instance, requiring a 100% match of the facial recognition in a scanner, the algorithm matches all the features of a face and produces an output; at the same time, it does not match each and every pixel since it might take much more time (Yu et al. 2020). In this situation, the attackers may know the feature set and generate a combination of matched image pixels and generate a new image that matches the original and then send a fake image for face recognition with a handful of correct pixels that satisfies the algorithm behind face recognition and produces a correct result.

The second model, with 95% probability, introduced a new parameter called T, the temperature to the softmax layer of the network. This value is raised to a large value in, say, the 40–50 range so that it produces the smooth probability vectors.

$$S_{max}\left(a,T\right)_n = \frac{e^{a_n/T}}{\sum_m e^{a_{m/T}}} \tag{16.10}$$

Where S_{max} is the distillation function of the network, a sample set of images, and T is the temperature of the softmax layer. This softmax label produces 95% probability for an image that matches with the record or database. This is an additional filter applied to gain robustness and improves spoofing efficiency. This method best defends against white box attacks like FGSM and JSMA. This aspect of defense distillation proves to be difficult for the attackers to generate adversaries.

16.3.1.4 MagNet

This method is another defense mechanism in DNNs against the adversarial examples without the knowledge of the framework behind the adversaries. Meng and Chen (2017) introduced this defense, which consists of two components:

***Detector*:** Identifies the normal and adversarial samples
***Reformer*:** Generates a variant of the input and feeds into the DNN.

The detector process can be carried out by approximating the manifold of the normal samples. The significance of the reformer is to move the adversarial images more closely to the manifold of the normal samples. This approach is effective to accurately categorize even the small perturbations available in the adversarial examples. Perhaps the reformer makes use of autoencoders and a combination of autoencoders to strengthen the defense. The task of the encoder is to reframe the input sample s and map with s'. These two input images are classified to determine the probability divergence to predict whether they are adversary images or normal images. This method works well upon gray box attacks, where the attackers are aware of the network and defense technique but not the parameters included in the mechanism.

16.3.1.5 Defense GAN

GAN (Goodfellow et al. 2020) was originally introduced by Goodfellow et al. and can detect both white box and black box attacks. This method of defense utilizes the power of GANs, which are trained to formalize the perturbation in the adversarial examples. GAN basically "cleans" images that are not previously involved in an attack. The generator is fed by the input images for classification. As discussed in the previous section, GAN has two components in its architecture: generators and discriminators. The generator takes some input images from the input training set and generates images with adversarial perturbations that are indistinguishable from natural images. Following this, the discriminator predicts whether that particular image originates from the natural data set or the adversarial images, i.e., those that were generated artificially.

$$min_G max_D \left(D,G\right) = E_{s\sim p_{data}} \log D\left(s\right) + E_{v\sim p_{data}\left(v\right)} \log\left(1 - D\left(G\left(v\right)\right)\right) \quad (16.11)$$

This GAN (Samangouei, Kabkab, and Chellappa 2018) is said to be optimal when the probability distribution of the generator reaches this condition: $p_g{=}{=}p_{data}$. Since GAN is computationally expensive and difficult to train for large applications, another enhanced version of GAN, called conditional GAN, was developed to compress the complex formulations of GAN by reducing the loss function.

$$min_G max_D F\left(D,G\right) = E_{s\sim p_{data}} D\left(s\right) - E_{v\sim p_{data}\left(v\right)} D\left(G\left(v\right)\right) \quad (16.12)$$

Conditional GAN is the natural extended version of GAN that adds the random noise function.

16.3.2 Summarizing Adversarial Attacks and Defense Systems

It is very difficult to design a defense system for a random attack, and it is also challenging to defend against adversarial examples by fluctuating the structure of the DNN models. At the same time, training the defense system for an attack is a tedious task, and it should be acceptable that a single defense mechanism could not solve all types of adversarial examples. Hence, the evolution of dynamic defense mechanisms in DL may solve the problems of dynamic cyberattacks, like adversarial examples. In the future, these DL models' architectures may be compatible with the architectures of evolving adversarial cases while still allowing for effective training.

16.4 CONCLUSION AND FUTURE SCOPE

In this chapter, we provided a general overview of the types of adversarial attacks and defense mechanisms in the field of deep learning in cybersecurity. With recent developments in the DNN model in cybersecurity, it has been proven that DNN models are more vulnerable against attacks. In this view, algorithms and methods in adversarial attacks and defense mechanisms have been investigated effectively. The fundamental workflow of the defense mechanism against the adversarial attacks are elaborated, and it is observed that there is still a need for advanced effective defense mechanisms to fight against the latest adversaries. It is also inferred that, if the efficiency of the defense mechanism tends to be increased to face the adversaries, the complexity and difficulty in the practical deployment of the system will tend to be more complex. In this regard, it is concluded that the hybrid version of more than one defense mechanism should be designed to demonstrate the vulnerability reduction of a specific attack or any combination of adversarial attacks. Hence, the future direction of this work shall focus on the effective combination of defense mechanisms with efficient computational power and practical adaptability to fight against cyberthreats in the network.

REFERENCES

Carlini, Nicholas, and David Wagner. 2017. "Towards Evaluating the Robustness of Neural Networks." In *Proceedings—IEEE Symposium on Security and Privacy*, 39–57. https://doi.org/10.1109/SP.2017.49.

Dong, Yinpeng, Fangzhou Liao, Tianyu Pang, Hang Su, Jun Zhu, Xiaolin Hu, and Jianguo Li. 2017. "Boosting Adversarial Attacks with Momentum." In *Proceedings of the IEEE Computer Society Conference on Computer Vision and Pattern Recognition*, October, 9185–9193. http://arxiv.org/abs/1710.06081.

Goodfellow, Ian, Jean Pouget-Abadie, Mehdi Mirza, Bing Xu, David Warde-Farley, Sherjil Ozair, Aaron Courville, and Yoshua Bengio. 2020. "Generative Adversarial Networks." *Communications of the ACM* 63 (11): 139–144. https://doi.org/10.1145/3422622.

Karatas, Gozde, Onder Demir, and Ozgur Koray Sahingoz. 2019. "Deep Learning in Intrusion Detection Systems." In *International Congress on Big Data, Deep Learning and Fighting Cyber Terrorism, IBIGDELFT 2018 — Proceedings*, 113–116. Institute of Electrical and Electronics Engineers Inc. https://doi.org/10.1109/IBIGDELFT.2018.8625278.

Khalid, Faiq, Hassan Ali, Hammad Tariq, Muhammad Abdullah Hanif, Semeen Rehman, Rehan Ahmed, and Muhammad Shafique. 2019. "QuSecNets: Quantization-Based Defense Mechanism for Securing Deep Neural Network against Adversarial Attacks." In *2019 IEEE 25th International Symposium on On-Line Testing and Robust System Design, IOLTS 2019*, 182–187. Institute of Electrical and Electronics Engineers Inc. https://doi.org/10.1109/IOLTS.2019.8854377.

Lansley, Merton, Nikolaos Polatidis, Stelios Kapetanakis, Kareem Amin, George Samakovitis, and Miltos Petridis. 2019. "Seen the Villains: Detecting Social Engineering Attacks Using Case-Based Reasoning and Deep Learning." Paper presented at Workshops Proceedings for the Twenty-seventh International Conference on Case-Based Reasoning.

Li, Jiao, Yang Liu, Tao Chen, Zhen Xiao, Zhenjiang Li, and Jianping Wang. 2020. "Adversarial Attacks and Defenses on Cyber-Physical Systems: A Survey." *IEEE Internet of Things Journal*. Institute of Electrical and Electronics Engineers Inc. https://doi.org/10.1109/JIOT.2020.2975654.

Liu, Hongyu, and Bo Lang. 2019. "Machine Learning and Deep Learning Methods for Intrusion Detection Systems: A Survey." *Applied Sciences (Switzerland)*. MDPI AG. https://doi.org/10.3390/app9204396.

Moosavi-Dezfooli, Seyed-Mohsen, Alhussein Fawzi, and Pascal Frossard. 2015. "DeepFool: A Simple and Accurate Method to Fool Deep Neural Networks." In *Proceedings of the IEEE Computer Society Conference on Computer Vision and Pattern Recognition 2016*, November/December, 2574–2582. http://arxiv.org/abs/1511.04599.

Meng, Dongyu and Hao Chen. 2017. "MagNet: A Two-Pronged Defense against Adversarial Examples", In *Proceedings of the 17th ACM conference on Computer and communications security (CCS)*, pp. 135–147.

Papernot, Nicolas, Patrick McDaniel, Somesh Jha, Matt Fredrikson, Z. Berkay Celik, and Ananthram Swami. 2015. "The Limitations of Deep Learning in Adversarial Settings." In *Proceedings—2016 IEEE European Symposium on Security and Privacy, EURO S and P 2016*, November, 372–387. http://arxiv.org/abs/1511.07528.

Samangouei, Pouya, Maya Kabkab, and Rama Chellappa. 2018. "Defense-Gan: Protecting Classifiers against Adversarial Attacks Using Generative Models." In *6th International Conference on Learning Representations, ICLR 2018 — Conference Track Proceedings*. International Conference on Learning Representations, ICLR. https://github.com/kabkabm/defensegan.

Short, Austin, Trevor La Pay, and Apurva Gandhi. 2019. *Defending Against Adversarial Examples*. Albuquerque, NM and Livermore, CA. https://doi.org/10.2172/1569514.

Thomas, Tony, Athira P. Vijayaraghavan, and Sabu Emmanuel. n.d. *Machine Learning Approaches in Cyber Security Analytics*. Springer, ISBN 978-981-15-1705-1.

Yu, Fangchao, Li Wang, Xianjin Fang, and Youwen Zhang. 2020. "The Defense of Adversarial Example with Conditional Generative Adversarial Networks." *Security and Communication Networks*. https://doi.org/10.1155/2020/3932584.

Index

L

M

N

O

P

R

S

T

U

V

W

Z

Printed in the United States
by Baker & Taylor Publisher Services

Printed in the United States
by Baker & Taylor Publisher Services